PROCEEDINGS OF THE

VIth INTERNATIONAL WORKSHOP ON

PHOTON-PHOTON COLLISIONS

PROCEEDINGS OF THE

VIth INTERNATIONAL WORKSHOP ON PHOTON-PHOTON COLLISIONS

GRANLIBAKKEN, LAKE TAHOE, CALIFORNIA

SEPTEMBER 10-13, 1984

Edited by
R.L. Lander

World Scientific

Published by

World Scientific Publishing Co. Pte. Ltd.
P. O. Box 128, Farrer Road, Singapore 9128

ISBN 9971-978-22-9

Printed in Singapore by Kim Hup Lee Printing Co Pte Ltd.

INTERNATIONAL ADVISORY COMMITTEE

G. Barbiellini (CERN)

Ch. Berger (DESY)

S. Brodsky (SLAC)

H. Harari (Weizmann Inst.)

P. Kessler (Collège de France)

P. Oddone (SLAC)

J.C. Sens (NIKHEF)

V. Sidorov (Novosibirsk)

K. Strauch (Harvard)

T. Walsh (U. Minnesota)

G. Wolf (DESY)

LOCAL ORGANIZING COMMITTEE

S. Brodsky (SLAC)

J. Gunion (U.C. Davis)

J. Kiskis (U.C. Davis)

Winston Ko (U.C. Davis)

R. Lander, Chair (U.C. Davis)

G. Masek (U.C. San Diego)

D. Pellett (U.C. Davis)

B. Shen (U.C. Riverside)

J. Smith (U.C. Davis)

W. Wagner (U.C. Davis)

Conference Secretary

Mary Schenck

SPONSORS

U.C. Davis, Dept. of Physics
Dean, College of Letters and Science
U.S. Department of Energy
National Science Foundation

God, whose law it is that he who learns must suffer,
and even in our sleep pain that cannot forget,
falls drop by drop upon the heart,
and in our own despite, against our will,
comes wisdom to us by the awful grace of God

                          Aeschylus

PREFACE

Our understanding of two-photon interactions has grown enormously in recent years. The first theoretical and experimental work goes back to the 1930's. It was the e^+e^- colliders, however, that provided the greatest acceleration of the flow of experimental data, which in turn justified an increase in theoretical work. The first major meeting on this topic was held in 1973. The second did not occur until 1979, at Lake Tahoe, California, shortly after PETRA operated. Then meetings were organized at Amiens (1980), Paris (1981), and ·Aachen (1983). There is now a regular series of meetings devoted to the two-photon interaction, of which the present one is the most recent. The next one is to be held at Paris in 1986.

Much has been learned about the subject, yet there is a considerable distance still to be travelled. The theoretical and experimental papers in this year's proceedings provide a review of the present status and point the way to future understanding. The authors are to be deeply thanked for their efforts. It is they, along with the other participants, who determine the degree of success of these meetings.

The Editor wishes gratefully to acknowledge the excellent and essential advice of the International Advisory Committee, as well as the help of the Local Organizing Committee and the Chairs of the various sessions. Special thanks are given to M. Schenck and A. Balics for their untiring service that continues to this moment. W. Parsons and the Granlibakken staff were superb hosts, as always.

The meeting was made possible by generous support from the University of California Physics Department, the Dean of the College of Letters and Science, the United States Department of Energy and the National Science Foundation.

Davis, California   R. Lander
December 1984

C O N T E N T S

Parallel Session (Monday) - Chair: G. Farrar:

Parallel Session (Monday) - Chair: J.C. Sens:

<u>Parallel Session (Tuesday) - Chair: D. Stork</u>:

<u>Parallel Session (Tuesday)- Chair: D. Pellett</u>:

PLENARY SESSIONS

A SURVEY OF TWO-PHOTON PHYSICS:
WHAT HAVE WE DONE, WHERE ARE WE GOING ?

Hermann Kolanoski

Physikalisches Institut der Universität Bonn, 53 Bonn
FED. REP. OF GERMANY

ABSTRACT

After a brief general introduction into two-photon
scattering processes the status of two-photon physics
as of the beginning of this workshop is reported.
Emphasis is given to those subjects which should get
more attention in the future.

1.   Introduction

In the last few years two-photon physics has enjoyed an intensive experimental activity at the electron-positron storage rings, in particular
in Hamburg and Stanford. A large body of experimental results has been obtained providing answers to many theoretical questions.

In this talk an attempt is made to scrutinize were we stand at the
beginning of this workshop. We shall discuss the achievements in the various fields of two-photon physics, such as the investigation of resonance
formation, of hard scattering processes and of the photon structure function. For each of these fields it was tried to point at the problems which
have still to be tackled, both theoretically and experimentally.

4

We will begin with a brief general introduction to two-photon scattering processes. What follows is in some sense a reminder of what we have learnt at the last workshop in Aachen in April 1983. It is not intended to give all the original references to the experimental results, rather we will often refer to the review talks at Aachen where more details can be found. Clearly, you will find the most recent developments in two-photon physics in the review talks presented at this 1984 meeting.

1.1 The Photon-Photon Scattering Process

The subject of this workshop is the production of particles in two-photon collisions (Fig. 1).

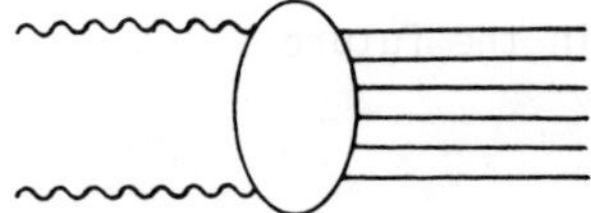

Figure 1

In Maxwell's classical theory electromagnetic waves do not scatter off one another because of the linear superposition principle, however quantum electrodynamics allows photon-photon scattering via the polarisation of the vacuum, i. e. the creation and absorption of fermion pairs[1] (Fig. 2).

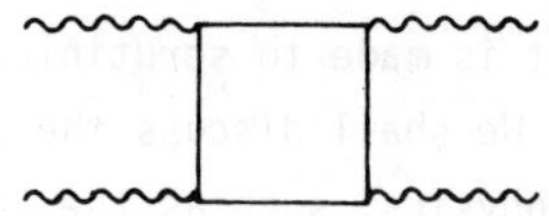

Figure 2

The cross section for scattering of visible light according to this diagram is only of the order of 10^{-60} cm^2. An early experiment[2] using the sun as a photon source and the "dark-adapted" eye as a detector (see Fig. 3) obtained an upper limit for the cross section of the photon

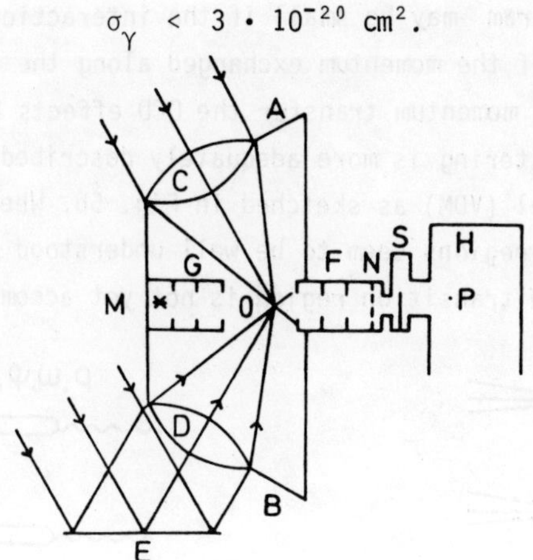

<u>Figure 3</u>   Apparatus for a light-light scattering experiment:
The two lenses C and D focus sun light on the same spot O in
a light-tight box AB. The dark-adapted eye of an observer at
the point P serves as the detector for scattered light.

This experiment was far too insensitive to measure the cross section as
calculated later in QED. But even the most powerful lasers today cannot
provide enough luminosity to make elastic light-light scattering observable.

At photon energies above the threshold for the production of particle pairs inelastic processes occur.[3] The lowest order QED process
is the lepton pair production as given by the diagram in Fig. 4.

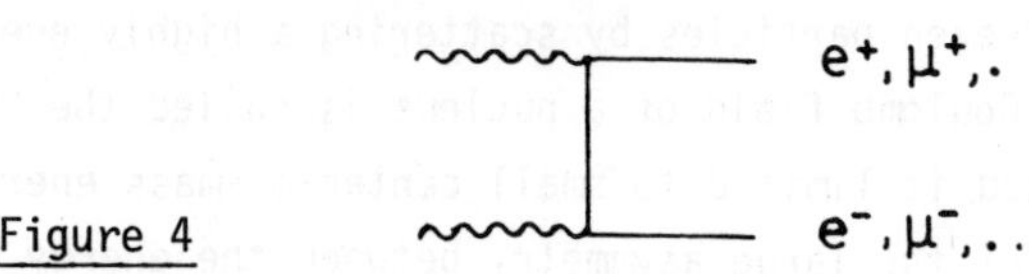

<u>Figure 4</u>

One may expect that two-photon production of hadrons can be described by the corresponding diagram for quark pair creation with the
quarks subsequently fragmenting into hadrons (Fig. 5a). The QCD correc-

tions to this diagram may be small if the interaction occurs at small distances, i. e. if the momentum exchanged along the inner quark line is large. At small momentum transfer the QCD effects become large and photon-photon scattering is more adequately described by the vector meson dominance model (VDM) as sketched in Fig. 5b. Whereas these two extreme kinematical regions seem to be well understood a satisfactory understanding of the transition region is not yet accomplished.

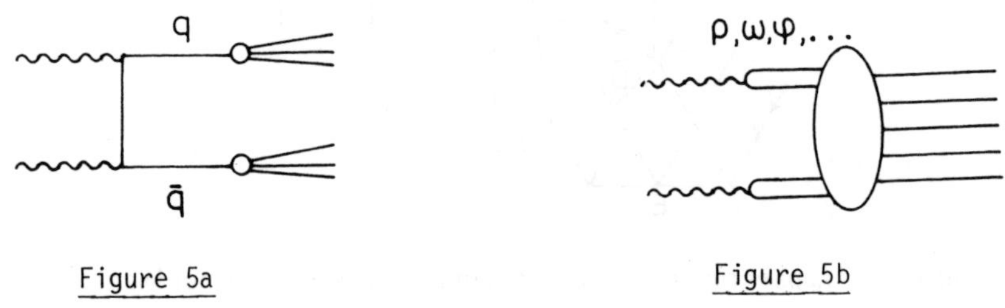

Figure 5a   Figure 5b

1.2 Photon-Photon Scattering Experiments

How can photon-photon scattering reactions be observed by experiments? The most obvious idea - using two beams of real photons - would yield in practice much too low luminosities. The two methods which have been successfully used, the Primakoff effect and e^+e^- scattering, both exploit the high photon densities in the electromagnetic fields surrounding charged particles.

The production of C-even particles by scattering a highly energetic photon beam off the Coulomb field of a nucleus is called the Primakoff effect.[4] This method is limited to small center-of-mass energies of the two photons because of the large asymmetry between the energy of the beam photons and the target photons. Up to now only the two-photon production of the π^0 and the η have been measured with the Primakoff effect.[5]

Most of the experimental results on two-photon scattering have been obtained at e^+e^- storage rings. Following a suggestion by Low[6] and by Calogero and Zemach[7] the clouds of virtual photons accompanying the electron and positron beams can be used to investigate photon-photon scattering as a subprocess of the reaction shown in Fig. 6:

$$e^+e^- \rightarrow e^+e^- X \tag{1}$$

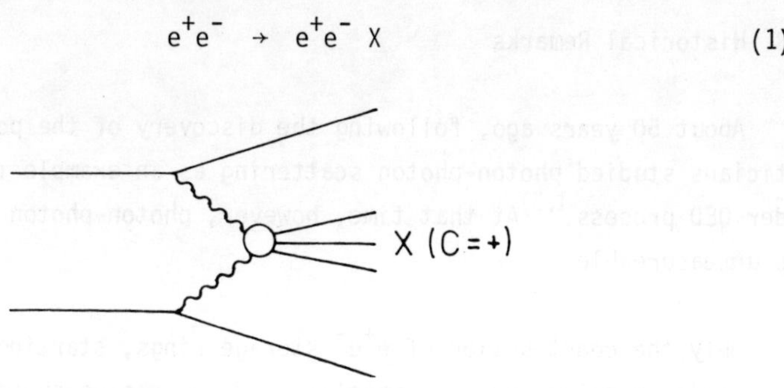

<u>Figure 6</u>

With this method one does not only achieve large $\gamma\gamma$ luminosities, but reaction (1) also offers the possibility to study $\gamma\gamma$ interactions with continuously tunable photon kinematics and photon polarisations. The two photons in Fig. 6 are linearly polarized with respect to the scattering planes of the radiating electrons. Tagging of the photons, i. e. detection of the scattered electrons, allows one to control the polarisation as well as the two-photon invariant mass W and the four momentum transfer Q_1^2, Q_2^2 of the photons. The cross section for $e\,e \rightarrow e\,e\,X$ can be decomposed into different cross section terms depending on the transverse (T) and longitudinal (L) polarisation of the photons (ϕ is the angle between the lepton scattering planes):

$$\sigma_{ee \rightarrow eeX}\,(W,q_1^2,q_2^2) = L_{TT}\,\sigma_{TT} + L_{TL}\,\sigma_{TL} + L_{LT}\,\sigma_{LT} + L_{LL}\,\sigma_{LL}$$

$$+ \, \mathcal{L}_{TT}\,\tau_{TT} \cdot \cos 2\phi + \mathcal{L}_{TL}\,\tau_{TL}\,\cos\phi$$

The two-photon luminosity functions L_{ij} depend only on the kinematics of the electrons and can be calculated within QED.[8,9] Due to the bremsstrahlung-type radiation process in Fig. 6 the bulk of the photons is quasi-real and the $\gamma\gamma$ luminosity drops with the $\gamma\gamma$ c. m. energy like $1/W^2$ (but increases logarithmically with the $e^{\pm}$ beam energy).

1.3 Historical Remarks

About 50 years ago, following the discovery of the positron, theoreticians studied photon-photon scattering as an example of a higher order QED process.[1,3] At that time, however, photon-photon scattering was unmeasureable.

Only the construction of e^+e^- storage rings, starting in the sixties, made an extensive experimental study of photon-photon reactions possible. We mentioned before that the use of storage rings as sources for photons was suggested in 1960.[6,7] Detailed theoretical studies of the two-photon process at e^+e^- storage rings

$$e^+ e^- \rightarrow e^+ e^- X$$

were started at the end of the sixties. Most of the basic concepts and formulae which are now standard for the physicists working in this field have been developed during that time. Valuable reviews are found in references 8, 9, 10. The first two-photon processes experimentally observed were the QED processes of lepton pair production reported by groups from Novosibirsk and Frascati.[11] However, the productive era of experimental two-photon physics, which we are right now experiencing, only started around 1979 with experiments at the storage rings SPEAR, PETRA, PEP, DCI and DORIS. The experiments cover subjects such as QED processes, resonance production, total cross section and structure function measurements as well as hard scattering reactions (for a review see ref. 12).

The theoretical concepts in two-photon physics were quite advanced even before any experimental results were available. Impor-

tant and, given todays knowledge, quite far-sighted, have been the ideas
that the pointlike coupling of the photon to quarks plays an important
role in photon-photon reactions, even for real photons. In 1971, Berman,
Bjorken and Kogut[13] pointed out that the pointlike coupling of the photons should become visible in the production of hadrons with large transverse momenta and in the formation of hadron jets. Walsh and Zerwas[14]
suggested in 1973 that the structure function of the (real!) photon is
dominated by the pointlike coupling of the photon to quarks. Quite a
brave statement at a time where most people would have bet that the
structure of the real photon is just that of a vector meson. The point-
like scattering phenomena, first derived in the framework of the quark-
parton model, are now understood as resulting from the asymptotically
free behaviour of the underlying dynamical theory, i. e. quantum chomo-
dynamics. The first QCD evaluation of the structure function of the photon was done by Witten in 1977 [15] and the production of hadrons with
large transverse momenta was treated extensively in the framework of QCD
by Brodsky, DeGrand, Gunion and Weis.[16]

Together with the study of resonance production, the test of pertu-
bative QCD predictions belongs to the most interesting subjects investigated in two-photon physics.

2.   Assessing the Achievements in Two-Photon Physics

In the following we will discuss what one had hoped to learn from
studying two-photon reactions and compare it to what was actually achieved. (Questions and problems which were still open at the beginning of the
workshop when this talk was given, may have been solved or at least cla-
rified during this workshop. It is an obvious problem to write down such
a talk after having absorbed all the new information presented at this
meeting.)

2.1 QED Reactions

The $O(\alpha^4)$ QED processes

$$e^+ e^- \to e^+ e^- \ell^+ \ell^- \quad (\ell = e, \mu)$$

have been measured in a wide range of invariant masses of the lepton pairs and Q^2 values of the photons.[17] The experimental results are in agreement with QED at a level of about 10 %. That means, for example, that no exotic objects decaying into lepton pairs have been seen to be produced in $\gamma\gamma$ reactions. Such objects have recently been suggested to be produced in radiative Z^0 decays. In some models they could have a quite substantial rate in the reaction

$$\gamma\gamma \to X \to \ell^+ \ell^-$$

(see the talk on "new particles" given by F. Schrempp at this workshop.) In this context it should be mentioned that τ-pair production by two photons has not yet been measured. It is conceivable, though not very likely, that we could miss, for example, an exotic object which can couple to $\gamma\gamma$ and preferentially decays into the heaviest leptons.

Even if one accepts the argument that the chance to find new physics in $\gamma\gamma$ QED processes is not particularly good (e. g., the $\gamma\gamma$ energy is always lower than the corresponding e^+e^- energy), there is another good reason to study QED processes: Lepton pair production, in particular $\mu^+\mu^-$ production, can be seen as a prototype reaction for quark pair production according to the diagram in Fig. 5a. The simultaneous measurement of both reactions allows testing the absolute normalisation of hadron production including the QED radiative corrections.

2.2 General Remarks on the Two-Photon Production of Hadrons

Hadrons are produced in two-photon collisions via the coupling of the photons to the charged constituents of hadrons, the quarks. Since the quarks are also strongly interacting via gluon exchange the bare $\gamma q\bar{q}$ vertex has to be corrected for gluon radiation corresponding to an infinite sum of the type

For real or nearly real photons, this sum is known to be dominated by poles in the timelike region, the vector mesons (vector meson dominance). However, it has been predicted that there are kinematical situations where the bare $\gamma q\bar{q}$ vertex dominates:

1) if hadrons are produced with large transverse momenta (p_T) with respect to the $\gamma\gamma$ direction,[13,16]

2) if at least one of the photons has a large four-momentum transfer Q^2, a kinematical situation usually described in terms of the structure functions of the photon.[14]

3) An "anomalous" pointlike component was also predicted to contribute to the total cross section for two-photon production of hadrons.[18]

The pointlike coupling of the photons to quarks is also seen in the two-photon production of resonances: The ratios of $\gamma\gamma$ widths of mesons belonging to the same SU(3) multiplet depend in good approximation only on the pointlike coupling of the photons to the quark charges. The dependence on the spatial wave functions, which contain the strong interaction effects, is supposed to cancel in the ratios.

2.3 Hadron Production at Low Q^2 and p_T

Hadron production by two real photons is usually described in terms of "old fashioned" physics using VDM and Regge model terminology. In these models photon-photon scattering is dominantly like peripheral hadron-hadron scattering leading to forward-backward production of hadrons with a typical transverse momentum distribution (Fig. 7):

$$\frac{d\sigma}{dp_T^2} \sim e^{-bp_T}$$

with $b \approx 6$ GeV^{-1}.

<u>Figure 7</u>

Regge estimates predict a total cross section:[19]

$$\sigma(\gamma\gamma \to \text{hadrons}) \approx (240 + 270 \text{ GeV/W}) \text{ nb}.$$

A pointlike contribution would show up as an additional $1/W^2$ term contributing most significantly in the low W region, in particular in the resonance region. The experiments had problems to establish definitely such a term.[20]

At energies above the resonance region the measured cross section has roughly the predicted size. The Q^2 dependence at not too large Q^2 can be described by a generalized vector meson dominance model. The event topology has the predicted exponential p_T behaviour at small p_T. Deviations from this behaviour already starting at p_T values as low as 1 GeV have been interpreted as hard scattering effects.

The lack of detector acceptance at small transverse particle momenta in combination with the steep p_T dependence of the cross section caused in the past severe problems for the determination of the total cross section. The systematic uncertainties become particularly large in the low W region. In the resonance region the techniques applied up to now to determine the total cross section may not be adequate (see e. g. ref. 20). The progress in this field will be reported at this conference by G. Knies.

2.4 Two-Photon Production of Resonances

Two-photon resonance physics has been started by the Mark II group in 1979 with the measurement of the $\gamma\gamma$ width of the η'.[21] Subsequently many more two-photon couplings have been determined (a review is given in ref. 12). Accounting for the large amount of information available on this subject, it will be covered by three speakers at this workshop: the current experimental status will be reported by A. Cordier and F. Erné, while M. Chanowitz will explain to us the theoretical implications.

The $\gamma\gamma$ widths of the pseudoscalar mesons (π^0, η, η') and the tensor mesons (A_2, f, f') have been measured allowing a detailed comparison with the predictions of the SU(3) quark model. Until recently, the scalar mesons (δ, ε, S^*) failed to show up in $\gamma\gamma$ reactions. (An experimental hint for $\gamma\gamma$ production of the $\delta(980)$ will be reported at this workshop.) also the two-photon production of $c\bar{c}$ states appears to be difficult to observe, presumable because of lack of luminosity.

Within the quark model, the $\gamma\gamma$ width is a measure of the charges and thus of the quark composition of a meson. According to the diagram in Fig. 8

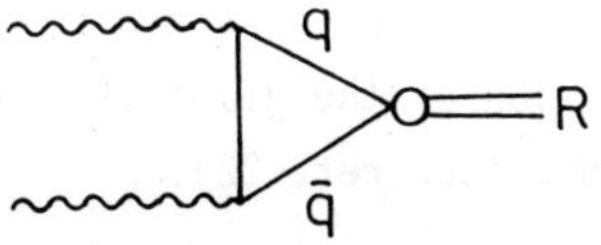

Figure 8

the $\gamma\gamma$ width is proportional to a coherent sum over the squares of the quark charges:

$$\Gamma_{\gamma\gamma} \sim \left| \sum_q c_q\, e_q^2 \right|^2$$

The coefficients c_q determine the quark content in a meson which is given by the SU(3) representation together with the singlet-octet mixing angle. Thus the $\gamma\gamma$ widths provide tests of SU(3) symmetry and allow the determination of the mixing angles.

Inconsistencies of the measured $\gamma\gamma$ widths with the SU(3) relations could be a hint for admixtures of exotic states, such as glueballs or four-quark states, to the normal $q\bar{q}$ states. Probably, only a very systematic investigation of the meson spectroscopy and the comparison of different resonance production processes will tell us if there is room for such additional states.

Concerning the search for glueballs, it appears to be particularly promising to compare the resonance spectra obtained in radiative J/ψ decays to those obtained in $\gamma\gamma$ interactions. The radiative decay

$$J/\psi \rightarrow \gamma\,X$$

is a socalled "glueball favoured" channel as the transition of the J/ψ to X is mediated by two gluons (Fig. 9a).

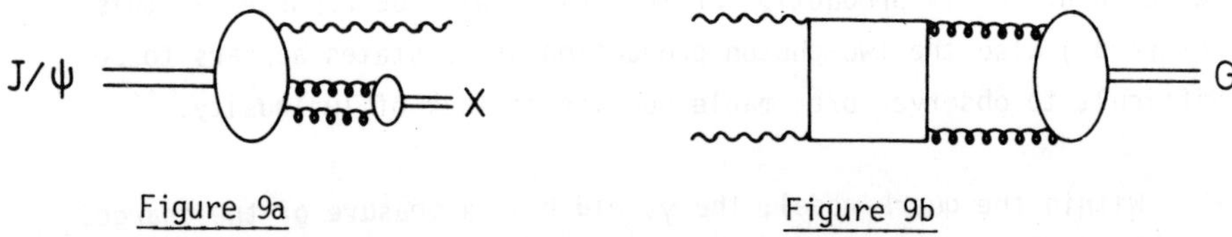

Figure 9a   Figure 9b

In contrast, two photons can only couple to a glueball, G, via an intermediate quark loop (Fig. 9b). Thus, this coupling is Zweig-suppressed and relatively small $\gamma\gamma$ widths are expected for glueballs. Upper limits on the two-photon production of the glueball candidates $\iota(1440)$ and $\Theta(1700)$ have been obtained (see ref. 22).

The strong threshold enhancement observed in $\gamma\gamma \rightarrow \rho^0\rho^0$ [23] has been interpreted to be due to four-quark states (see references in ref. 20). Degenerate four-quark states decaying mainly into $\rho^0\rho^0$ were predicted in this region, one with isospin 0 and the other with isospin 2. The superposition of both yields a suppression of the $\rho^+\rho^-$ channel, which was actually observed (Fig. 10). A simple non-exotic resonance, instead, would decay twice as often to $\rho^+\rho^-$ than to $\rho^0\rho^0$. The various models explaining $\rho^0\rho^0$ production near threshold can be tested by measuring the production of other vector meson pairs:

$$\gamma\,\gamma \;\to\; V\,V' \quad (V,\,V' = \rho,\,\omega,\,\phi,\,J/\psi,\,\ldots).$$

Except for $\rho\rho$ production, all other searches for vector meson pair
production yielded only upper limits.

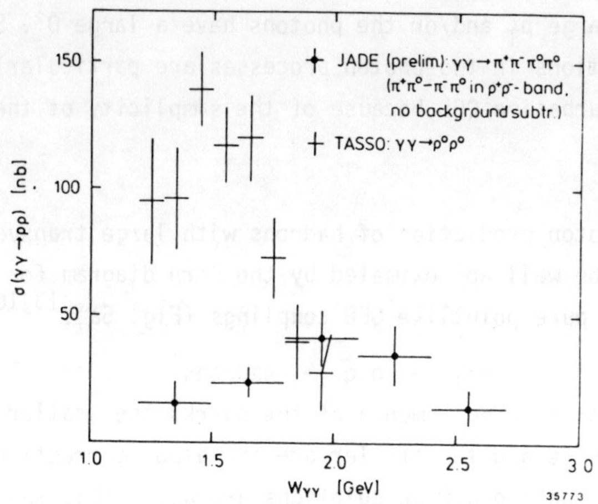

<u>Figure 10</u>   Cross section for $\gamma\gamma \to \rho^0\rho^0$ from TASSO and for
$\gamma\gamma \to \rho^+\rho^-$ from JADE (prelim.). The latter cross section has
to be taken as an upper bound since no background has been
subtracted.

Assessing the achievements of two-photon resonance physics, one has to
say that this was one of the most prolific fields of two-photon physics. Yet, more luminosity is desireable to extend the investigation
beyond the study of pseudoscalar and tensor mesons to the search for
scalars, $c\bar{c}$ states, four-quark states and glueballs. A significant increase of the accumulated luminosity, say by a factor of 10, will add a
new quality to two-photon resonance physics and will almost certainly
lead to new surprises in meson spectroscopy.

2.5 Hard Scattering Reactions in Two-Photon Processes

The study of the direct pointlike coupling of photons to quarks is one of the most challenging tasks in two-photon physics. As pointed out in Sect. 2.2, the pointlike coupling is expected to become observable in reactions involving a hard momentum transfer, i. e. if the hadrons are produced with large p_T and/or the photons have a large Q^2. Such hard scattering reactions in two-photon processes are particularly well suited to test perturbative QCD because of the simplicity of the initial state.

The two-photon production of hadrons with large transverse momenta is expected to be well approximated by the Born diagram for quark pair production with pure pointlike QED couplings (Fig. 5a):[13,16]

$$\gamma\,\gamma \;\rightarrow\; q\,\bar{q} \;\rightarrow\; \text{hadrons}$$

The larger the transverse momenta of the quarks the smaller are the interaction distances and the smaller are the gluon corrections. The region where perturbative QCD calculations are applicable may extend down to p_T values as low as 1 GeV. The p_T distribution below about 1 GeV is characterized by peripheral scattering of vector mesons leading to an exponential p_T dependence, while the Born term (Fig. 5a) leads to a scale invariant differential cross section:

$$\frac{d\sigma}{dp_T^2} \sim \frac{1}{p_T^4}$$

A flattening of the p_T distribution above about 1 GeV has in fact been experimentally observed, approaching a slope close to p_T^{-4} (Fig. 11)[24]. If the high p_T tail is dominantly due to the Born process, the events with high p_T hadrons should exhibit 2-jet topologies. The data were found to be consistent with this expectation.

The absolute normalisation of hadron production via the Born process is given by $R_{\gamma\gamma}$, which is defined as the cross section ratio for quark and muon pair production via the Born diagram. $R_{\gamma\gamma}$ depends on the

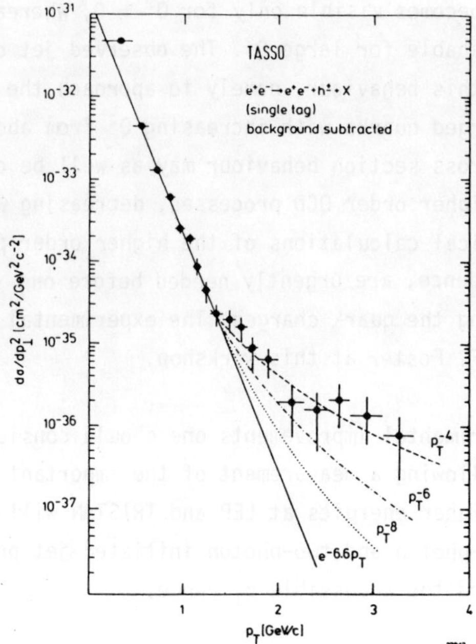

<u>Figure 11</u>   Transverse momentum distribution for inclusive
hadron production by two photons (TASSO)

fourth power of the quark charges:

$$R_{\gamma\gamma} = 3 \sum_q e_q^4 = 34/27 \quad \text{for} \quad q = u, d, s, c.$$

In the Han-Nambur model of integrally charged quarks $R_{\gamma\gamma}$ is about three
times larger than the above value for fractionally charged quarks:

$$R_{\gamma\gamma}^{H-N} = 10/3 \quad \text{for} \quad q = u, d, s, c.$$

This value has already been ruled out by the first experiments looking
for two-photon production of jets,[25] although the data were, in the
covered p_T range, still higher than expected for fractionally charged
quarks. Since these experiments have been done in the single tag mode,
i. e. $Q^2 \neq 0$ for one of the photons, it has been pointed out that in a
gauge invariant formulation of the integrally charged quark model $R_{\gamma\gamma}$
is Q^2 dependent (see references in ref. 24). The full difference

between both models becomes visible only for $Q^2 = 0$, whereas both models are indistinguishable for large Q^2. The observed jet cross sections indeed seem to have this behaviour, namely to approach the prediction for fractionally charged quarks with increasing Q^2 from above. On the other hand, such a cross section behaviour may as well be caused by contributions from higher order QCD processes, decreasing with Q^2 and p_T. Detailed theoretical calculations of the higher order processes, including the Q^2 dependence, are urgently needed before one can draw conclusions concerning the quark charges. The experimental situation will be reported by F. Foster at this workshop.

For future experimental improvements one should consider building 0^0 tagging devices allowing a measurement of the important $Q^2 = 0$ kinematical point. The higher energies at LEP and TRISTAN will make the separation between one-photon and two-photon initiated jet production easier and will extend the accessible p_T range.

2.6 Hadron Pair Production

Following a suggestion of Brodsky and Lepage[26] perturbative QCD can also be tested in exclusive processes, such as two-photon production of hadron pairs at large transverse momenta. This model is based on the assumption that, if a large momentum transfer is involved in the process, the matrix element factorizes into a hard scattering amplitude T_H and "soft" distribution functions ϕ (see Fig. 12):

$$M \sim \phi \, T_H \, \phi^*$$

ϕ cannot be calculated, but has to be inferred from related processes containing the same distribution function (e. g., $\gamma\gamma \to \pi^+\pi^-$ is related to the pion form factor). Such calculations then yield
- the absolute normalisation,
- the angular dependence,

- and the energy dependence
of the cross section.

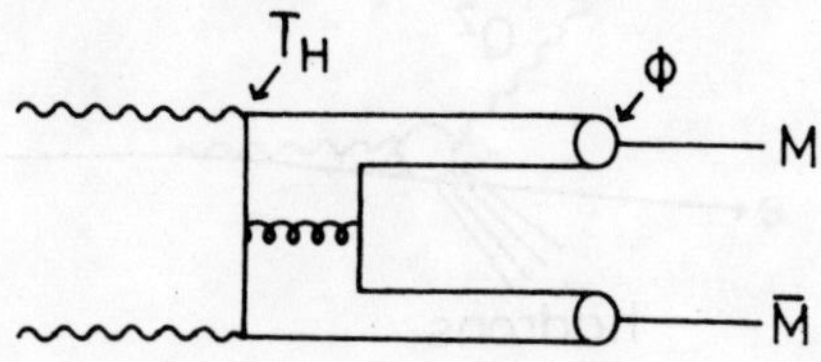

<u>Figure 12</u>

Measurements of two-photon production of hadron pairs are presumably
well suited to test this concept. Particularly interesting is the question if such a model, which is probably correct in the asymptotic limit, can be applied at the relatively moderate energies available in
two-photon reactions.

The experimental situation before this workshop showed no consistent picture. Whereas meson pair production ($\pi^+\pi^-$, K^+K^-) seems to be
in quite reasonable agreement with the QCD calculations,[27] baryon
pair production ($p\bar{p}$) in the invariant mass range below 3 GeV [28] seems
to be far off the predictions. However, in the latter case two theoretical predictions exist which are inconsistent with one another. These
problems will be discussed from a theoretical point of view by G. Farrar; the experimental results will be reported by H. Kück. Experimental
improvements, as extending the measurements to other channels and
higher energies and testing the predicted energy dependence   ($\sim 1/s^4$
for mesons, $\sim 1/s^6$ for baryons) need considerably more luminosity.

2.7 The Structure Functions of the Photon

The structure of a photon can be probed by scattering an electron
off a quasi-real target photon[29]. Detecting the scattered electron at
large angles ensures that the probing photon has a high Q^2 (Fig. 13).

20

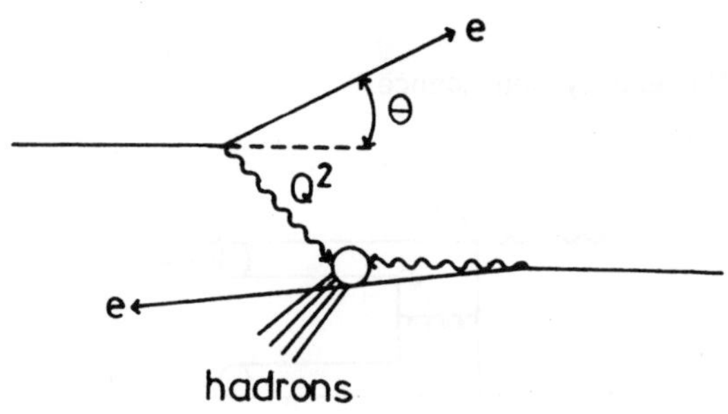

Figure 13

If the photon is a vector meson most of the time, one expects to
measure a typical hadronic structure function, i. e. falling with increasing Bjorken variable x and independent of Q^2 (apart from logarithmic gluon corrections). However, it has been predicted[14] that the
pointlike coupling of the photons to quarks (Fig. 5a) should contribute
and should even become dominant at Q^2 values of a few GeV^2. In the
quark-parton model the pointlike part is given by the Born diagram for
quark pair production yielding for the structure function F_2 (m_q is the
quark mass):

$$F_2 \sim \ln \frac{Q^2}{m_q^2}$$

$F_2(x,Q^2)$ has a maximum at large x, quite in contrast to hadronic structure functions, and increases with Q^2 due to the increasing phase space
of the produced quark pairs. That means that the photon structure function does not scale, even not in the quark-parton model. (The scaling
of the hadronic structure functions is understood as being due to the
limited transverse momenta of the quarks in a hadron.) Fig. 14 shows
$F_2(x,Q^2)$ for the hadronic component, estimated by ρ dominance, and for
the pointlike component. The latter clearly dominates at large Q^2 and
not too small x.

The first experimental results from the PLUTO collaboration[30] indeed show that the structure function is large as x approaches 1 (Fig.
15). The data can be well described by a sum of a pointlike and a VDM

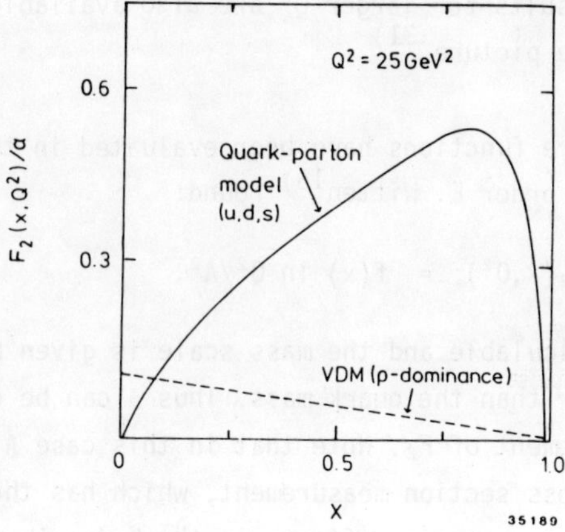

Figure 14   The photon structure function $F_2(x,Q^2)$ predicted by the quark-parton model and by the vector meson dominance model

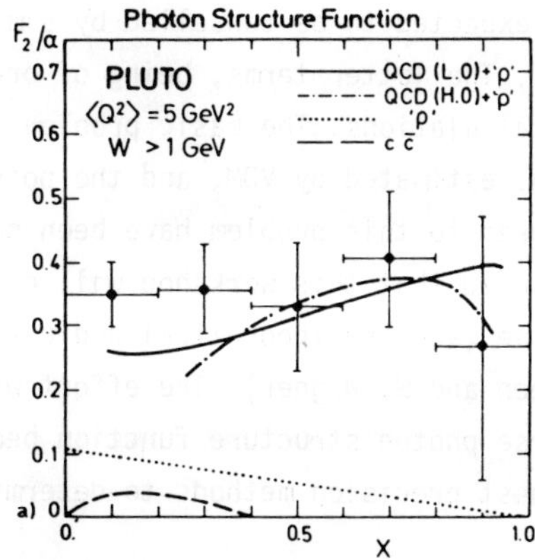

Figure 15   The structure function $F_2(x,Q^2)$ for $\langle Q^2 \rangle = 5$ GeV^2 as measured by PLUTO. The data are compared to QCD calculations for the pointlike contribution and a VDM ('ρ') estimate for the hadronic part. The charm quark contribution ($c\bar{c}$) is still small at this Q^2 value.

component. Meanwhile results for larger Q^2 are also available which all fit nicely into the same picture.[31]

The photon structure functions have been evaluated in the framework of QCD. In leading order E. Witten[15] found:

$$F_2(x,Q^2) \; = \; f(x) \; \ln Q^2/\Lambda^2 .$$

The function $f(x)$ is calculable and the mass scale is given by the QCD scale parameter Λ rather than the quark mass. Thus Λ can be determined from an absolute measurement of F_2. Note that in this case Λ can be extracted from a total cross section measurement, which has the virtue of not being affected by fragmentation effects in the hadronic final state.

The next-to-leading order QCD corrections for the photon structure functions have also been calculated and were found to be small in the medium x range.[32] However, at small x the "pointlike" terms in F_2 develop singularities which are expected to be cancelled by similar singularities in the "hadronic" pieces. The latter terms, being of order $(\ln Q^2/\Lambda^2)^{-1}$, are neglected in these calculations. The basic problem is how to connect the hadronic piece, e. g. estimated by VDM, and the pointlike piece as calculated using QCD. Cures to this problem have been suggested and are heavily debated. It is hoped that this workshop will contribute to the clarification of the issue (see the theoretical and experimental talks on this subject by W. Bardeen and W. Wagner). The effort will certainly be worthwhile: the transverse photon structure function bears the potential of being one of the highest precision methods to determine Λ.

3. Conclusions

It would be neither wise nor true to conclude already in this first talk of the workshop that everything has been done in two-photon physics. But in fact, the last about 5 years of experimental work have been very productive. Nearly all the physics questions which had been raised before the experiments started have been tackled, such as tests of QED and SU(3) symmetry, studies of the characteristics of hadron production at low and high Q^2 and p_T, including the investigation of jet production and of the photon structure. To many of the questions answers were found, most of them in agreement with the theoretical predictions.

These achievements owe a lot to the good performance of the storage ring machines. At PETRA, for example, most of the experiments have only become possible with the dramatic increase of the luminosity after the insertion of the mini-β quadrupoles in 1981. Equally important as high luminosities are smooth running conditions which allow triggering on the low energy and low multiplicity two-photon events. Clearly, the higher the beam energy, the more the average energies and multiplicities of two-photon and one-photon initiated processes differ and thus the trigger requirements change as well.

Where we are going in two-photon physics depends partly on general decisions on the physics priorities at the various machines. Here we want to give some arguments why and how the two-photon physics programs should be extended. The two-photon experiments so far not only answered physics questions, but also raised new ones as the statistics of the data improved and the experiments became more elaborate.

In two-photon resonance physics the program was extended from testing SU(3) symmetry to searching for possible exotic states. Since these states may not have very distinct signatures to tell them apart from normal $q\bar{q}$ mesons, it seems that one type of experiment by itself will not be able to discover and verify such a state. One rather has to combine the information from various resonance production mechanisms. The radiative coupling of a meson, being sensitive to the quark charges, is a particularly important piece of information. To put the two-photon spectroscopy on equal footing with, e. g., the spectroscopy observed in radiative J/ψ decays, we need a lot more luminosity, desireably some order of magnitude more. Such a program should be possible at the existing storage ring machines.

Other physics questions may not be solvable by just increasing the luminosity, such as the study of the various aspects of perturbative QCD in two-photon physics. Testing the application of perturbative QCD to two-photon physics poses a particular challenge, since two-photon scattering reactions have often been emphasized to be clean testing grounds because of their simple initial state. In particular the transition between the perturbative and the non-perturbative regime, the latter classically described by the vector meson dominance model, is very little understood. To make progress here requires a lot of interaction between theory and experiment. More luminosity could be helpful as it would allow us to study the details of hadron fragmentation, but it seems to be less urgently needed than in the case of the investigation of exclusive channels. Instead, machines with higher energies, like LEP and TRISTAN, will be advantageous in that they extend the accessible kinematical range in p_T and Q^2 for measurements of jet production and the photon structure functions. In addition, the separation of one-photon and two-photon processes may become easier at higher energies.

Thus there are two different areas of two-photon physics with somewhat different experimental requirements: two-photon resonance physics at moderate storage ring energies but high luminosities and on the other hand tests of QCD in hard or deep-inelastic scattering reactions at preferably higher energies. Depending on the capabilities of the specific machine one may concentrate more on one or the other subject. As we have tried to demonstrate there are many interesting and challenging physics questions for which the study of two-photon reactions provides unique possibilities for a solution.

I would like to thank Professor R. Lander and the Organizing Committee
for an interesting and stimulating workshop in a very pleasant atmosphere.

REFERENCES

1) H. Euler and B. Kockel, Nat. Wiss. 23 (1935) 246;
   H. Euler, Ann. Physik 26 (1936) 398;
   R. Karplus and M. Neuman, Phys. Rev. 83 (1951) 776

2) A.L. Hughes and G.E.M. Jauncey, Phys. Rev. 36 (1930) 773

3) G. Breit and J.A. Wheeler, Phys. Rev. 46 (1934) 1087

4) H. Primakoff, Phys. Rev. 81 (1951) 899

5) Particle Data Group, Rev. Mod. Phys. 56 (1984)

6) F. Low, Phys. Rev. 120 (1960) 582

7) F. Calogero and C. Zemach, Phys. Rev. 120 (1960) 1860

8) V.M. Budnev, I.F. Ginzburg, G.V. Meledin and V.G. Serbo, Phys.
   Rep. 15 (1975) 181

9) G. Bonneau, M. Gourdin and F. Martin, Nucl. Phys. B54 (1973) 573

10) A. Jaccarini, N. Arteaga-Romero, J. Parisi and P. Kessler, Compt.
    Rend. 269B (1969) 153, 1129; Nuovo Cim. 4 (1970) 933;
    V.E. Balakin, F.M. Budnev and I.F. Ginzburg, Zh.E.T.F. Pis'ma 11
    (1970) 559 (JETP Lett.II, 388);
    S.J. Brodsky, T. Kinoshita and H. Terazawa, Phys. Rev. D4 (1971)
    1532;
    H. Terazawa, Rev. Mod. Phys. 45 (1973) 615

11) V.E. Balakin et al., Phys. Lett. 34B (1971) 99;

C. Bacci et al., Nuovo Cim. Lett. 3 (1972) 709

12) H. Kolanoski, Springer Tracts in Modern Physics, Vol. 105 (1984)

13) S.M. Berman, J.D. Bjorken and J.B. Kogut, Phys. Rev. D4 (1971) 3388

14) T.F. Walsh and P.M. Zerwas, Phys. Lett. 44B (1973) 195

15) E. Witten, Nucl. Phys. B120 (1977) 189

16) S. Brodsky, T.A. DeGrand, J.F. Gunion and J.H. Weis, Phys. Rev. D19 (1979) 1418

17) M. Pohl, Proceedings of the 5th International Colloquium on $\gamma\gamma$-Interactions, Aachen, ed. Ch. Berger, Lecture Notes in Physics, Vol. 191, Springer Verlag (1983)

18) S.J. Brodsky, F.E. Close and J.F. Gunion, Phys. Rev. D6 (1972) 177

19) J.L. Rosner, Brookhaven report CRISP 71 26 (1971)

20) H. Kolanoski, Proceedings of the 5th International Colloquium on $\gamma\gamma$-Interactions, Aachen, ed. Ch. Berger, Lecture Notes in Physics, Vol. 191, Springer Verlag (1983)

21) G.S. Abrams et al., Phys. Rev. Lett. 43 (1979) 477

22) J. Olsson, Proceedings of the 5th International Colloquium on $\gamma\gamma$-Interactions, Aachen, ed. Ch. Berger, Lecture Notes in Physics, Vol. 191, Springer Verlag (1983)

23) TASSO Collaboration, R. Brandelik et al., Phys. Lett. 97B (1980) 448

24) N. Wermes, Proceedings of the 5th International Colloquium on $\gamma\gamma$-Interactions, Aachen, ed. Ch. Berger, Lecture Notes in Physics, Vol. 191, Springer Verlag (1983)

25) TASSO Collaboration, R. Brandelik et al., Phys. Lett. 107B (1981) 290;
   JADE Collaboration. W. Bartel et al., Phys. Lett. 107B (1981) 163

26) G.P. Lepage and S.J. Brodsky, Phys. Rev. D22 (1980) 2157;
   S.J. Brodsky and G.P. Lepage, Phys. Rev. D24 (1981) 1808

27) PLUTO Collaboration, Ch. Berger et al., Phys. Lett. 137B (1984) 267

28) TASSO Collaboration, R. Brandelik et al., Phys. Lett. 108B (1982) 67

29) S.J. Brodsky, T. Kinoshita and H. Terazawa, Phys. Rev. Lett. 27 (1971) 280;
   T.F. Walsh, Phys. Lett. 36B (1971) 121

30) PLUTO Collaboration, Ch. Berger et al., Phys. Lett. 107B (1981) 168

31) W. Wagner, Proceedings of the 5th International Colloquium on $\gamma\gamma$-Interactions, Aachen, ed. Ch. Berger, Lecture Notes in Physics, Vol. 191, Springer Verlag (1983)

32) W.A. Bardeen and A.J. Buras, Phys. Rev. D20 (1979) 166;
   D.W. Duke and J.F. Owens, Phys. Rev. D22 (1980) 2280

Questions and Discussion

G. KNIES (DESY): How true is the transition from QED → QCD, from 1934 to 1984? In other words: If we had only $\gamma\gamma$ → hadrons data up to now - i.e. no e^+e^- → hadrons, etc. - would we invent QCD on these grounds?

Question answered by S. BRODSKY (SLAC): It is amusing to consider whether QCD would have been discovered on the basis of present knowledge of $\gamma\gamma$ events. The scaling behavior and normalization of the photon structure function and $\gamma\gamma$ → jets indicates pointlike spin-$\frac{1}{2}$ carriers of the electromagnetic current; on the other hand the measurements of $\gamma\gamma \rightarrow \pi^+\pi^-$, K^+K^- show that the mesons are not pointlike, but scale as if the mesons are composite with underlying scale-invariant interactions. The only theory which predicts these phenomena is QCD.

GAMMA GAMMA ANNIHILATION TO EXCLUSIVE FINAL STATES

Glennys R. Farrar

Department of Physics
Rutgers University
New Brunswick, NJ

I.   Introduction

It is by now well-known how the asymptotic freedom of QCD can be exploited to permit perturbative calculation of many features of deep-inelastic and other large-momentum-transfer inclusive scattering processes. The situation is quite similar for large-momentum-transfer exclusive scattering processes: an amplitude is the convolution of a perturbatively calculable "hard" part with non-perturbatively-calculable hadronic wavefunctions, which can be taken from experiment, as is done for the structure functions of inclusive scattering.

Exclusive reactions in general however present several new features [1] which add to their experimental and theoretical interest: a) The "hard" calculations really are hard, involving tens to hundreds of thousands of Feynman diagrams even in Born approximation. Issues of the convergence of perturbation theory may be experimentally accessible. b) Integration of the Born approximation hard amplitude over the hadron wavefunctions in general involves kinematic configurations in which intermediate quarks and/or gluons of the hard amplitude go on mass shell. In these regions perturbation theory is not reliable, since freely propagating quarks and gluons are confined by non-perturbative effects. Nonetheless, it has been conjectured that contributions of non-perturbative kinematic regions are suppressed by so-called "Sudakov factors", possibly yielding a final result which is still perturbatively calculable. c) There are many more reactions with related amplitudes than in the inclusive case, permitting a much more detailed investigation of the predictions.

II. Applicability of Perturbative QCD to Exclusive Scattering

The most persuasive evidence which we have so far of the applicability of perturbative QCD to large momentum transfer exclusive scattering is the remarkably good agreement for $t>5$ GeV^2 of the predicted power-law energy dependence [2], $d\sigma/dt \sim s^{2-n}$ (where n is the minimum number of quanta scattering), with experimental observations of the pion and proton form factors, Compton scattering, photoproduction, and meson-nucleon and nucleon-nucleon scattering. That $t>5$ GeV^2 is a reasonable value at which it should be possible to neglect higher twist terms in the amplitude can be seen by looking at the minimum invariant mass of the internal quarks in the hard scattering amplitude, which for a nucleon is typically $\sim 2/9\, t$. This invariant mass should be large compared to the confinement and quark mass scales of ~ 300 MeV, as it is for $t>5$ GeV^2. A more definitive assessment of the applicability of perturbative QCD to exclusive scattering will have to wait for more data [3].

III. Gamma Gamma Annihilation to Baryon-Antibaryon Pairs

The prediction [4,5] of the cross-sections for gamma gamma $\rightarrow$ baryon anti-baryon requires evaluating the six fundamental quark amplitudes shown in Figure 1. At large t a minimum of two gluons is required to accomplish the scattering. By gauge invariance, the gluons and photons must be inserted in all possible ways, so that in Born approximation the amplitudes, A_0, A_3, and A_4, and A_2 and A_5, are the sum of 48 and 68 diagrams, respectively. The quark amplitudes must be convoluted with the wavefunctions of the baryons. Asymptotically these are predicted by perturbative QCD [6], up to the overall normalization, just as the structure functions of deep-inelastic scattering are predicted asymptotically to approach a delta function at x=0. Fortunately, for many reactions of interest the predicted cross-sections are rather weakly dependent on the detailed form of the wavefunction. They also depend at the factor-of-two level on the behavior assumed for the QCD running coupling constant. For a given ansatz for the form of the wavefunction and coupling constant behavior, one computes the ratio $br(\psi \to p\bar{p})/br(\psi \to e^+e^-)$, and fixes the overall normalization by requiring it to have its measured value, .22%. The results [4] of such a computation are shown for non-running coupling constant $\alpha_s = .2$, and the equipartition and asymptotic wavefunctions, in Figure 2. Detailed tables of the predictions for other choices are given in ref. 4. These predictions are substantially different than those for gamma gamma $\to$ $p\bar{p}$ given previously by Damgaard [5], and cannot be attributed to differences in normalization procedure, choice of wavefunction and coupling constant, etc.: one of the calculations must be in error. Since the computer-calculation of ref. 4 incorporates many checks, such as U(1) and SU(3) gauge invariance, we are quite confident of its correctness. Final resolution of the issue awaits a third calculation.

As will be discussed in the lecture of Kuch (elsewhere in these proceedings), the experimental results on gamma gamma $\to$ baryon antibaryon at large t consist [8] of a measurement of gamma gamma $\to$ pp at $W_{\gamma\gamma}$ up to 2.8 GeV (i.e., t up to ~ 3 GeV^2) and upper limits on gamma gamma $\to$ $\Delta^{++} \bar{\Delta}^{++}$. The cross section for gamma gamma $\to$ $p\bar{p}$ is measured to be roughly isotropic. In the bin of largest $W_{\gamma\gamma}$, 2.4 GeV $< W_{\gamma\gamma} <$ 2.8 GeV, $d\sigma/d\cos\theta = 1 + .5$ nb for $|\cos \theta| < .6$. From our discussion in Section II above, it is clear that one should not expect the leading twist predictions of perturbative QCD to be very accurate at such a low value of t. However 1 nb is about a factor of ten larger than the QCD predictions shown in Figure 2. Can we account for this discrepancy?

One possibility, suggested in ref. 4, is that one or more of the 0^{-+} resonances seen in $N\bar{N}$ reactions, e.g., at 2190 and 2350 MeV, are being produced on top of the non-resonant perturbative QCD contribution. In addition to accounting for the observed isotropic angular distribution, such 0^{-+} states could not decay to $\pi^{+}\pi^{-}$, thereby explaining why the perturbative QCD prediction works reasonably well for $\gamma\gamma \to \pi^{+}\pi^{-}$. These resonances are below threshold for $\Delta\bar{\Delta}$ production, so that if this is the explanation for the large $\gamma\gamma \to p\bar{p}$ cross section, $\gamma\gamma \to \Delta\bar{\Delta}$ production should be at the perturbative QCD level ($.34$ nb for $|\cos\theta|<0.6$ and $W_{\gamma\gamma} = 2.6$ GeV) unless even higher-mass resonances decaying to $\Delta\bar{\Delta}$ are produced. Present experimental bounds of $\sigma_{.6} < 2$ nb therefore do not exclude this possibility. (Note however that the bound $\sigma(\gamma\gamma \to \Delta^{++}\bar{\Delta}^{++})/\sigma(\gamma\gamma \to p\bar{p}) < 3$ does exclude the possibility that the prediction of Figure 2 is correct except for an overall scaling factor.) Since a 0^{-+} resonance can in general decay to a pair of vector mesons, checking experimentally the perturbative QCD prediction for $\gamma\gamma \to \rho\rho$ [7] is another crucial means of deciding whether resonances are obscuring the perturbative component of gamma gamma exclusive annihilation.

Other than obtaining data extending to larger values of W and t, with finer resolution in W and study of the $\rho\rho$ final state to resolve the resonance question, what means do we have for determining whether perturbative QCD is applicable to gamma annihilation experiments at present energies? One important indicator is to observe the predicted s^{-5} energy dependence of the cross section for fixed angle. There are other means as well: if higher twist effects are negligible the baryon and antibaryon must have opposite helicities, and the amplitudes for producing a helicity $\pm 3/2$ baryon-antibyron pair should vanish for photons of like helicities [9]. If the flavor wavefunctions have their asymptotic SU(6) form even at modest W, then the fact that all gamma gamma $\to$ baryon anti-baryon reactions depend on the same set of six amplitudes means that there will be relations among the amplitudes for reactions of different flavors beyond those predicted by U-spin conservation alone [9]:

$$\text{Amp}[\gamma\gamma \to p\bar{\Delta}^{-}) = \sqrt{2}\,\text{Amp}[\gamma\gamma \to (p\bar{p} + n\bar{n} - 2\Lambda\bar{\Lambda} - \Sigma^{-}\bar{\Sigma}^{-})]$$

$$\text{Amp}[\gamma\gamma \to n\bar{\Delta}^{\circ}] = 2\sqrt{2}\,\text{Amp}[\gamma\gamma \to (n\bar{n} - \Lambda\bar{\Lambda})]$$

The above relations should hold for each helicity combination separately; they depend on the validity of neglecting components of the baryon wavefunctions which have "extra" $q\bar{q}$ pairs. However their validity is independent of the values of the amplitudes themselves. Thus for instance if the absolute theoretical predictions are incorrect (say because the guessed x-dependence of the wavefunctions is grossly incorrect, or the wrong form of the running coupling is being used, or even because the higher order corrections to the Born amplitudes are important), yet if leading twist perturbative QCD is in fact applicable and if the flavor wavefunctions have their asymptotic SU(6) form, these relations will be correct and can be (in principle) used to discriminate between lack of applicability of the theory, and poor calculational ability of the theorists [10]. The relations among reactions in gamma gamma $\rightarrow$ meson meson [7] provide a less powerful testing ground for the applicability of perturbative QCD. Nevertheless they ought to be examined experimentally because they can be very useful in deciding whether the apparent success of perturbative QCD in describing gamma gamma $\rightarrow \pi^{+}\pi^{-}$ is fortuitous.

For gamma gamma $\rightarrow$ baryon anti-baryon, unlike large t Compton scattering, photoproduction, and purely hadronic exclusive scattering, non-zero phases must come from higher order perturbative corrections to the Born amplitudes; thus relative phases provide a direct measure of the importance of higher order corrections. [9] If perturbative QCD with SU(6) for the baryon wavefunctions is determined to be applicable at practical energies, and experiments measuring initial and final polarizations are done, the above relations can be used to determine the relative phases of the six fundamental quark amplitudes. Alternatively, in principle, comparisons of cross-sections using linearly and circularly polarized photons could be used to determine some of the relative phases.

IV. Post-Conference Result on Gamma Gamma $\to$ p$\bar{\text{p}}$

Since the Workshop, I have explored another possible explanation for the "large" gamma gamma $\to$ p$\bar{\text{p}}$ cross section, namely its sensitivity to the choice of proton wavefunction. [11] Three ansatze for the x-dependence of the wavefunction were taken in ref. 4, xyz, $(xyz)^2$, and $\delta(1/3-x)\ \delta(1/3-y)\ \delta(1/3-z)$. While they differ substantially in the spread of x, y, and z about their mean, they are all symmetric in x, y, and z, and none of them gives the observed signs and magnitudes of both the proton and neutron form factors. On the other hand Cernyak and Zhitnitsky [3] have proposed a wavefunction, including normalization, which they derived from QCD sum rules; using $\alpha_s = .3$, they claim to get good agreement with the data for $\psi \to$ p$\bar{\text{p}}$ and the nucleon magnetic form factors. I have taken their wavefunctions and convoluted them with the QCD Born amplitudes. The resulting predict-ions for gamma gamma $\to$ p$\bar{\text{p}}$ are shown in Figure 3 for a variety of choices of coupling constant dependence. Defining $c1 = 4\Lambda^2/s$ ($c1 \equiv c$ in the notation of ref. 4) and $c2 = \alpha_{max}$, we have $c1 = 0.016$ for $\Lambda_{QCD} = 0.2$ GeV and $s = 10$ GeV^2. To select the most reasonable choice of coupling constant dependence, and to assess the credibility of the CZ wavefunction, I calculated, also in Born approximation, $br(\psi \to p\bar{\text{p}})/br(\psi$-$e^+e^-)$ and G_M^P and G_M^n. The choice $c1 = 0.016$, $c2 = 0.5$ gives good agreement with the experimental values of these quantities; it is better than the fixed $\alpha_s = .3$ used by Cernyak and Zhitnitsky (presumably for simplicity). Therefore, in the following, results quoted for the CZ wavefunction are for the choice $c1 = 0.016$, $c2 = 0.5$; this is the solid curve of Figure 3.

Integrating the CZ prediction for $d\sigma/d\cos\theta(\gamma\gamma\to$p$\bar{\text{p}})$ over $|\cos\theta|<0.6$ gives $\sigma = 0.17$ nb at $W_{\gamma\gamma} = 2.4$ GeV. This is much closer to the observed cross-section than the result of the symmetric wavefunctions of Figure 2, and within two standard deviations of the experimental number, due to the large error bars. It should be emphasized that the improved agreement between theory and the experimental results is not due to having "cranked up" the coupling constant enough to suitably increase the prediction: in the method of refs. 4 and 5, where $\psi \to p\bar{p}$ is used to normalize the wavefunction, increasing the coupling constant actually decreases the prediction for $\gamma\gamma \to p\bar{p}$, because $\sigma(\gamma\gamma \to p\bar{p})/\Gamma(\psi \to p\bar{p}) \sim \alpha_s^{-2}$. Note that the prediction of the CZ wave-function with fixed $\alpha_s = .3$ is smaller by a factor of six or more than is obtained with running α, taking $c1 = 0.016$ and $c2 = 0.5$. This indicates that with the Cernyak-Zhitnitsky wavefunction, quarks carrying small momentum fractions make an important contribution. We cannot yet say on purely theoretical grounds whether the Born approximation should be adequate in this case.

Unfortunately Cernyak and Zhitnitsky have not given a sum-rule wavefunction for the Δ^{++}, so that at present I am unable to give the "CZ" prediction for $\gamma\gamma \to \Delta^{++}\Delta^{++}$. If the ratio of this cross-section to that for $\gamma\gamma \to p\bar{p}$ is 16, as naively obtained from the ratio of the charges to the fourth power, the cross-section at $W_{\gamma\gamma} = 3$ GeV, integrated over $|\cos\theta| < 0.6$, is predicted to be 1.7 nb. This is just at the upper limit of ref. 4. Therefore measuring this cross-section to compare to the theoretical prediction with the sum-rule wave-function for the Δ^{++} will provide another important test of whether perturbative QCD applies to exclusive gamma gamma annihilation to baryon anti-baryon pairs.

Given the impressive success of the CZ wavefunction for the form factors and $\psi \to p\bar{p}$, I think the present result should be taken very seriously. Possibly perturbative QCD actually is accounting for the measurements, even at $t < 3$ GeV^2! Improving the experimental error bars on the $\gamma\gamma \to p\bar{p}$ cross-section by even a factor of two could prove decisive.

References

1. See refs. 4 and 9 for a detailed set of references.

2. S.J. Brodsky and G.R. Farrar, Phys. Rev. Lett. 31,1153 (1973) and Phys. Rev. D11,1309 (1975); V.A. Matveev, R.M. Muradyan, and A.V. Tavkhelidze, Lett. Nuovo Cimento 7,719 (1973).

3. Recently N. Isgur and C.H. Llewelyn-Smith, Phys. Rev. Lett. 52, 1080 (1984), have questioned the applicability of perturbative QCD to the proton form factor for $t < 25$ GeV, by attempting to place an upper bound on the perturbative QCD contribution which they find to be 1-2 orders of magnitude below experiment. The proton form-factor is a difficult case to predict reliably, because at the Born-approximation there is an exact cancellation between two terms when the asymptotic wavefunction is used, giving a net zero result. This cancellation disappears when any of the corrections are included, such as permitting the coupling constants to "run", using a different form for the wavefunctions, or keeping the one-loop corrections to be Born amplitude. Thus the Isgur-Llewelyn-Smith "bound" is highly model-dependent. Using QCD sum rules V.L. Cernyak and I.R. Zhitnitsky, Nucl. Phys. B246,52 (1984), have determined the proton wavefunction appropriate for experimentally relevant values of t; with that wavefunction both the proton and neutron form factors are in quite good agreement with experiment.

4. G.R. Farrar, E. Maina, and F. Neri, Rutgers University preprint RU-84-13.

5. P. Damgaard, Nucl. Phys. B211,435 (1983).

6. S.J. Brodsky and G.P. Lepage, Phys. Rev. D22,2157 (1980).

7. S.J. Brodsky and G.P. Lepage, Phys. Rev. D24,1808 (1981).

8. TASSO collaboration, M. Althoff et al., Phys. Lett. 130B,449 (1983) and DESY preprint 84-015.

9. G.R. Farrar, Phys. Rev. Lett. 53,28 (1984).

10. Since the C-Z wavefunction does not have the asymptotic SU(6) form, it will be interesting to see how much the relations above are modified when baryon wavefunctions are taken from QCD sum rules.

11. Details of this and further work on this problem will be published elsewhere.

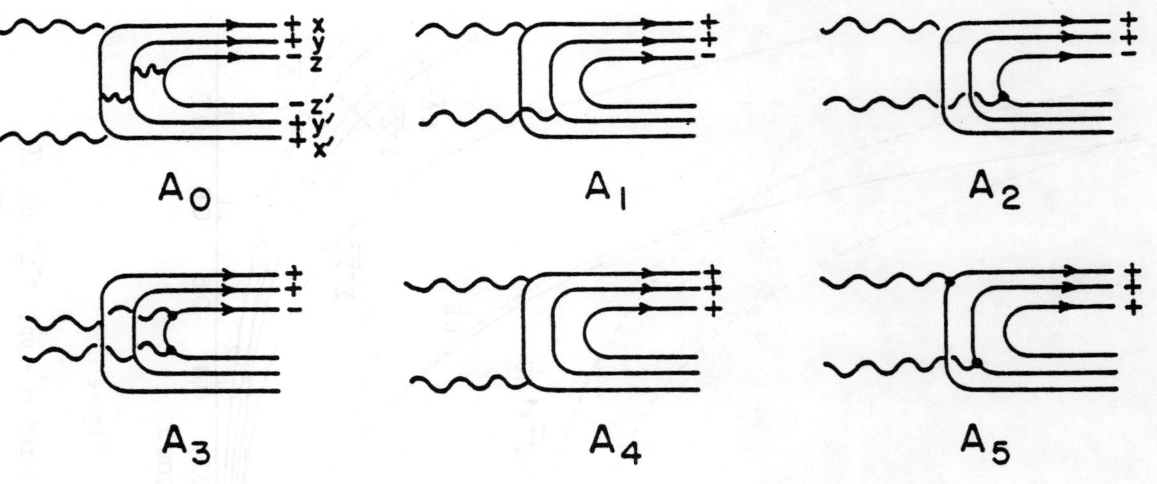

Figure 1

Independent quark amplitudes necessary to completely determine

$\gamma\gamma \to B_8\overline{B}_8$, $B_8\overline{B}_{10}$, and $B_{10}\overline{B}_{10}$, with photon symmetrization and

gluon connections understood. The +(-) labels the chirality of

the fermion line.

38

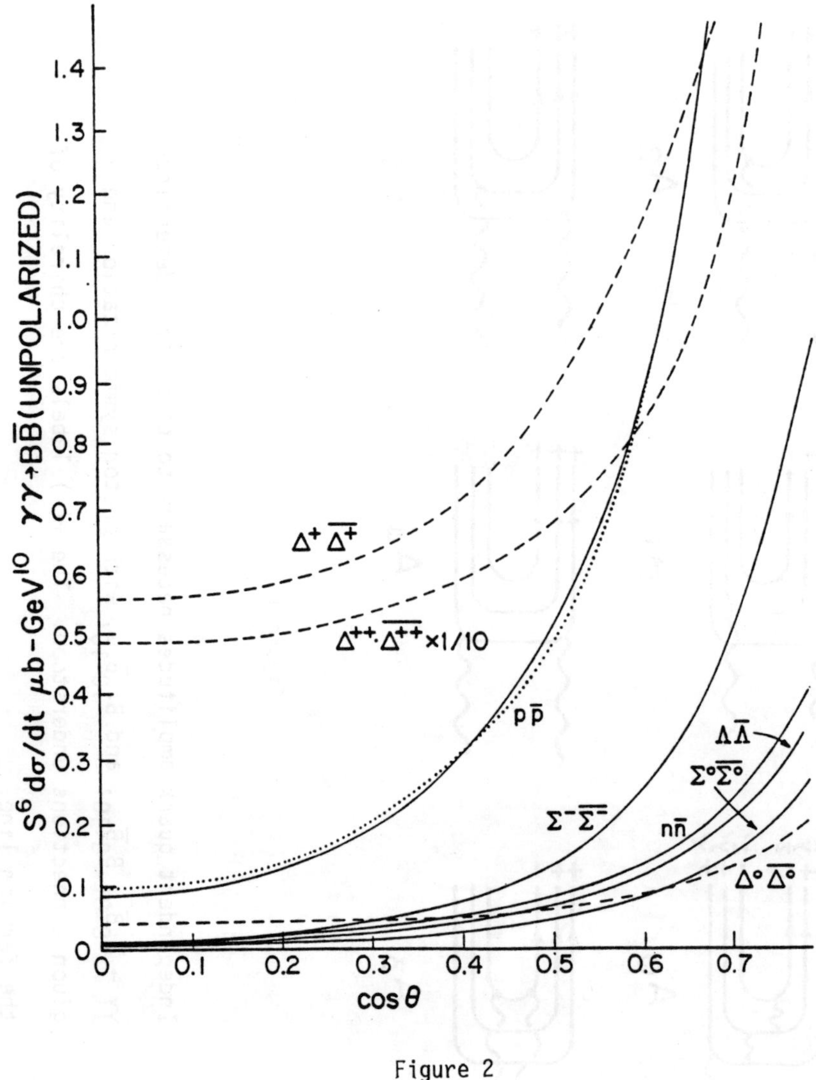

Figure 2

$s^6 \frac{d\sigma}{dt}$ in μb-GeV^{10} for unpolarized $\gamma\gamma \to p\bar{p}$, $\Sigma^-\overline{\Sigma^-}$, $n\bar{n}$, $\Lambda\bar{\Lambda}$, $\Sigma^0\overline{\Sigma^0}$, $\Delta^0\overline{\Delta^0}$, $\Delta^+\overline{\Delta^+}$ and (reduced by factor of 10) $\Delta^{++}\overline{\Delta^{++}}$, using the "equipartition" wave function.[f5] The prediction using the "asymptotic" wavefunction is shown for $p\bar{p}$, with a dotted line.

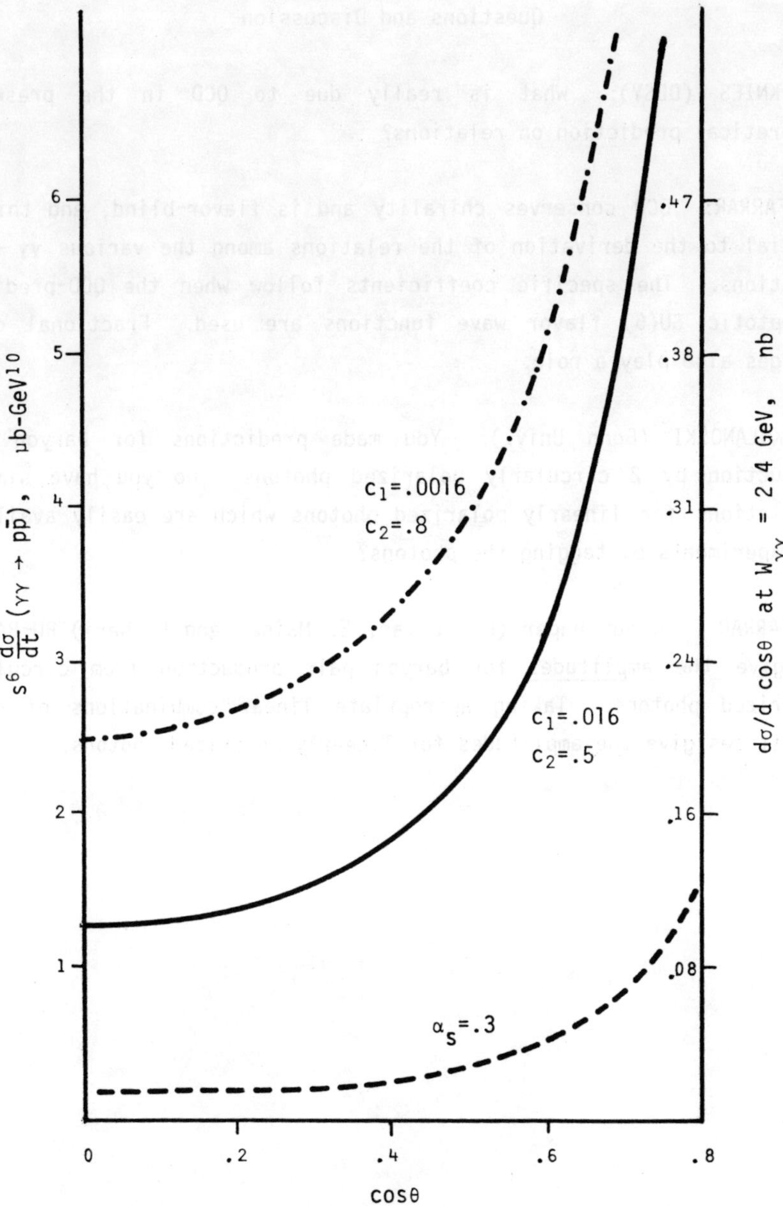

$s^6 \frac{d\sigma}{dt}$ ($\gamma\gamma \to p\bar{p}$), $\mu b\text{-GeV}^{10}$
6
5
4
3
2
1
c_1=.0016
c_2=.8
c_1=.016
c_2=.5
α_s=.3
.47
.38
.31
.24
.16
.08
$d\sigma/d\cos\theta$ at $W_{\gamma\gamma}$ = 2.4 GeV, nb
0   .2   .4   .6   .8
$\cos\theta$

Figure 3

Questions and Discussion

G. KNIES (DESY): What is really due to QCD in the presented theoretical prediction on relations?

G. FARRAR: QCD conserves chirality and is flavor-blind, and this is crucial to the derivation of the relations among the various $\gamma\gamma \to B\bar{B}'$ reactions. The specific coefficients follow when the QCD-predicted asymptotic SU(6) flavor wave functions are used. Fractional quark charges also play a role.

H. KOLANOSKI (Bonn Univ.): You made predictions for baryon pair production by 2 circularly polarized photons. Do you have similar predictions for linearly polarized photons which are easily available in experiments by tagging the photons?

G. FARRAR: In our paper (G. Farrar, E. Maina, and F. Neri) RU-84-13, we give the _amplitudes_ for baryon pair production from circularly polarized photons. Taking appropriate linear combinations of these amplitudes give the amplitudes for linearly polarized photons.

EXPERIMENTAL RESULTS ON HADRON PAIR PRODUCTION BY TWO PHOTONS

Heinz Kück

Physikalisches Institut der Universität Bonn, FRG

ABSTRACT

This report reviews the experimental results on non-resonant exclusive production of hadron pairs in photon-photon collisions. The cross section for charged pion pairs in the invariant mass range below the f resonance is discussed. The measured cross sections show no evidence for scalar resonances. At momentum transfer $|t| \gtrsim 1.6$ GeV^2 the sum of the cross sections for charged pion and kaon pairs are found to be in remarkable agreement with a theoretical model based on perturbative QCD. Cross sections for proton-antiproton production and upper limits on $\Lambda\bar{\Lambda}, \Delta\bar{\Delta}$ production are discussed.

Introduction

This report reviews the experimental results on non-resonant production of hadron pairs in photon-photon collisions. By now there is a lot of experimental data on the exclusive production of pseudoscalar mesons, vector mesons and baryons. Most of the results have been worked out in order to study the production or formation of resonances in photon-photon collisions. These aspects of hadron production and the physical implications are discussed by A.Cordier and F.Erné at this conference. The intention of this report is to review the non-resonant production of hadron pairs. Three topics will be covered:

- production of charged pions at low invariant mass
- production of pions and kaons at large $|t|$
- production of baryon pairs.

I. Production of charged pion pairs at low invariant mass

The production of pion pairs near threshold plays an important role for the search for scalar resonances which are predicted from quark models[1] and suggested by $\pi\pi$-phase shift analyses[2]. However the search for resonances, especially if they are broad, is complicated by possible interference of the continuum with the resonant amplitude. Therefore good knowledge of the non-resonant part of the amplitudes is crucial for the separation of the resonance contribution to the observed cross section.

The simplest physical model describing charged pion production is given by the QED Born diagrams of Fig.1. The corresponding cross section is shown in Fig.2 for the two possible helicity contributions ($\lambda = 0, 2$ corresponding to photons with antiparallel, parallel spins). The $\lambda = 0$ amplitude, where scalar resonances can contribute, dominates the cross section at threshold and decreases rapidly with increasing c.m. energy $W_{\gamma\gamma}$. The $\lambda = 2$ contribution dominates the cross section above ~ 0.5 GeV and makes up roughly 90 % of the cross section at 0.65 GeV.

This simple model must be considered as an approximation because strong interaction has not been taken into account. Attempts have been made to extend the theoretical description beyond the Born approximation in order to include resonances properly[3,4]. Mennessier[4] has done a coupled channel analysis of $\pi\pi$ and $K\bar{K}$ production as visualized in Fig.3. In this model vector meson exchange is taken into account and the amplitudes are unitarized by including $\pi\pi$ and $K\bar{K}$ final state interactions as given by the phase shifts. In addition the direct production of resonances, which is proportional to their $\gamma\gamma$-width, is taken into account. The outcome of the calculation is that the modification of the pure Born approximation is small for invariant masses below the f resonance. This is because the $J \geq 2$ $\pi\pi$-waves are small below the f and because the $\lambda = 0$ Born term dies out rapidly with increasing mass. Furthermore no f peak at all is expected without the direct resonance contribution, i.e. for vanishing $\gamma\gamma$-width. The resonance peak is predicted to be shifted due to the interference of the direct contribution with the $\lambda = 2$ continuum amplitude.

There is by now plenty of experimental data on the reaction $\gamma\gamma \rightarrow \pi^+\pi^-$ [5] in the invariant mass range around the f resonance. The experiments have detected the $\pi^+\pi^-$ final state in the central detectors and the trigger conditions restrict the c.m. energy to values typically larger than 0.7 GeV. The large background from lepton pair production, i.e. $\gamma\gamma \rightarrow e^+e^-$, $\gamma\gamma \rightarrow \mu^+\mu^-$, is subtracted on a statistical basis in most cases. Latest results are described in detail in the report by A.Cordier. The common features are: there is direct f production corresponding to a $\gamma\gamma$-width of about 3 keV; the Mennessier model fits the data well; the Born approximation with an f contribution interfering with the $\lambda = 2$ amplitude also fits the data well; there is no need for scalar resonances. As an example the invariant mass distribution obtained by the CELLO experiment together with a fit according to Mennessiers model is shown in Fig. 4.

In the analysis of single and double tag events for $\pi^+\pi^-$ invariant masses in the range between threshold and ~0.7 GeV, the DM1 group[6] has found a cross section, which is roughly twice as large as the Born term. They have confirmed this result using data taken with the DM2 detector[7]. They can explain the $\pi^+\pi^-$ yield by the model of Mennessier including the contribution of a scalar resonance ε with the preliminary parameters: $M_\varepsilon \sim 0.82$ GeV, $\Gamma_\varepsilon \sim 0.34$ GeV and $\Gamma_\varepsilon^{\gamma\gamma} = 15 \pm 5$ keV.

The PLUTO group has published a new measurement of $\gamma\gamma \to \pi^+\pi^-$ for invariant masses as far down as 0.36 GeV [8]. The $\pi^+\pi^-$ system was detected in the forward spectrometer of the PLUTO detector. In the forward direction the particle momenta are large due to the Lorentz boost of the $\gamma\gamma$-system allowing the invariant mass to be small. Because of the high momenta they can use the information of the tracking device, the shower counters and the muon chambers to identify electron and muon pairs and exclude them from the two prong data sample. The contribution of K^+K^- pairs to the final data sample is estimated to be smaller than 5 %. The background subtracted invariant mass distribution for $\pi^+\pi^-$ pairs is shown as the shaded area in Fig. 5. Due to the finite mass resolution of the forward spectrometer the f resonance appears as a broad shoulder around 1.25 GeV. The background from beam gas events and inclusive one photon annihilation events is given by the histogram on top and shows a signal of ρ^0 electroproduction. The double shaded area represents the calculated background from two photon processes with additional undetected particles. To extract a cross section the PLUTO group has refitted the invariant mass constraining the net transverse momentum of the event to be less than 0.015 GeV. The $\pi^+\pi^-$ cross section is obtained by normalzing to the measured rate of $\mu^+\mu^-$ pairs thus minimizing systematic uncertainties. The cross section as plotted in Fig. 6 has been extrapolated to $\cos\Theta = 0$, where Θ is the scattering angle in the $\gamma\gamma$ c.m. system. After refitting the data show a clear f signal. Curve (1) in Fig. 6 shows the prediction of the Mennessier model while curve (2) corresponds to the Born approximation with the f resonance interfering with the $\lambda = 2$ amplitude. Both models give an acceptable fit to the data for $m_{\pi\pi} \gtrsim 0.8$ GeV. The systematic shift of

the data points in the f region towards larger $m_{\pi\pi}$ values is attributed to the mass reconstruction with the forward spectrometers.

At roughly ~ 0.6 GeV the data have a minimum, i.e. the average
cross section in the $m_{\pi\pi}$ range from 0.52 to 0.76 GeV is lower by $\approx 7\sigma$
than the prediction of either model (~115 nb). Though not yet
understood the minimum cannot be explained by a destructive
interference with a scalar resonance. A scalar resonance can only
contribute to the $\lambda = 0$ amplitude which makes up only ~10 % of the
Born cross section at these mass values. The minimum is in clear
contradiction to the DM1 results and the preliminary DM2 results.
The first two data points in Fig. 6 lie far above the theoretical
curves (~130 nb). Combining both points and adding the statistical
and systematic errors the access is ~2σ.

The experimental results on $\gamma\gamma \to \pi^+\pi^-$ in the mass range below
the f resonance might point to unexpected phenomena. The question
of scalar resonances is not yet settled. An independent measurement
with good statistics is needed.

II. Production of pions and kaons at large $|t|$

At high momentum transfer squared $|t|$ between the photons and
the hadrons, i.e. at high c.m. energies and large scattering angles,
the internal structure of the hadrons should become apparent. In
the model of Brodsky and Lepage[9] the amplitude for the exclusive
production of meson pairs by two photons is factorized into two
parts (see Fig. 7): two quark-antiquark pairs are generated in a
hard scattering process and subsequently fragment into physical
mesons. The fragmentation process is described by the distribution
function $\Phi(x,Q^2)$ corresponding to the probability amplitude that
the quark and antiquark carry the momentum fraction x and 1-x of
the meson. The distribution function is assumed to be independent
of the process under consideration and to depend only logarithmically on the c.m. energy. The absolute normalization of $\Phi(x,Q^2)$ is
related to the measured pion form factor. The hard scattering

measure the sum of the pion pair rate and the kaon pair rate. The PLUTO group found a total of 15 events in the invariant mass range above 2.0 GeV. The measured the cross section at $\cos\Theta = 0$ for $\gamma\gamma \to h^+h^-$ assuming a π : K-ratio of 1:1. The cross section is plotted in Fig. 10 versus the particle momentum in the c.m. system, p_{cm}, rather than $W_{\gamma\gamma}$, in order to minimize the uncertainties due to the unknown particle masses. The full curve in Fig. 10 is the QCD prediction of Brodsky and Lepage. The dashed curve is the best fit of the QCD model shape to the data. The difference is roughly 40 %. Keeping in mind that the average momentum transfer is only $|t| = 2.1$ GeV^2, the model is in rough agreement with the data.

The MARK II group gives the cross section for $\gamma\gamma \to \pi^+\pi^- + K^+K^-$ integrated over $|\cos\Theta| < 0.3$ as a function of the invariant mass assuming pion masses for all particles. The data correspond roughly to 50 events with $m_{\pi^+\pi^-}$ between 1.6 and 2.5 GeV after background subtraction. The cross section is shown in Fig. 11 together with the prediction of the QCD model for $\sigma(\gamma\gamma \to \pi^+\pi^- + K^+K^-)$. The model is in very good agreement with the data although the momentum transfer is even smaller on average than for the PLUTO data. The MARK II[12] group has confirmed their earlier result with increased statistics by roughly a factor of 7 and by extending the mass range up to $m_{\pi^+\pi^-} = 3.5$ GeV. Their preliminary data are shown in Fig. 12. The QCD model is a remarkably good description of the absolute normalization and s dependence of the integrated cross section down to $m_{\pi^+\pi^-}$ values as small as ~ 1.8 GeV, which means momentum transfers as small as ~ 1.6 GeV^2. This result is another indication that in photon induced reactions hard scattering shows up at significantly lower values of momentum transfer than in hadron induced processes.

III. Production of baryon pairs

Given the presently available luminosities, the study of baryon pairs by two photons is restricted to $W_{\gamma\gamma} \lesssim 4$ GeV. This restriction is mainly due to the decrease of the $\gamma\gamma$-flux with increasing $W_{\gamma\gamma}$ and to the need for identifying protons. Fig. 13 shows the effective four momentum transfer ($|t_{eff}| = |t - t (\cos\Theta = 1)|$) as a function of $W_{\gamma\gamma}$ for proton-antiproton production at $\cos\Theta = 0$. The curve crosses the line $|t_{eff}| = 1.0$ GeV at $W_{\gamma\gamma} \approx 2.4$ GeV. In the $W_{\gamma\gamma}$ range from 2.5 to 3.0 GeV the momentum transfer is of the same magnitude as in the measurements of $\gamma\gamma \rightarrow \pi^+\pi^- + K^+K^-$ discussed above, although in the case of $p\bar{p}$ production this region is close to threshold. Therefore the QED Born term (Fig. 14) cannot be expected to be a reasonable description of $\gamma\gamma \rightarrow p\bar{p}$ in this region, because an extended object far off mass shell is exchanged. One is led to compare to QCD calculations similar to those for $\pi^+\pi^-$, K^+K^- production, which described the measured cross section quite well. As discussed above, the cross section for an exclusive reaction involving two photons and two protons should scale with s^{-6}. The first experimental indication for this scaling law to hold came from data on Proton Compton Scattering, the crossed channel of $\gamma\gamma \rightarrow p\bar{p}$. The differential cross section for $\gamma p \rightarrow \gamma p$ measured by three experiments[13] is given in Fig. 15. Plotted is $s^6 \, d\sigma/dt$ versus $\cos\Theta$ for s values between 3.9 and 12.1 GeV^2. While the quantity s^6 varies by roughly 3 orders of magnitude the experimental data are spread in a band which is at most a factor of 3 wide.

The TASSO collaboration[14] published angular distributions for $\gamma\gamma \rightarrow p\bar{p}$ for $W_{\gamma\gamma}$ between 2.0 and 2.8 GeV (see Fig. 16). The measured cross section is compatible with a flat angular distribution. The data are about one order of magnitude lower than the QED Born approximation with pointlike couplings (dashed curve). This discrepancy becomes even larger, if the anomalous magnetic moment of the proton is included in the calculation[15]. The shaded band in Fig. 16 gives the absolute prediction of a QCD model calculation done by Damgaard[16]. In the bins of largest momentum transfer the prediction is below the data by a factor of roughly 4. Taking into account an overall uncertainty of a factor of two for the amplitude[17], this can be interpreted as rough agreement with the data. Farrar and collaborators[18] subsequently carried out a more elaborate calculation for the production of octet and decuplet baryon pairs by two photons based on the same QCD model. The results are given in Fig. 17. The two curves for $\gamma\gamma \to p\bar{p}$ corrspond to two different quark distribution functions and demonstrate that the cross section depends only weakly on the choice of the distribution function. The calculated cross section for $\gamma\gamma \to p\bar{p}$ of Farrar et al. is a factor of 20 to 100 lower than that computed by Damgaard. There are good reasons for the calculation of Ref. 18 to be reliable as discussed in detail in the report by Farrar at this conference. However, the cross section for $p\bar{p}$ production below 3.0 GeV predicted by these authors is smaller than the data by roughly two orders of magnitude.

New experimental results on proton-antiproton production have been presented to this conference by the JADE[19] and the PEP4-PEP9[20] groups. The Jade experiment identified protons via dE/dx measurements in their jet chamber in the $W_{\gamma\gamma}$ range from threshold to 2.6 GeV. The preliminary cross section values integrated over $|\cos\Theta| < 0.6$ are given in Fig. 18 (full points). The data must be considered as lower bounds of the cross section, since the systematic uncertainties introduced by antiproton annihilation in the JADE detector are still under study. The final corrections could bring up the data points by as much as 50 %. For comparison the TASSO data are also shown in Fig. 18 (open points). The data are consistent with one another.

The PEP4-PEP9 group[20] has done a single tag measurement of $\gamma\gamma \to p\bar{p}$ in the Q^2-range from 0.1 to 7.0 GeV^2. They identify protons either by a dE/dx measurement in the TPC or by a time-of-flight measurement in the forward spectrometers. One particle is required to be a sure proton and the other must be compatible with the proton hypothesis. They find 15 events in the $W_{\gamma\gamma}$ range between threshold and 4.0 GeV. The uncorrected angular distribution of the detected $p\bar{p}$ events is shown for two $W_{\gamma\gamma}$ bins in Fig. 19. The full curve is the detection efficiency for tagged $p\bar{p}$ events (i.e. not including the tagging efficiency). The detector acceptance is reasonably large over the full angular range $|\cos\Theta| \leq 1.0$. The corresponding cross section is shown in Fig. 20. Figure 21a shows the Q^2 dependence of the cross section for the whole W range. In Fig. 21b the Q^2 dependence for the $W_{\gamma\gamma}$ range from 2.0 to 2.9 GeV is plotted including the $Q^2 \approx 0$ point of TASSO. The full curve corresponds to a pole form factor and the dashed curve is the QED expectation for Dirac protons. Both curves are normalized to the TASSO point. Within the limited statistics the measured Q^2 dependence appears to be extremely flat. However, before drawing conclusions from this result, the flat Q^2 dependence has to be confirmed with higher statistics.

The TASSO collaboration has searched for two photon production of $\Lambda\bar{\Lambda}$, $\Delta^{++}\overline{\Delta^{++}}$, $\Delta^0\overline{\Delta^0}$ via the reaction $\gamma\gamma \rightarrow pp\pi^+\pi^-$ [21]. Particle identification is done by a time-of-flight measurement. They find 15 events with an estimated background of one event. Fig. 22 (23) shows the correlation of the invariant masses of the two neutral (doubly charged) pion proton systems. No $\Lambda\bar{\Lambda}$ candidate is observed and there is no evidence for $\Delta^{++}\overline{\Delta^{++}}$ production. The corresponding 95 % c.l. upper limits for the integrated cross section $(|\cos\Theta| < 0.6)$ are given in Fig. 24. The most restrictive upper limit for a comparison with the QCD model prediction [18] is for $\Delta^{++}\overline{\Delta^{++}}$ production in the $W_{\gamma\gamma}$ range from 2.5 to 3.0 GeV. In this bin the measured upper limit is larger than the theoretical cross section by roughly a factor of 30. Thus the measured upper limits for $\Lambda\bar{\Lambda}$, $\Delta^{++}\overline{\Delta^{++}}$, $\Delta^0\overline{\Delta^0}$ production are not in conflict with QCD model results.

On the other hand the measured $p\bar{p}$ rate is large compared to the QCD model cross sections. It is also large when compared to the upper limit on the $\Delta^{++}\overline{\Delta^{++}}$ event yield taking the $\Delta^{++}\overline{\Delta^{++}}$ to $p\bar{p}$ ratio from the QCD model. Therefore one might suspect the large cross section for $p\bar{p}$ to be due to resonance production. Fig. 25 shows the invariant mass distribution of the TASSO $p\bar{p}$ events. The chosen bin width roughly corresponds to the experimental mass resolution. The data are compared to Breit-Wigner shapes with parameters as observed for the so called T,U regions in $p\bar{p}$ scattering [22]. Within the present experimental statistics the question if the $p\bar{p}$ cross section is dominated by resonances cannot be settled.

Summary

In the mass range around the f resonance the measured cross sections for $\gamma\gamma \to \pi^+\pi^-$ are well described by the model of Mennesier. An equally good description is however given by the Born approximation with an f resonance contribution interfering with the $\lambda = 2$ amplitude. New data for masses down to 0.36 GeV show a minimum of the cross section at $m_{\pi\pi} \sim 0.6$ GeV. No evidence is found for the formation of a scalar resonance in the mass range from 0.5 to 0.8 GeV, in disagreement with earlier published data. In the energy range above threshold further measurements are highly desirable.

In the mass range above 1.8 GeV, i.e. for momentum transfer $|t| \gtrsim 1.6$ GeV^2, the production rate of charged pion pairs and charged kaon pairs has been measured with good statistics. The measured cross sections show the power law dependence on s and the absolute magnitude as predicted by the QCD model of Brodsky and Lepage. Thus pion and kaon pair production by two photons is in remarkable agreement with the hard scattering picture for momentum transfers as low as ~ 2.0 GeV^2.

The $p\bar{p}$ production rate between threshold and 4.0 GeV exceeds by far the values calculated in the QCD model by Farrar et al., although the momentum transfer is of the same magnitude as for $\gamma\gamma \to \pi^+\pi^- + K^+K^-$. The early $p\bar{p}$ cross section measurements are confirmed by two recent experiments. There has been a first look at the Q^2 dependence of $p\bar{p}$ production. Within the limited experimental statistics the cross section shows an extremely flat Q^2 dependence. No evidence for $\Lambda\bar{\Lambda}$, $\Delta^{++}\bar{\Delta}^{++}$, $\Delta^0\bar{\Delta}^0$ production has been found in the channel $p\bar{p}\pi^+\pi^-$. The $p\bar{p}$ cross section appears to be large as compared to the limits on the $\Delta^{++}\bar{\Delta}^{++}$ yield. The question whether $p\bar{p}$ resonances play a role cannot be answered with the present statistics.

Exclusive hadron pair production by two photons at high momentum transfer is a comparatively young field of photon-photon physics. Many interesting physics questions wait for an answer. The experimentalists are prepared to give most of the answers if the luminosities at PETRA and PEP could be increased by about an order of magnitude.

Acknowledgement

I am indebted to many colleagues for valuable discussions and
support in preparing this talk. In particular I want to mention here:
P. Bussey, E.H.Hilger, R.Kellog, H.Kolanoski, J.Olsson, D.Pandoulas,
B.Schrempp, F.Schrempp, R.Wedemeyer and my colleagues from F12.
I am very grateful to R.Lander and his staff for organizing this
interesting and enjoyable conference.

process is calculated in leading order QCD. The energy dependence is given by the number of elementary fields (quarks and photons) involved in the scattering process. According to the dimensional counting rule the cross section is

$$\frac{d\sigma}{dt} = \frac{1}{s^k} \cdot f(\Theta)$$

where k = 4 for meson pairs and k = 6 for baryon pairs. The function f gives the angular dependence. As the asymptotic distribution functions might not be applicable at finite values of $|t|$ the authors calculated the cross sections for $\gamma\gamma \to \pi^+\pi^-$, $\gamma\gamma \to \pi^0\pi^0$ for three different x-dependences of Φ (see Fig. 8). The calculated cross section for charged pions is nearly independent of the choice of Φ, while the cross section for neutral pions strongly depends on Φ. This suggests to check the absolute normalization as well as the Θ and s dependence of the model prediction using data on charged pion pairs and then decide on the distribution function, i.e. on the dynamics of the quarks in the mesons, by measuring the cross section for $\gamma\gamma \to \pi^0\pi^0$. Assuming that the quark content of mesons is given by flavour symmetry and that Φ is flavour independent the cross sections for the various pseudoscalar and vector mesons are related to the cross section for pion pairs. The predicted cross sections for the various meson pairs, integrated over $|\cos\Theta| < 0.3$, are shown in Fig. 9 as a function of $W_{\gamma\gamma}$. The solid line indicates the cross section that would result in 10 events observed in a $\Delta W_{\gamma\gamma} = 0.5$ GeV bin for beam energies, luminosities and efficiencies typical for PETRA/PEP experiments (17 GeV, 100 pb^{-1}, 5%). For a measurement of $\pi^0\pi^0$ and K^0K^0 production above 2.5 GeV an unrealistic increase in the integrated luminosity is necessary, while data on $\pi^+\pi^-$ and K^+K^- pairs are available by now.

The PLUTO group[10] and the MARK II group[11] have published data corresponding to L = 42 pb^{-1} and 35 pb^{-1}, respectively. The MARK II group has presented preliminary data from 230 pb^{-1} at this conference[12]. Both experiments use the central detectors, shower counters and the muon systems to exclude electron and muon pairs from their samples of hadron pairs. The $p\bar{p}$ event rate is estimated to be negligible. As they do not identify pions and kaons, the experiments

References

1) L.Montanet, Rep.Prog.Phys. 46 (1983) 337 and references therein

2) Particle Data Group, 1984 edition

3) D.H.Lyth, Nucl.Phys. B30 (1971) 195;
    G.Schierholz and K.Sundermeyer, Nucl.Phys. B40 (1972) 125;
    O.Babelon et al., Nucl.Phys. B113 (1976) 445

4) G.Mennessier, Z.Phys. C16 (1983) 241

5) PLUTO Collaboration, Ch.Berger et al., Phys.Lett. 94B (1980) 254;
    TASSO Collaboration, R.Brandelik et al., Z.Phys. C10 (1981) 117;
    A.Roussarie et al., Phys.Lett. 105B (1981) 304;
    J.Smith et al., SLAC-PUB-3205 (1983);
    CELLO Collaboration, H.J.Behrend et al., Z.Phys. C23 (1983) 223;
    A.Courau et al., SLAC-PUB-3362 (1984);
    PEP-4/PEP-9 Two Gamma Collaboration, Contribution to the XXII
    International Conference on High Energy Physics, Leipzig,
    July 19-25, 1984

6) A.Courau et al., Phys.Lett. 96B (1980) 402;
    R.Wedemeyer, Proceedings of the 10th Intern. Symposium on Lepton
    and Photon Interactions at High Energies, Bonn (1981),
    ed. W.Pfeil

7) A.Courau et al., paper contributed to this conference

8) PLUTO Collaboration, Ch.Berger et al., DESY 84-074 (1984)

9) S.J.Brodsky, G.P.Lepage, Phys.Rev.D24 (1981) 1808

10) PLUTO Collaboration, Ch.Berger et al., Phys.Lett. 137D (1984) 267

11) J.Smith et al., SLAC-PUB-3205 (1983)

12) MARK II Collaboration, paper contributed to this conference

13) M.Deutsch et al., Phys.Rev. D8 (1973) 3828
    M.Shupe et al., Phys.Rev. D19 (1979) 1921
    J.Duda et al., Z.Phys. C17 (1983) 319

14) TASSO Collaboration, M.Althoff et al., Phys.Lett. 130B (1983) 449

15) N.Arteaga-Romero, Seminar on $\gamma\gamma$-Physics, LPC 182-14, Paris (1982)

16) P.H.Damgaard, Nucl.Phys. B211 (1983) 435

17) P.H.Damgaard, private communication

18) G.R.Farrar et al., Rutgers University RU-83-3 and RU-84-13
19) JADE Collaboration, paper contributed to this conference
20) PEP-4/PEP-9 Two Gamma Collaboration, paper contributed to this
     conference
21) TASSO Collaboration, M.Althoff et al., Phys.Lett. 142 (1984) 135
22) For a list of references see Particle Data Group

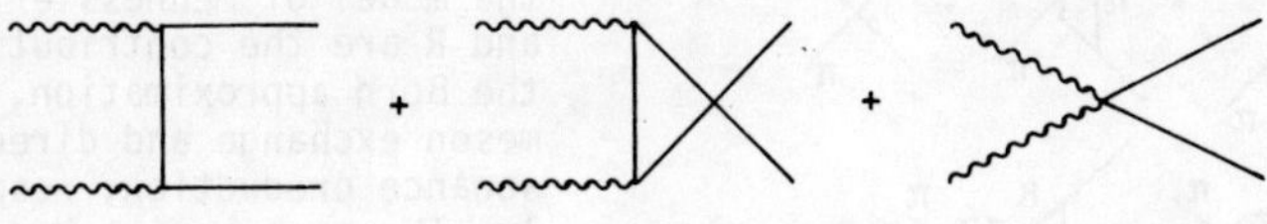

Fig. 1 Born diagrams for two-photon production of a pair
of charged scalar particles.

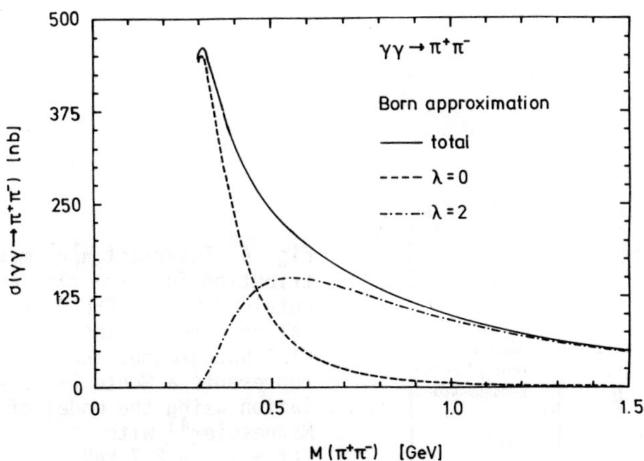

Fig. 2 Cross section for $\gamma\gamma \to \pi^+\pi^-$ in Born approximation. The curves show the cross sections for $\gamma\gamma$ helicities
$\lambda = 0, 2$ and the sum of both.

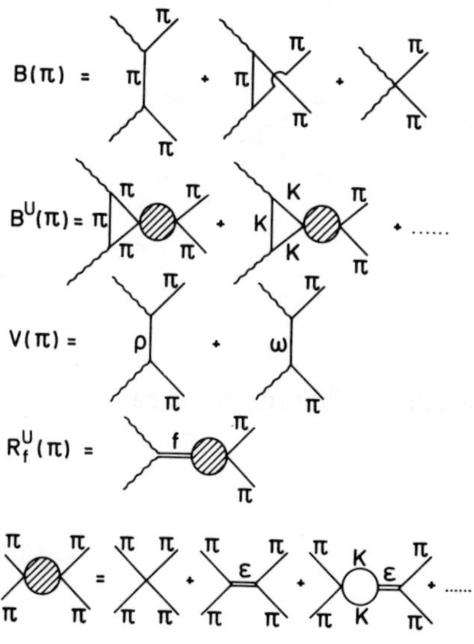

Fig. 3 The diagrams contributing to $\gamma\gamma \to \pi^+\pi^-$ according to the model of Mennessier[4]. B, V and R are the contributions from the Born approximation, vector meson exchange and direct resonance production, respectively. The superscript U denotes unitarized amplitudes.

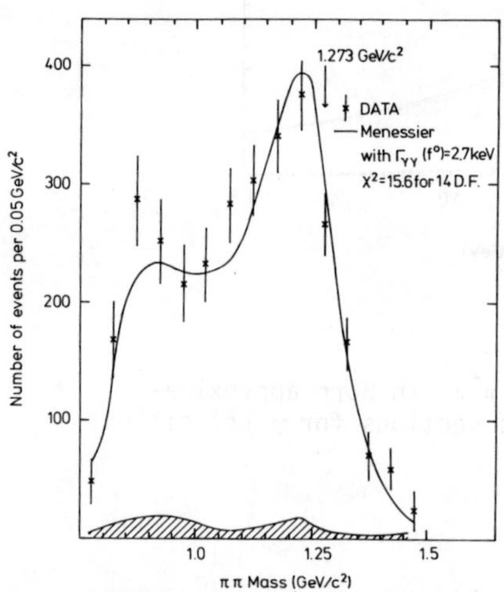

Fig. 4 Invariant $\pi^+\pi^-$ mass distribution for $\gamma\gamma \to \pi^+\pi^-$ as obtained by the CELLO group. The hatched area is the estimated K^+K^- background. The solid curve represents a Monte Carlo simulation using the model of Mennessier[4] with $\Gamma(f \to \gamma\gamma) = 2.7$ keV.

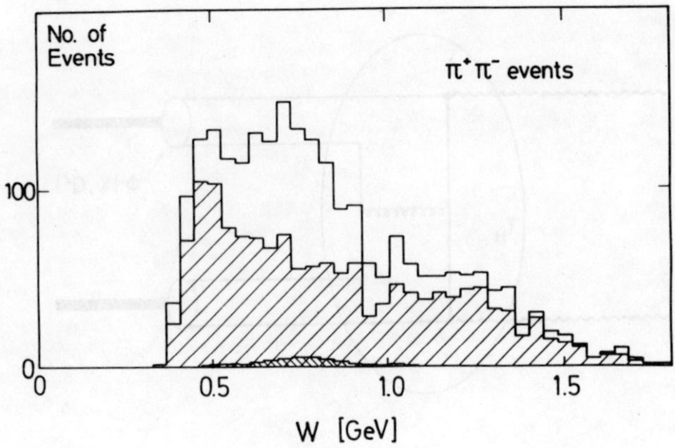

Fig.5 The shaded area shows the invariant $\pi^+\pi^-$ mass distribution for $\gamma\gamma \to \pi^+\pi^-$ events corrected for misidentified $\mu^+\mu^-$ and e^+e^- pairs of the PLUTO group. The histogram on top gives the background from beam gas events and inclusive one photon annihilation events. The doubly shaded area represents the calculated background from two photon processes having undetected final state particles.

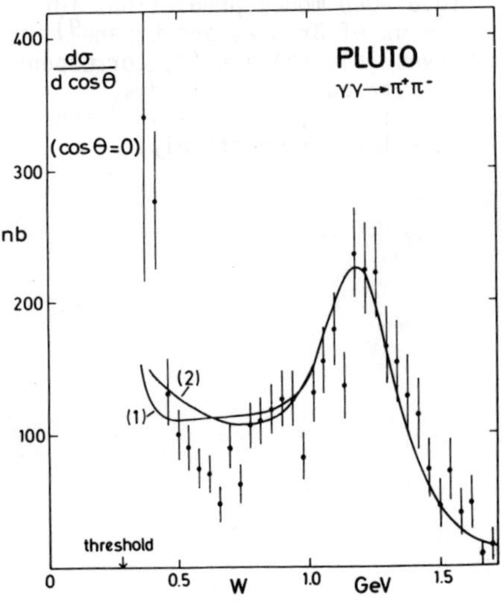

Fig.6 The cross section for $\gamma\gamma \to \pi^+\pi^-$ extrapolated to $\cos\Theta = 0$ as obtained by the PLUTO group. Curve (1) shows the prediction of the model of Mennessier[4]. Curve (2) gives the Born approximation with an f resonance interfering with the $\lambda = 2$ amplitude.

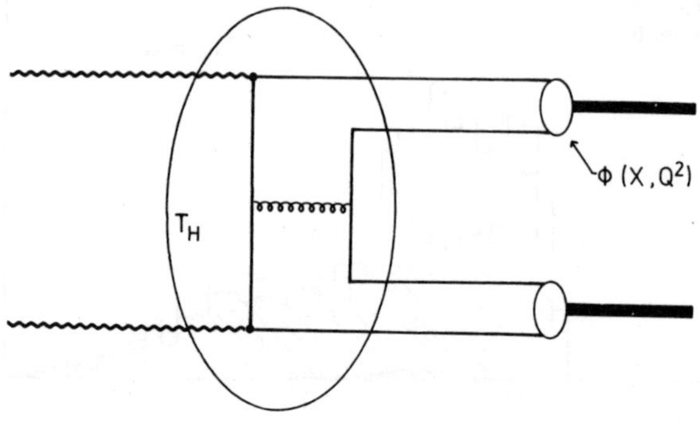

Fig.7 Illustration of the QCD model of Brodsky and Lepage[9] for two-photon production of meson pairs.

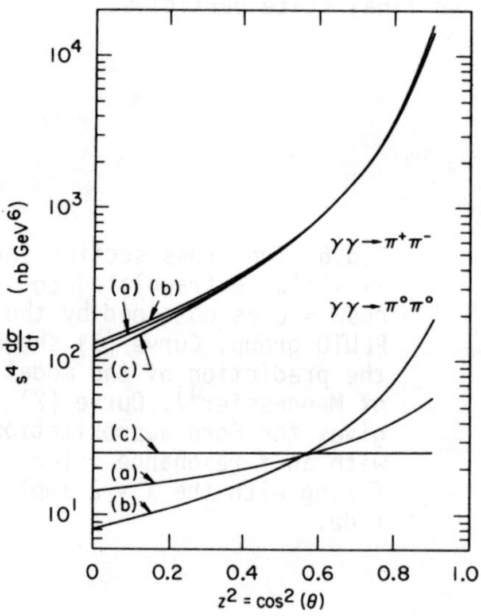

Fig.8 QCD model predictions for $\gamma\gamma \to \pi\pi$ of Brodsky and Lepage[9]. Curves (a), (b) and (c) correspond to distribution amplitudes $\Phi = x(1 - x)$, $(x(1 - x))^{1/4}$ and $\delta(x - 1/2)$, respectively.

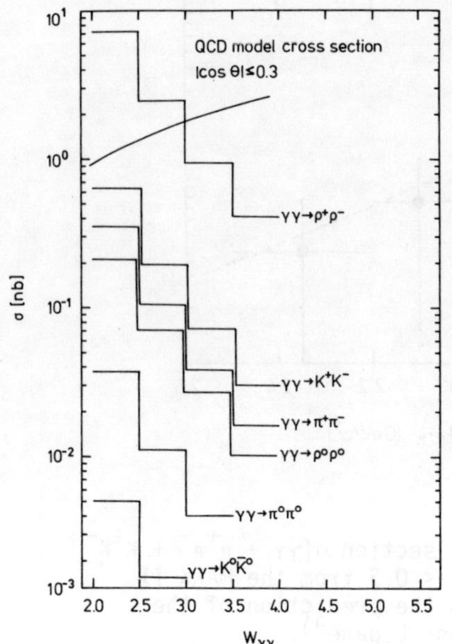

Fig.9 QCD model cross sections for two-photon production of meson pairs[9] integrated over $|\cos\Theta| < 0.3$. The solid line gives the cross section that would result in 10 events observed in a $\Delta W_{\gamma\gamma} = 0.5$ GeV bin for $E_{beam} = 17$ GeV, $\int Ldt = 100$ pb^{-1} and a detection efficiency of 5 %.

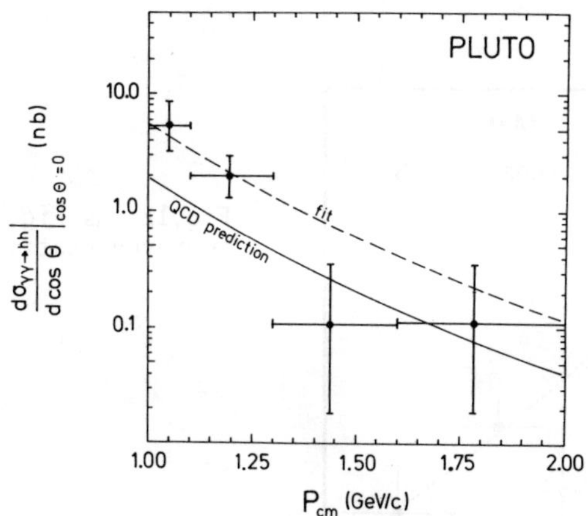

Fig.10 The cross section $d\sigma/d\cos\Theta$ at $\cos\Theta = 0$ for $\gamma\gamma \to h^+h^-$ as obtained by the PLUTO group[10]. The solid line is the prediction of the model of Brodsky and Lepage[9]. The dashed curve is the same prediction with the normalization adjusted to the data.

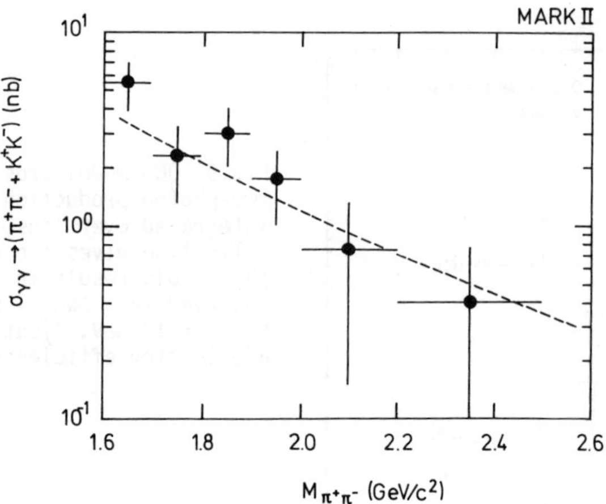

Fig.11 Measured cross section $\sigma(\gamma\gamma \to \pi^+\pi^- + K^+K^-$
integrated over $|\cos\Theta| < 0.3$ from the MARK II
group[11]. The curve is the prediction of the
QCD model of Brodsky and Lepage[9].

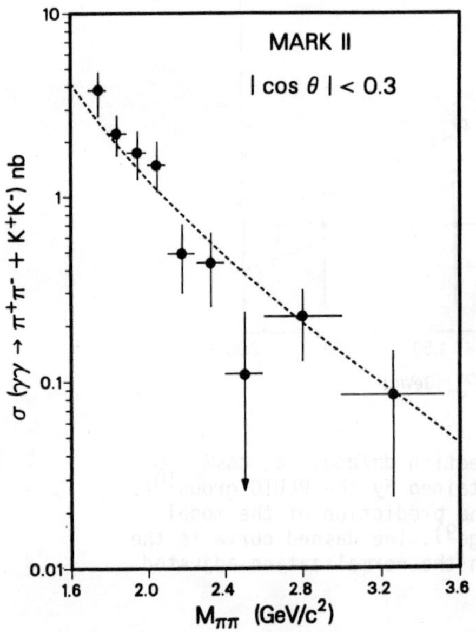

Fig.12 As Fig. 11, but with
improved statistics[12].

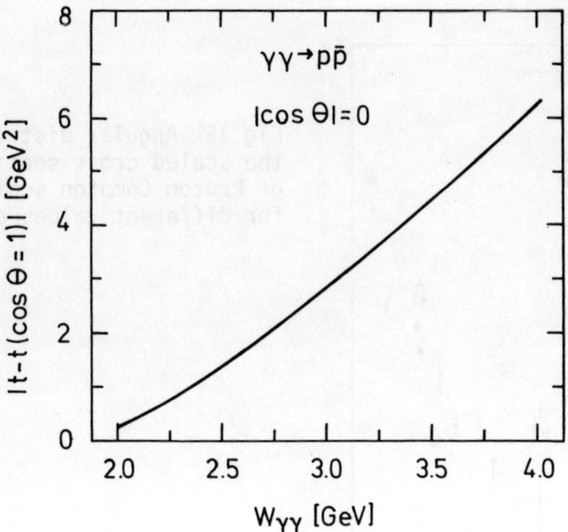

Fig.13 Effective four momentum transfer squared $|t - t(\cos\Theta = 1)|$ versus $W_{\gamma\gamma}$ for $\gamma\gamma \to p\bar{p}$ at $\cos\Theta = 0$.

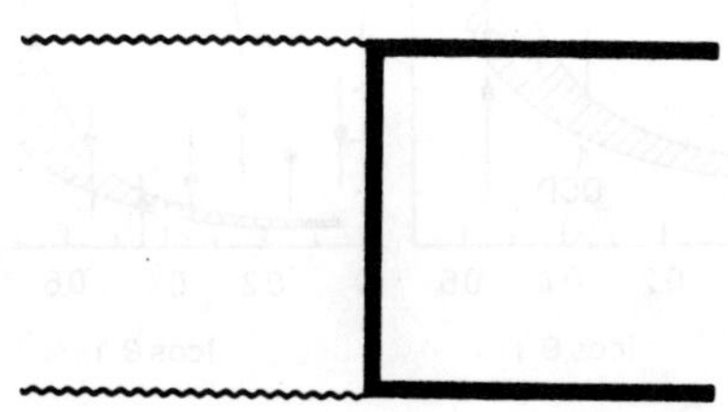

Fig.14 Born diagram for the two-photon production of a pair of Dirac particles.

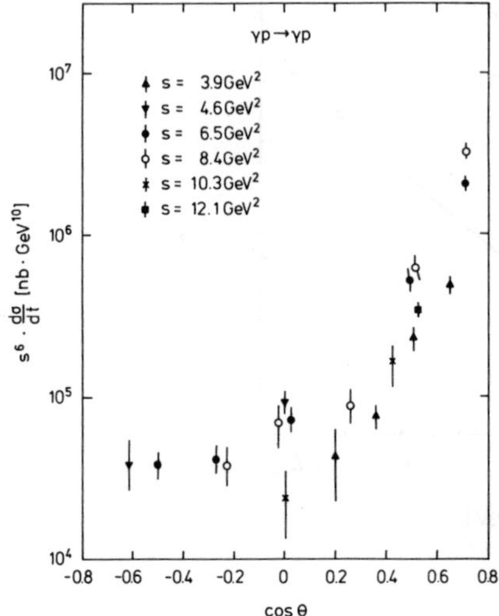

Fig.15 Angular distribution of the scaled cross section $s^6 * d\sigma/dt$ of Proton Compton scattering[13] for different values of s.

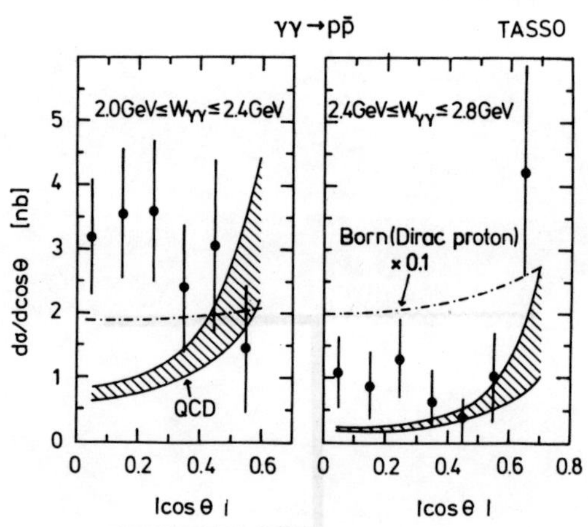

Fig.16 Angular distributions for γγ → pp̄ measured by the TASSO group[14].

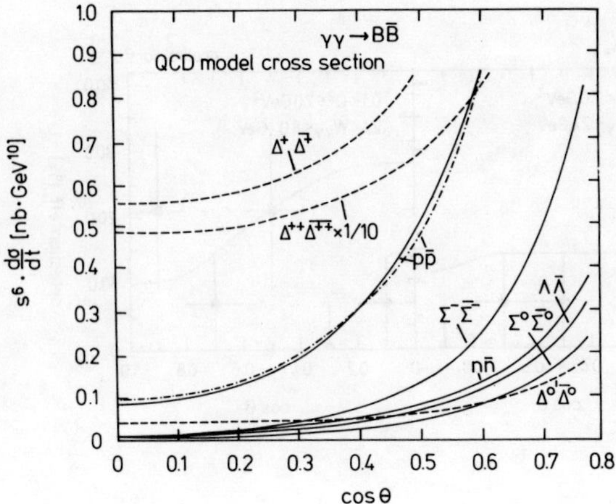

Fig.17 Differential cross sections for two-photon production of baryon pairs calculated by Farrar et al.[18] according to the QCD model. The scaled cross section $s^6*d\sigma/dt$ is plotted versus $\cos\Theta$.

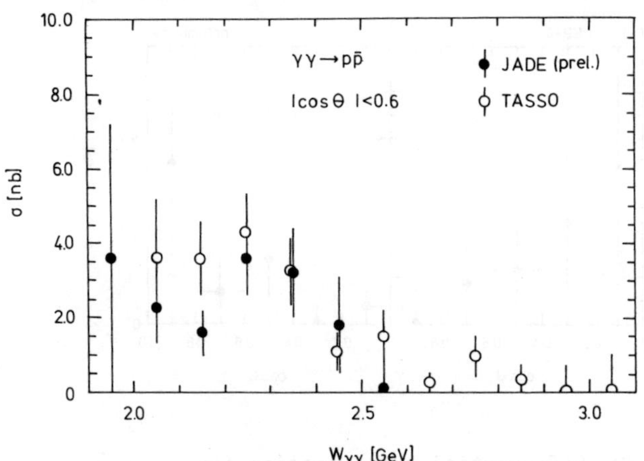

Fig.18 Measured cross sections for $\gamma\gamma \to p\bar{p}$ integrated over $|\cos\Theta| < 0.6$ from the JADE collaboration[19] and the TASSO collaboration[14].

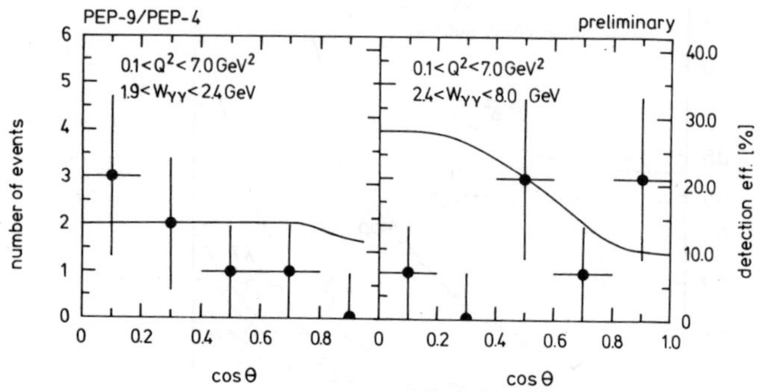

Fig.19   Uncorrected angular distribution of tagged $\gamma\gamma \to p\bar{p}$ events ($0.1 < Q^2 < 7.0\,GeV^2$) for two $W_{\gamma\gamma}$ bins from the PEP-4/PEP-9 collaboration[20]. The full curve is the detection efficiency for tagged $\gamma\gamma \to p\bar{p}$ events (scale on right side).

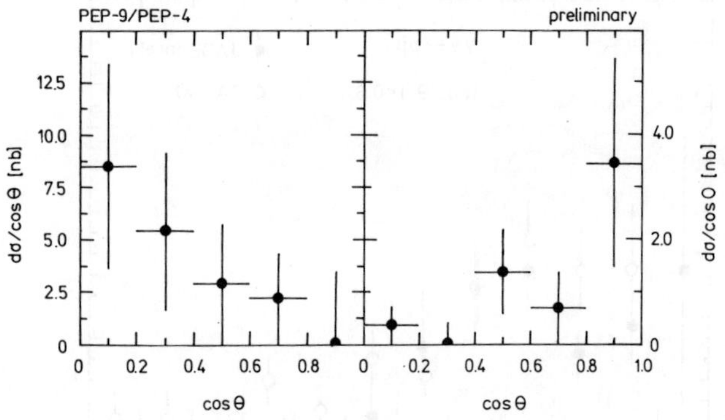

Fig.20   Differential cross section for $\gamma\gamma \to p\bar{p}$ corresponding to the event distribution in Fig.19 measured by the PEP-4/PEP-9 collaboration.

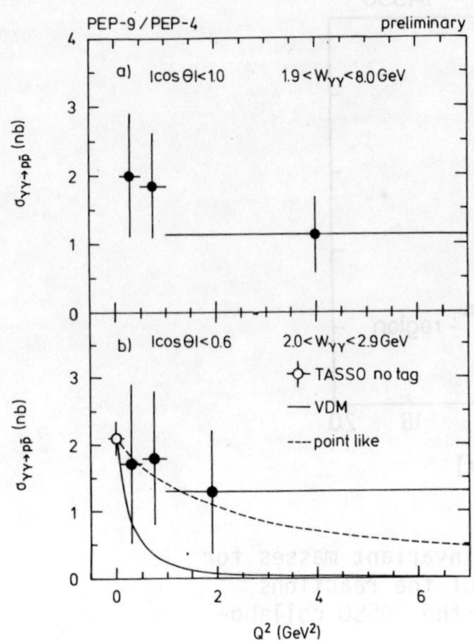

Fig.21 Q^2-dependence of the cross section for $\gamma\gamma \to p\bar{p}$ from the PEP-9/PEP-4 collaboration for the whole data sample (a) and for $2.0 \leq W_{\gamma\gamma} \leq 2.9$ GeV (b). The full curve corresponds to a ρ-pole form factor. The dashed curve is the expectation for Dirac protons. Both curves are normalized to the TASSO $Q^2 \approx 0$ point.

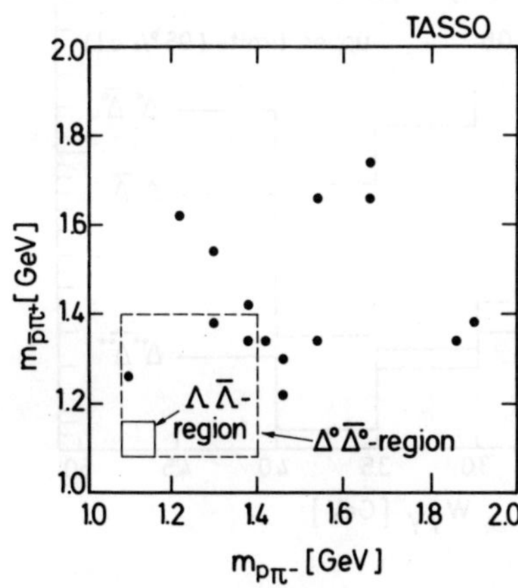

Fig.22 Correlation of invariant masses for $p\pi^-$ and $\bar{p}\pi^+$ for events of the reaction $\gamma\gamma \to p\bar{p}\pi^+\pi^-$ measured by the TASSO collaboration[21].

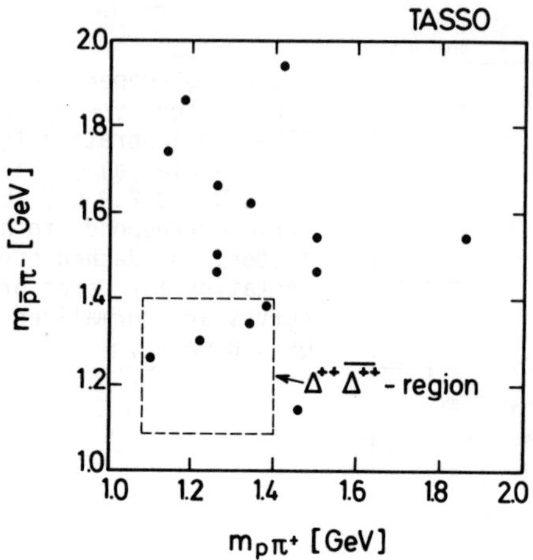

Fig.23 Correlation of invariant masses for
$p\pi^+$ and $\bar{p}\pi^-$ for events of the reactions
$\gamma\gamma \rightarrow p\bar{p}\pi^+\pi^-$ measured by the TASSO collaboration[21].

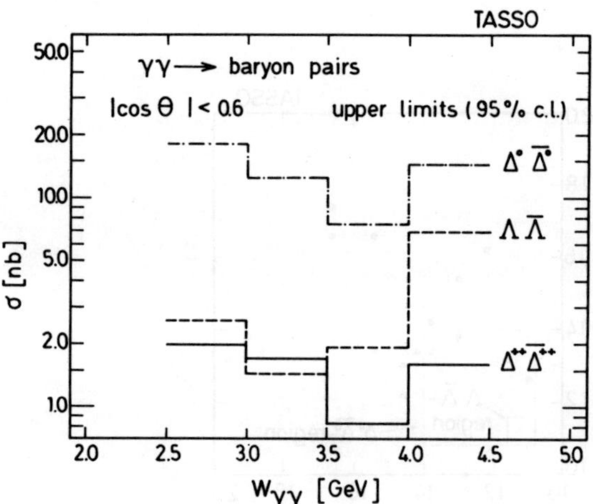

Fig.24 Upper limits (95 % c.l.) on the cross
sections ($|\cos\Theta| < 0.6$) for baryon pair produc-
tion by two photons as measured by the TASSO
collaboration[21].

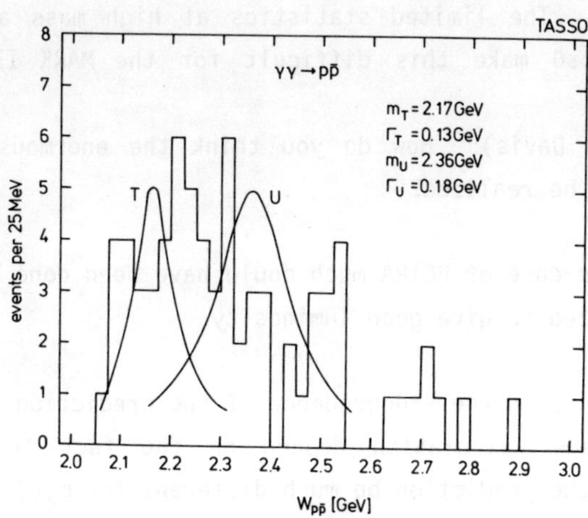

Fig.25 Uncorrected invariant mass distribution for γγ → pp̄ events as measured by the TASSO collaboration. The curves show resonance shapes generated with parameters typical for the T,U resonances observed in pp̄ scattering.

Questions and Discussion

L. KÖPKE (UC Santa Cruz): Did any experiment look at the angular distribution in the channels $\gamma\gamma \to \pi^+\pi^-$ and $\gamma\gamma \to K^+K^-$ at high $|t|$? Are these consistent with the QCD predictions by Brodsky and Lepage?

G. GIDAL (LBL): The limited statistics at high mass and the small lever arm in $\cos\theta$ make this difficult for the MARK II experiment.

J.F. GUNION (UC Davis): How do you think the enormous increase in luminosity could be realized?

H. KÜCK: For the case of PETRA much could have been done by running at an energy optimized to give good luminosity.

G. KNIES (DESY): Does the s-dependence of the prediction on $\gamma\gamma \to \pi\pi$ in the Brodsky-Lepage calculation depend on the fact that $\alpha_s(Q^2)$ is running? Would the prediction be much different for $\alpha_s(Q^2)$ = constant? (The question goes to S.J. Brodsky).

S.J. BRODSKY (SLAC): The $\gamma\gamma \to M\bar{M}$ is proportional to $\alpha_s^2(Q^2)$, but the resulting logarithmic correction to the power scaling $s^4 d\sigma/dt = f(\theta)$ is small for $\Lambda_{QCD}^{\overline{MS}} \lesssim 200$ MeV. Since the prediction can be normalized to $e^+e^- \to \pi^+\pi^-$ pion form factor data there is little uncertainty in the prediction. The $\gamma\gamma \to \pi^\circ\pi^\circ$ cross section, however, depends in detail on the form of the pion distribution amplitude.

G.R. FARRAR (Rutgers Univ.): What is the outlook for improved statistics at large s and t, e.g. for $\Delta^{++}\bar{\Delta}^{++}$ say within the next year?

H. KÜCK: Within the next year the increase in luminosity necessary to proof your calculation will not be provided.

HIGH TRANSVERSE MOMENTUM JETS

Frank Foster

University of Lancaster
LANCASTER LA1 4YB
U.K.

ABSTRACT

Four recent experimental studies of high transverse momentum
jets in photon-photon collisions are discussed. The results give
strong evidence for the existence of pointlike scattering in the
interaction and show that the quark parton model gives a good
description of two-jet production at high transverse momentum. There
is some indication that multi-jet processes predicted by Q.C.D. may
also be involved.

1. Introduction

In this talk we will be concerned with the observation of hadron
jets with high transverse momentum, $p_\perp^{jet}$, relative to the γ-γ direction
measured in the hadronic centre of mass system. The description "high"
will be taken rather arbitrarily to mean $p_\perp^{jet} \gtrsim 2$ GeV. The motivation
for searching for such events is of course that high transverse momentum
jets are the most obvious consequences of hard, pointlike scattering
behaviour. Further it is confidently expected that the dominant
contribution to jet production in photon-photon collisions must come
from the lowest order "Born" diagram when $p_\perp^{jet}$ is large[1,2]. This
diagram or the Quark Parton Model (Q.P.M.) yields the simplest two jet
structure and its contribution to the jet cross section may be calculated using the electromagnetic coupling of the photons to the quark
line. All the experiments discussed here use the Vermaseren event
generator[3] to evaluate the cross section and then a fragmentation
scheme[4] is invoked before tracking the resultant hadrons through the
detector simulation. Clearly the higher order gluon radiative

corrections to the Born term are neglected but when such corrections are evaluated it is found that they may be small[5]. Also these corrections will depend on the particular definition of a "jet" used by the experimenter and the statistical significance of the data is now sufficient for these refinements to be seriously considered in the future. It is as well to remember that the V.D.M. mechanism will also produce two-jet events for $W_{\gamma\gamma} \gtrsim 4$ GeV but the jet axis will have $P_\perp$ limited with respect to the γ-γ direction. The quark counting rule, however, predicts that the Born term cross section goes like $(P_\perp^{jet})^{-4}$ so the "hard scattering" jets are expected to dominate over the soft jets for $P_\perp^{jet} \gtrsim 2$ GeV. Q.C.D. makes firm predictions for the existence of gluon mediated events with more than two jets[2]. These events arise when one or both photons fragments into its quark constituents before the hard scattering takes place so they will be suppressed relative to the Born term although the cross section is still $(P_\perp^{jet})^{-4}$. One or more of the photon quarks will be unaffected by the hard scatter and so produce one or two "beam pipe" jets. With the limited acceptance of today's detectors in the forward and backward directions it is expected that many three jet events will be classified as two-jet. An even greater limiting factor is the photon-photon luminosity which decreases rapidly with hadron centre of mass energy $W_{\gamma\gamma}$, at an energy of 20 GeV per beam there are very few two photon high $P_\perp$ jet events with $W_{\gamma\gamma} > 15$ GeV. Experience with annihilation jets suggests that $W_{\gamma\gamma} \gtrsim$ 30 GeV is a realistic minimum energy for the separation of 3 jet from 2 jet events.

The so-called "higher twist" jet like events i.e. those having a $(P_\perp^{jet})^{-n}$ dependence with $n \geq 6$ are also expected to make significant contributions to the cross section[2]. Here one or more of the photons converts into a meson which then participates in the hard scattering process. One signature for such events is a single meson observed at high $P_\perp$ with a balancing quark jet and a low $P_\perp$ beam line jet; frequently however such events are suppressed by the experimenters' definition of a "jet" containing some minimum number of particles.

One expects then that the jet like events will be dominated for $P_\perp^{jet} \gtrsim 2$ GeV by the "leading twist" $(P_\perp^{jet})^{-4}$ processes producing two, three, four or more jets. Only the two jet events are experimentally observable and the multijet events will either be misinterpreted as two-jet, when one jet disappears down the beam pipe, or be seen as a rather "spherical" non-jet-like event. Following Stirling[2] we can define the scale invariant ratio $X_T = 2P_\perp/\sqrt{s}$ and show the expected ratios of 3-jet, 4-jet to 2-jet production in pointlike $\gamma\gamma$ interactions (fig.1). Only when $X_T \gtrsim 0.2$ does the 2-jet process dominate over the multi-jets and as we shall see the experiments are limited to the range $0.1 \lesssim X_T \lesssim 0.3$.

In summary then for a selection of two photon events with $W_{\gamma\gamma} \gtrsim$ 4 GeV one expects the cross section to be dominated by the V.D.M. process producing low $P_\perp^{jet}$, 2-jet events. For $P_\perp^{jet} \gtrsim 2$ GeV the hard scattering dominates with perhaps one half of the events arising from the two jet Born diagram and the remaining events unresolved three or four jet structures.

If the Q^2 of one of the photons is increased i.e. events are selected by tagging a scattered electron at large angles then the soft V.D.M. background is strongly suppressed since the V.D.M. contribution will decrease like $1/Q^4$ and the Born diagram decreases like $1/Q^2$.

We will discuss the following recent experiments which have been done at the PETRA accelerator DESY (Table I). They include both "single

TABLE I

Group	Method	$\int \mathcal{L} dt$	Q^2 range
TASSO[8]	No tag	71 pb^{-1}	~ 0
CELLO[9]	No tag	7.9 pb^{-1}	~ 0
JADE[10]	Single Tag	66 pb^{-1}	$0.2 < Q^2 < 1.75$ GeV^2
PLUTO[11]	Single Tag		
	S.A.T.	29.3 pb^{-1}	$0.1 < Q^2 < 1.0$ GeV^2
	L.A.T.	39.4 pb^{-1}	$1.0 < Q^2 < 18$ GeV^2
	End Cap	40.1 pb^{-1}	$15 < Q^2 < 120$ GeV^2

tag" and "no tag" procedures and a variety of analysis techniques. But we are beginning to get a common message from all the results. The earlier published results from JADE[6] and TASSO[7] will not be discussed.

72

2. <u>TASSO No-Tag Analysis</u>

This experiment[8] presented by the TASSO group at DESY demonstrates
that it is possible to overcome the horrendous background problems
caused by the annihilation channel when one turns off the single tag
trigger. The experiment was motivated by two factors 1) There is
interest in the physics when Q^2 of both photons is close to zero, 2)
the number of events is increased by roughly a factor of ten over the
single tag sample for the same luminosity. It was relatively easy to
eliminate the beam gas background by rejecting events with a p or $\bar{p}$
and by a z-vertex cut; also to eliminate the tau pairs from annihilation
and two photon events using topology and Monte Carlo simulations. Subtraction of the annihilation hadronic background was done after first
suppressing these events by rejecting events with total longitudinal
momentum greater than 0.3 of the beam momentum. The final subtraction
requires a good annihilation Monte Carlo including Q.E.D. radiative
effects[12] and a good knowledge of the detector "fringe" acceptances.
In fig.2 we show the distribution of $\Sigma|\underline{P}|/P_b$ for events with at least
one hadron with $P_t > 1.5$ GeV. The V.D.M. contribution has also been
subtracted from this distribution. The TASSO version of the V.D.M. is
a P_t limited phase space model which fits the low P_t inclusive spectra:

$$\frac{d\sigma}{dP_t^2} = a\ \exp(-\ bP_t)$$

$$n_{ch} = c_1 + c_2\ \exp c_3\sqrt{\ell n\ 4W_{\gamma\gamma}^2}$$

$$\text{where} \qquad b = 4\ \text{GeV}^{-1}$$

$$c_1 = 3.96$$

$$c_2 = 0.0029$$

$$c_3 = 2.85$$

Also shown in fig.2 is the Q.P.M. expectation from the Vermaseren
generator[3] and Field-Feynman fragmentation   The authors point out
that whereas the Q.P.M. cross section varies greatly with the assumed
quark masses the number of accepted events does not and the largest
contribution to the systematic error is the uncertainty in the low
energy fragmentation parameters. The Born term calculations are given
as bands showing the maximal variation that can be obtained by varying

these parameters.

Figure 3 shows the inclusive particle distribution, $\frac{dN}{dP_t^2}$ – remember this is the high P_t "tail" remaining after V.D.M. subtraction – and the Q.P.M. expectation. We see a factor of approximately four in the ratio of data to Q.P.M. expectation over the range $1.5 < P_t < 2.3$ GeV. The authors conclude that there is plenty of room for Q.C.D. corrections and other hard scattering multi-jet production and higher twist effects but quantitative statements cannot be made because theoretical calculations of the additional terms do not exist in a form which can be compared with the data.

A jet analysis was also done by searching for clusters in a plane perpendicular to the beams using the same events as in figs. 2 and 3. Figure 4 shows the distribution of the angle ϕ between the highest P_t track in an event and all the other tracks weighted by $P_t(P_t\frac{dN}{d\phi})$. The curves represent the shapes of the distribution expected from the Q.P.M. and their V.D.M. (both normalised to the observed number of events). Focussing on the low ϕ region the data clearly favours the shape of the two jet Q.P.M. suggesting that the observed large rate of high P_t track is correlated with particle clusters or jets.

3. CELLO No Tag Analysis

The data selection and background subtraction for this experiment[9] were similar to those employed by the TASSO group but the total integrated luminosity was much smaller. Events were obtained with the visible centre of mass energy W^{vis} greater than 4 GeV and the scale invariant ratio $X_T = \frac{2P_\perp^{jet}}{\sqrt{s}}$ between 0.12 and 0.35. Again the inclusive particle distribution at high P_t shows the ratio of Data to Q.P.M. expectation between two and four although no explicit subtraction of a (mainly low P_t) V.D.M. contribution has been made.

The jet structure of the events was investigated by making a cluster search in the γ–γ centre of mass system. First, preclusters are found with tracks, both charged and neutral, less than 30° apart and then clusters found with preclusters less than 45° apart. Clusters must then have at least two particles and at least 2 GeV energy to be defined as a "jet". Figure 5 shows the distribution of $(P_\perp^{jet})^2$ for events with two or more jets, there is a clear excess of data over the Q.P.M. expectation at all values of $P_\perp^{jet}$. A quantitative check on the results is given in Table II which has the ratio of events with two jets and $P_\perp$ greater than some minimum value ($P_\perp$(min)) to the Q.P.M. expectation. The ratios are given with and without a subtraction of the expected V.D.M. contribution.

TABLE II

$P_\perp$(min) GeV	Data/Q.P.M.	(Data–V.D.M.)/Q.P.M.
2	2.4±0.4	2.0±0.5
3	2.6±1.3	2.0±1.3

4. JADE Single Tag Analysis

This experiment[10] uses lead-glass small angle tagging system covering the angular range 43 to 75 mrad. The JADE detector has particularly good charged and neutral particle acceptance which allows a generally large ratio of the visible energy W^{vis} to the true centre of mass energy $W_{\gamma\gamma}$. This ratio is shown in fig.6 for the single tagged events. Only those events with W^{vis} greater than 4 GeV and measured Q^2 in the range 0.2 to 1.75 GeV^2 were used in the analysis. Significant background contamination came from beam gas scattering and this could be eliminated as in the CELLO experiment. The contribution of tau pairs is negligible in the final data because of the definition of jets. However the annihilation background is potentially serious because the tagging system lacks tracking information and hard radiated photons can trigger the system. Detailed Monte Carlo simulations showed that the contamination is roughly 20% for W^{vis} greater than 12 GeV and $P_\perp^{jet}$ greater than 3 GeV. The $P_\perp^{jet}$ distribution shown has been corrected for this source of background.

The jet search was made in the hadron centre of mass system using two different cluster algorithms:

    A) The minimum spanning tree (Dorfan[13])

    B) Angle clusters (Daum, Meyer, Burger[14]).

[Algorithm B was optimised to the energy range between 4 GeV and 15 GeV by modifying the precluster and cluster angles to $\alpha = 45^\circ$ and $\beta = 65^\circ$.] For both algorithms a "jet" was required to have three or more particles of which two or more were charged and the total energy of the cluster was greater than 2 GeV. This definition probably rules out observation of high P_t mesons occurring via some higher twist mechanism. All events were then classified as having 0,1,2,3.... jets so, unlike the "thrust" analysis of earlier TASSO work[7] and the present PLUTO analysis, events are not forced into a two-jet category. In order to emphasise the "jet" characteristics of the selected JADE events we can show that thrust distributions for the "high P_t^{jet}" two-jet events, Fig. 7, and the distribution of individual particle P_t relative to the jet axis for these events, fig.8. In both these figures the Q.P.M. expectation is also shown with absolute normalisation.

Finally we show the $(P_t^{jet})^2$ distribution for the sample of two-jet events compared with both the Q.P.M. and V.D.M. expectations, fig.9. The JADE version of the V.D.M. is normalised to fit the inclusive particle spectra particularly at values of P_t less than 1 GeV but it does include an artificial tail to fit the observed high P_t inclusive distribution also. Nevertheless it is clear from fig.9 that the V.D.M. expectation falls below the jet distribution for P_t^{jet} greater than 2 GeV and is negligible for P_t^{jet} greater than 3 GeV. The Q.P.M. however is approached by the data from above as P_t^{jet} increases and the data is saturated by the naive Q.P.M. for P_t^{jet} greater than 3 GeV.

5.   The PLUTO Single Tag Analysis

In the results of this experiment we see the full power of the dedicated two photon experiment. It is possible for the first time to follow both the Q^2 and P_t^2 evolution of the data. The small angle tagger (S.A.T.) covers the angular range 30 to 55 mrad, the large angle tagger (L.A.T.) the range 87 to 260 mrad and the end cap (E.C.) 349 to 681 mrad. One has therefore a fairly continuous distribution of Q^2 measurements from 0.1 to 120 GeV². Various tracking devices allow electrons to be distinguished from photons over the whole angular range, thus reducing the important background from radiative annihilation events. Good momentum resolution in the forward direction from the L.A.T. ensures an optimum ratio of the visible and true invariant mass W^{vis}/W, in fact two thirds of the data have this ratio between 0.5 and 1.0.

After the usual annihilation and tau pair background subtraction events were selected with four or more charged tracks and with W^{vis} between 4 and 13 GeV. This yielded a total of 846 events from the S.A.T. with $<Q^2>$ of 0.35 GeV , 381 events from the L.A.T. with $<Q^2>$ of 5.3 GeV^2 and 50 events from the E.C. having $<Q^2>$ of 49 GeV . (The relevant integrated luminosities are given in Table I.)

The search for jets was done using a "thrust" analysis in the two photon centre of mass system. Defining thrust, T, as:

$$T = \text{Max} \frac{\Sigma |P_L|}{\Sigma |\bar{P}|} \quad ,$$

the jet axis is the event axis relative to which the net longitudinal momentum is maximised. The procedure forces all events into a two-jet configuration but there is ample evidence in the data that the events are jet-like in the accepted sense. Whereas the mean longitudinal momentum increases with W^{vis} from 400 to 800 MeV, the mean transverse momentum about the thrust axis remains close to 300 MeV for all W^{vis}. The observations are in agreement with any jet model, either Q.P.M. or V.D.M., and totally inconsistent with particles distributed according to phase space. Naturally the differences between Q.P.M. and V.D.M. predictions appear in the angular distribution of the thrust (jet) axis relative to the photon-photon axis. The PLUTO methods of implementing the limited transverse momentum distribution of the V.D.M. and the hard scattering Q.P.M. are detailed in their preprint[11]. The jet momentum $\underline{P}^{jet}$ is defined as the net momentum in a hemisphere along the thrust axis and the transverse momentum $P_\perp^{jet}$ defined as $\underline{P}^{jet}\sin\theta$ where θ is the angle between the thrust axis and the γ-γ axis. Figure 10 shows the distributions of $(P_\perp^{jet})^2$ in the three intervals of Q^2 together with the predictions of the Q.P.M. and V.D.M. There is a significant excess of events at high $P_\perp^{jet}$ over the expectation of V.D.M. and this is true for all Q^2. Also the data approaches the Q.P.M. prediction from above as $P_\perp^{jet}$ increases. However it is quite clear that the higher Q^2 becomes the lower the value of $P_\perp^{jet}$ at which the data becomes consistent with the Q.P.M. expectation. These observations are consistent with the view that there is a significant pointlike component in two photon interactions and that the hadronic part of the photon decreases much more rapidly than the pointlike part as both Q^2 and $p_\perp^{jet}$ increase. The PLUTO group also notes that the data is well described by an incoherent sum of Q.P.M. and V.D.M. contributions over almost the whole kinematic range.

It is instructive to look in detail at the thrust distributions in regions where the Q.P.M. should dominate i.e. $p_\perp^{jet} > 2$ GeV. Figure 11 shows these distributions in three intervals of Q^2 and it is apparent that the events with $Q^2 < 1$ GeV^2 are significantly more "spherical" than expected for a pure two-jet configuration. This is an indication that a significant number of events at low Q^2 and high $p_\perp^{jet}$ do not have the two-jet configuration and may be unresolved multi-jet events produced by the Q.C.D. process discussed in the introduction.

Further evidence for these interpretations may be had from a study of the quantity $\tilde{R}$ defined by the PLUTO group:

$$\tilde{R} = \frac{\text{observed number of events at a measured } p_\perp^{jet}}{\text{number of events predicted by the Q.P.M. at that } p_\perp^{jet}}$$

$\tilde{R}$ is also very useful in making comparisons with the data from other experiments since the experimental acceptance factors are partly divided out and the relation between $p_\perp^{jet}$ measured and the "true" $p_\perp^{jet}$ is similar for most experiments. The PLUTO data on $\tilde{R}$ is shown in fig. 12 and we see that in the range of $p_\perp^{jet}$ between 2 and 4 GeV at low Q^2 values the data is significantly in excess of the sum of the Q.P.M. and V.D.M. This is the same region where the thrust distributions are significantly more spherical than the two-jet expectation.

6.   <u>Summary</u>

In order to make some useful comparisons between the single tagged data and the untagged ($Q^2 \tilde{\sim} 0$) data we plot the ratio $\tilde{R}$ in fig. 13 for the JADE and CELLO experiments in the same manner as for the PLUTO data (fig. 12). The trend of the JADE data at $\langle Q^2 \rangle = 0.8$ GeV^2 towards the Q.P.M. expectation ($\tilde{R} = 1$) as $p_\perp^{jet}$ increases is similar to the trends in the low $\langle Q^2 \rangle$ PLUTO data. However in the JADE experiment there is no significant excess of the data over the sum of the Q.P.M. and V.D.M. expectations – and this is for a good reason: Referring back to the JADE thrust distributions, fig. 7, we see that the mean thrust agrees

with the two-jet expectation and this is merely because of the two
cluster selection procedure. Any excess due to the multi-jet Q.C.D.
processes – largely unresolved – would only be seen as an excess of
JADE "single-jet" events over the Q.P.M. expectation at high $P_\perp^{jet}$.

Is there really any hard evidence for the multi-jet events, so
firmly predicted by Q.C.D., to exist in this region of X_T between 0.1
and 0.3? The answer must be "no" but the circumstantial evidence is
strong. It would certainly help the experimenters if the theoretical
predictions for three and four-jet events could be set in a helpful
form – e.g. angle and momentum distributions of the quark and gluon
four-vectors. A clear resolution of even three-jet events can only be
done with a significant number of events at higher values of the invariant mass W and this we can only achieve with much higher luminosity
or at L.E.P. energies.

For the immediate future we can look forward to high statistics
data from the PEP4/9 collaboration and to analyses of the hadronic
final states of the end-cap tagged events from CELLO, TASSO and JADE.
It may also be interesting to see the results of particle identification
studies in the jet-like events.

References

1)   e.g. S.J. Brodsky et al., Phys. Rev. <u>D19</u> (1979) 1418.

2)   W.J. Stirling, Proceedings International Workshop on γ-γ Collisions,
     Aachen (1983) 142.

3)   J.A.M. Vermaseren, Proceedings International Workshop on γ-γ
     Collisions, Amiens (1980) 35.

4)   R.D. Field and R.P. Feynman, Nucl. Phys. <u>B136</u> (1978) 1.

5)   F.A. Berends et al., Phys. Lett. <u>92B</u> (1980) 186.
     I. Kang, Oxford Preprint 50/82 (1982).
     See also discussion by H. Kolanoski, Bonn preprint HE 84/06 (1984)
     section 8.2.5.

6)   W. Bartel et al., Phys. Lett. <u>107B</u> (1981) 163.

7)   R. Brandelik et al., Phys. Lett. <u>107B</u> (1981) 1290.

8)   M. Althoff et al., DESY 83-115 (1983).

9)   Private communication, J.H. Field (CELLO collaboration).

10)   Private communication, A.J. Finch (JADE collaboration).

11)   Ch. Berger et al., DESY 84-070 (1984).

12)   P. Hoyer et al., Nucl. Phys. B161 (1979) 349.

13)   J. Dorfan, SLAC-PUB-2623 (1980).

14)   H.J. Daum, H. Meyer and J. Bürger, Z. Phys. C $\underline{C8}$ (1981) 167.

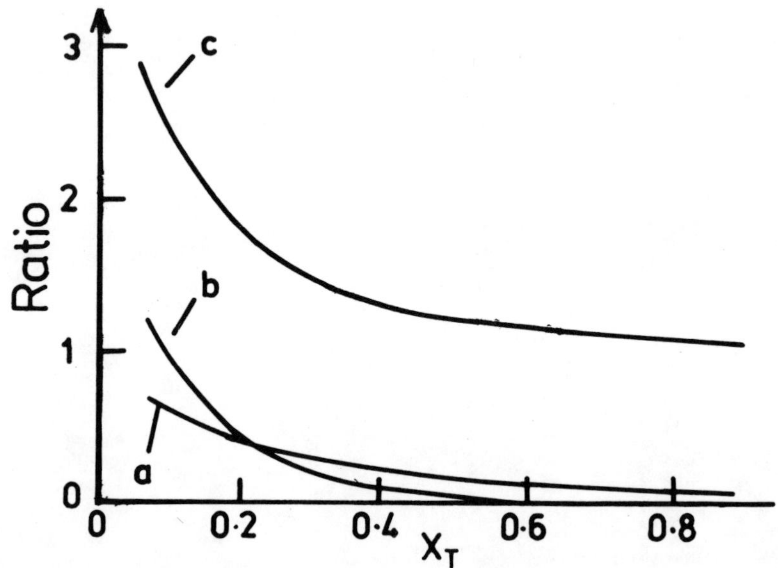

Fig.1   Relative contributions
        to inclusive jet
        cross-section:

        a) 3-jet/2-jet,
        b) 4-jet/2-jet,
        c) (2+3+4)-jet/2-jet

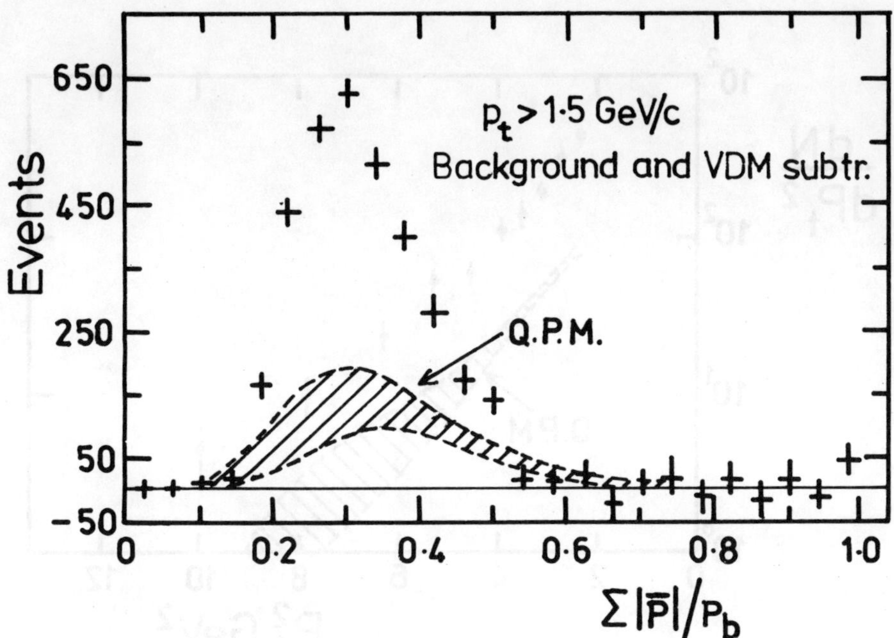

Fig.2 TASSO distribution of
$\Sigma|\underline{P}|/P_b$. Annihilation
background and V.D.M.
are subtracted. Q.P.M.
prediction shown as a
band

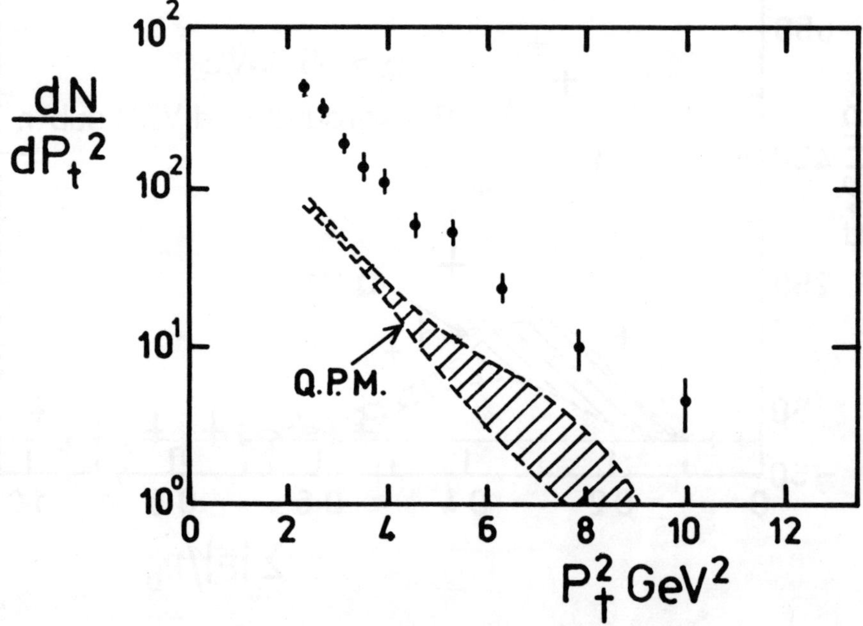

Fig.3 TASSO inclusive particle distribution, $\frac{dN}{dP_t^2}$, and the Q.P.M. expectation

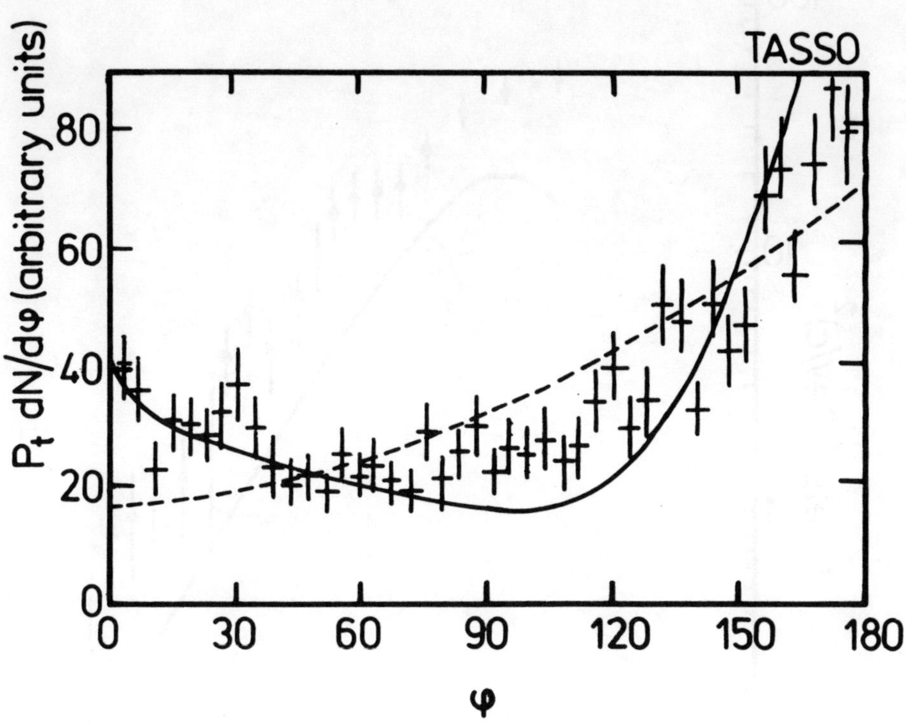

Fig.4   TASSO distribution of
$$P_t\frac{dN}{d\phi}$$

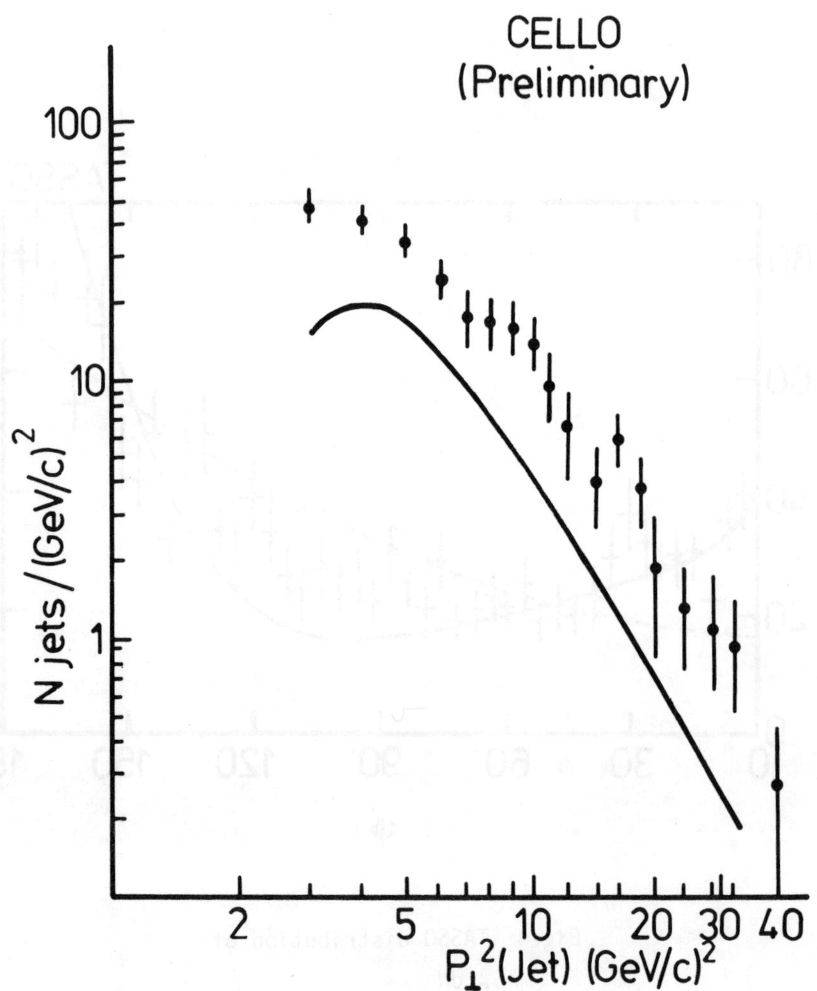

Fig.5   CELLO distribution of $dN/d(P_\perp^{jet})^2$

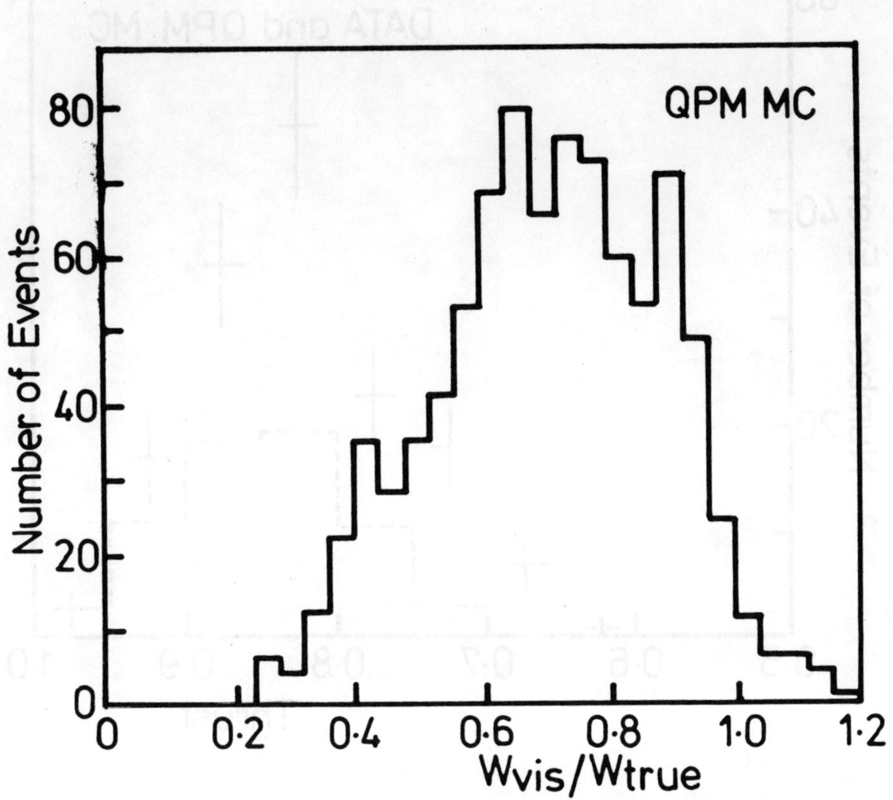

Fig.6   JADE distribution of
W^{vis}/W

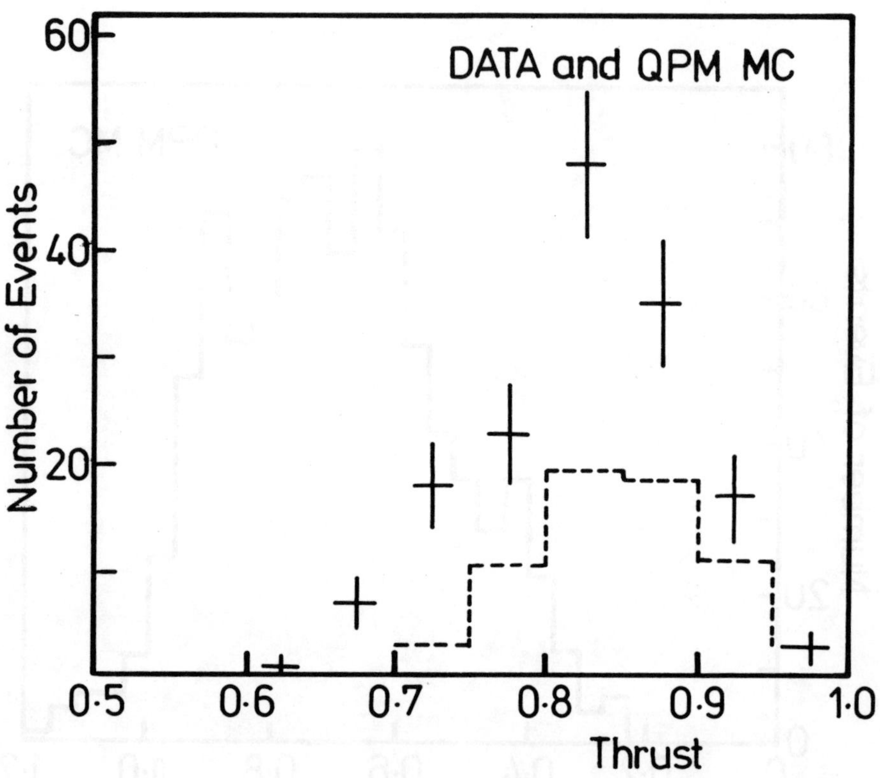

Fig.7   JADE distribution of
thrust for events with
$p_\perp^{jet} > 2$ GeV

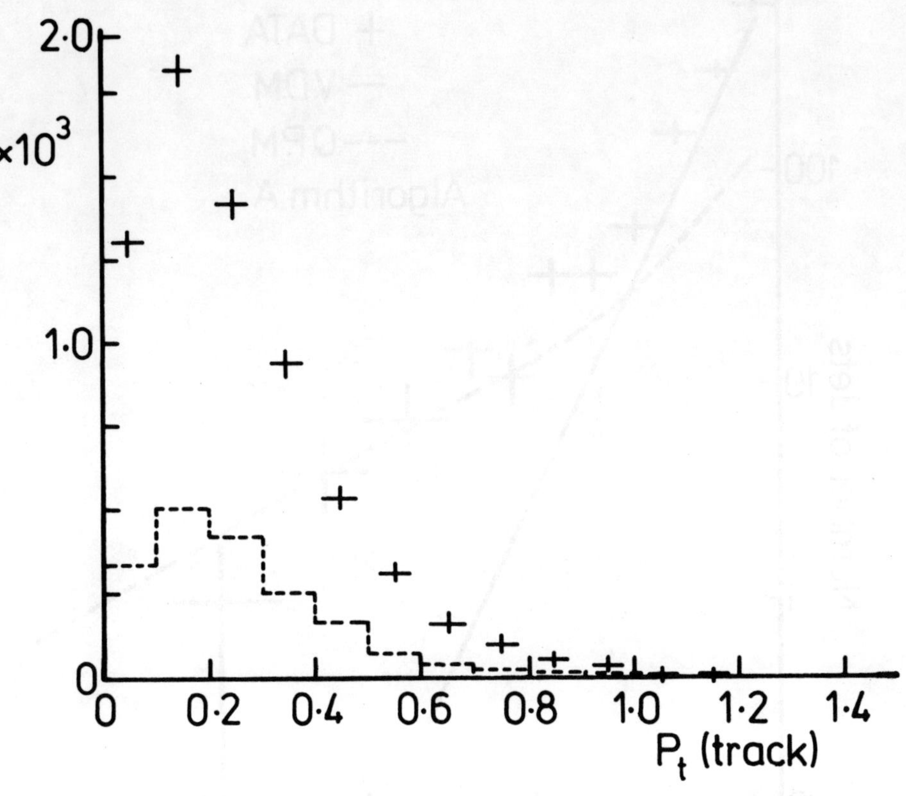

Fig.8 JADE distribution of
particle P_t relative
to jet axis for events
with $P_\perp^{jet} > 2$ GeV

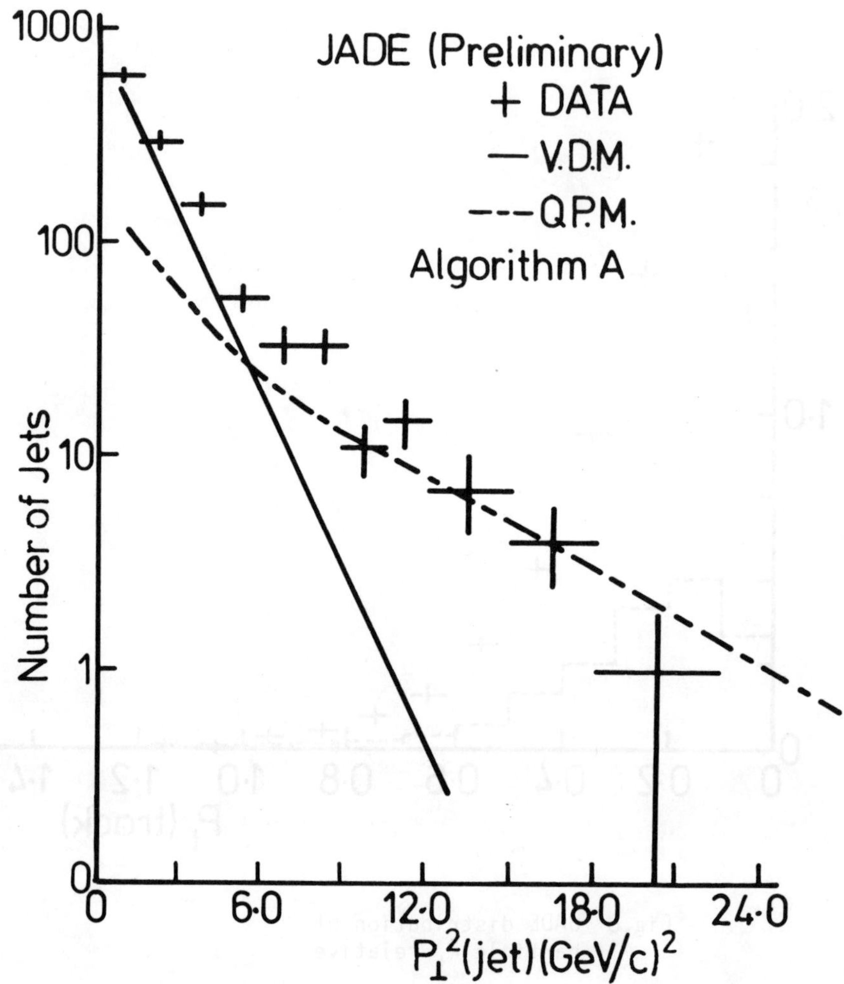

Fig.9 JADE distribution $dN/d(P_\perp^{jet})^2$ with the Q.P.M. and V.D.M. expectations

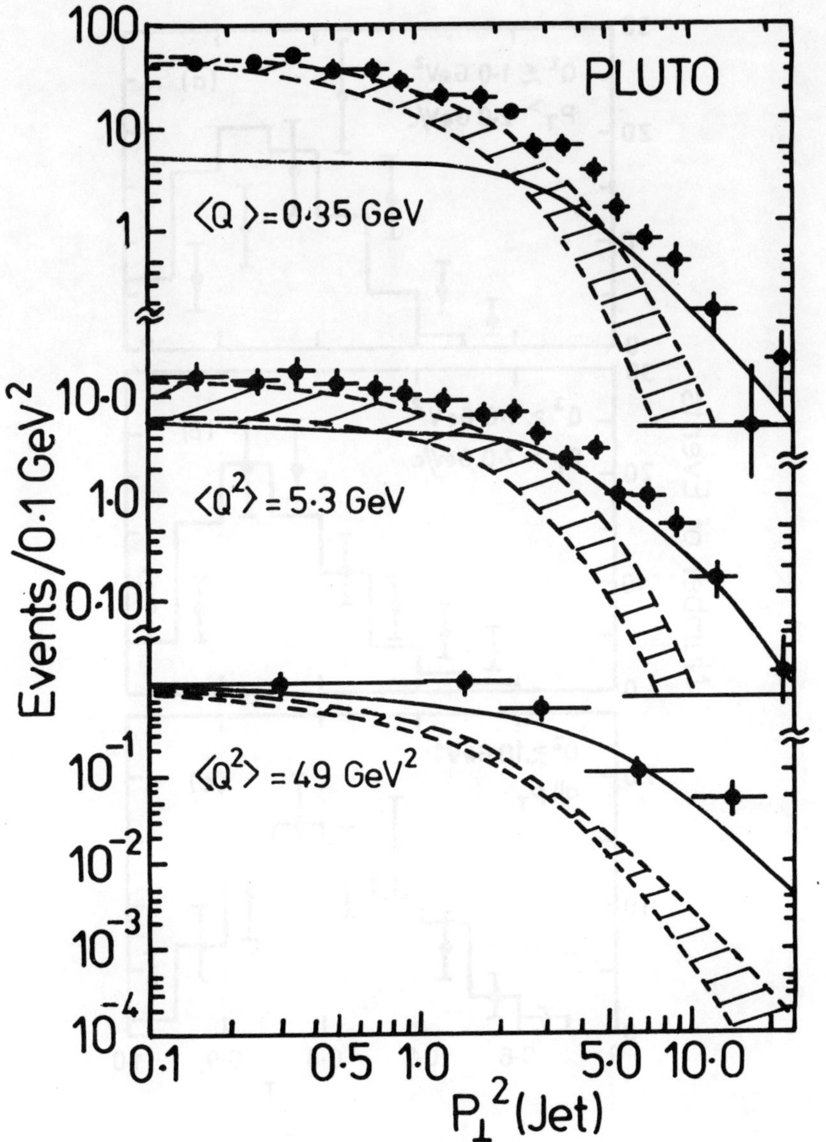

Fig.10 PLUTO distribution $dN/d(P_\perp^{jet})^2$ with Q.P.M. and V.D.M. expectations in three intervals of Q^2

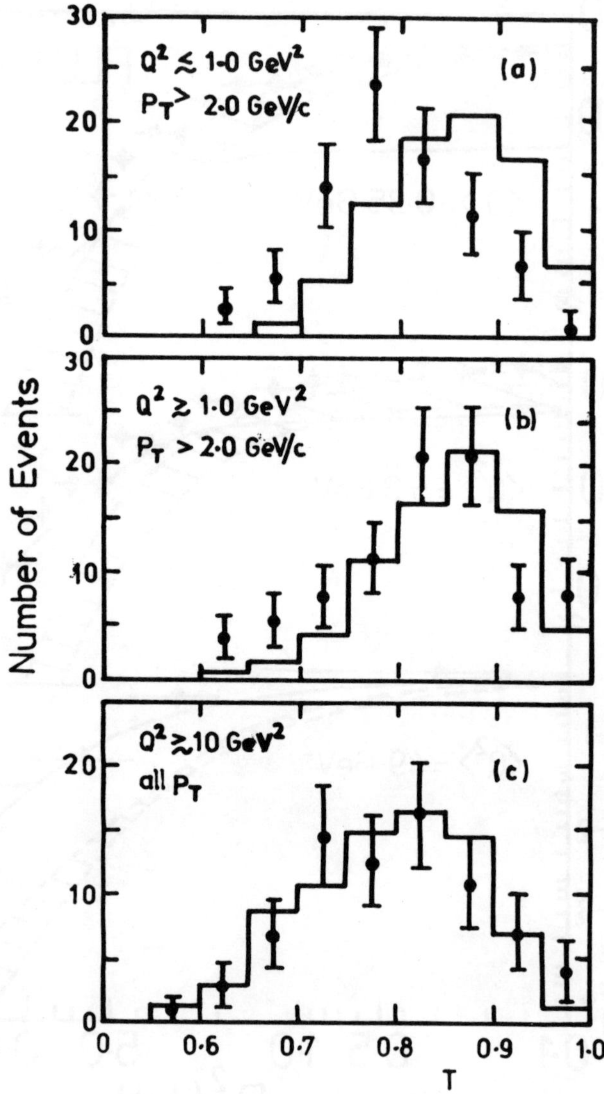

Fig.11 PLUTO distributions of thrust for events with $P_\perp^{jet} > 2$ GeV and the two-jet expectations

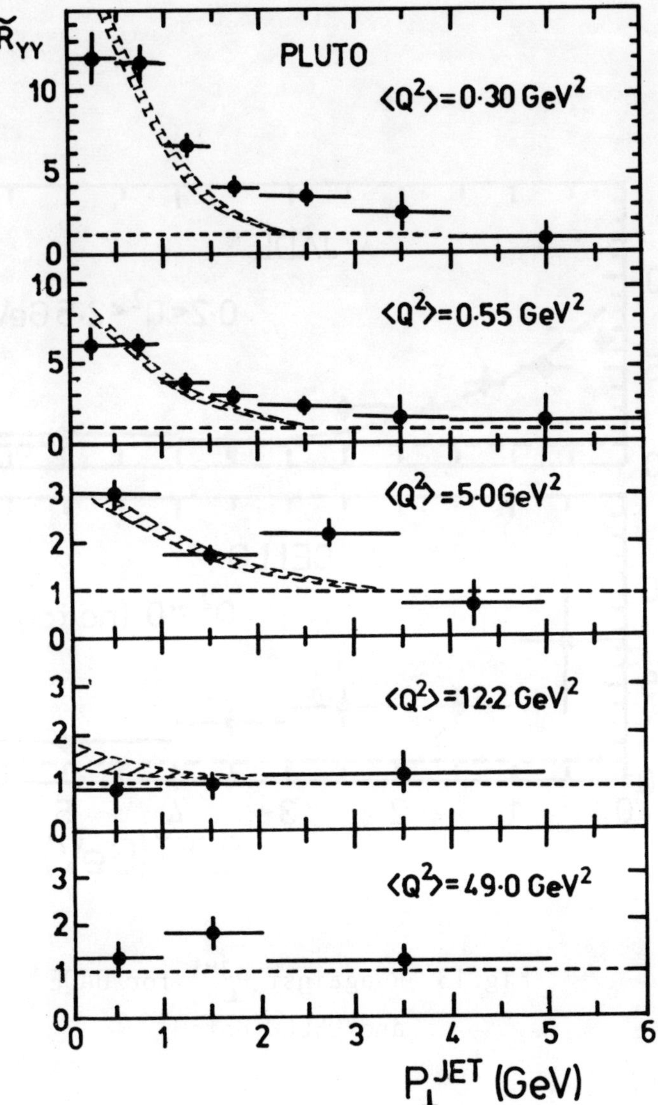

Fig.12 $\tilde{R}$ against $P_\perp^{jet}$ for
PLUTO data in five Q^2
intervals. V.D.M.
as a shaded band.

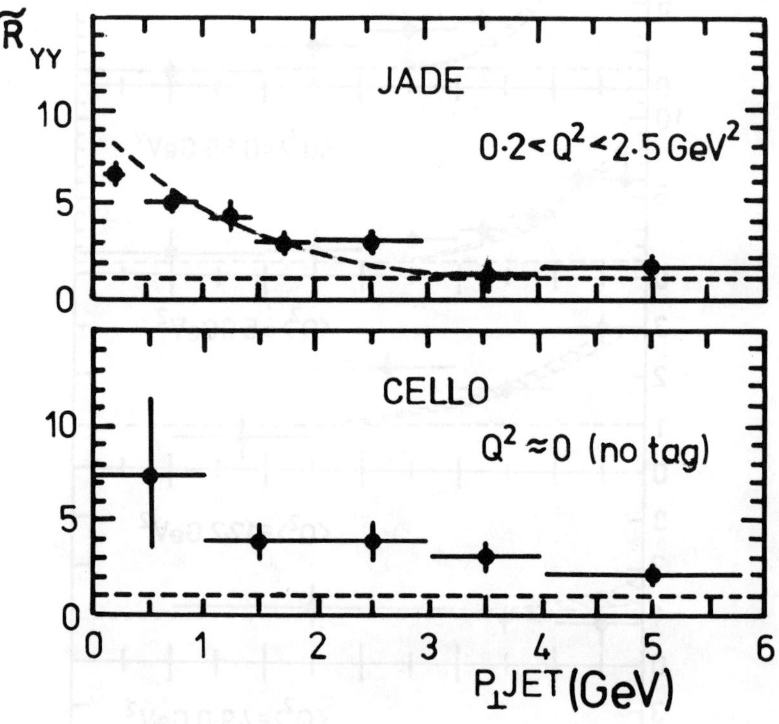

Fig.13 $\widetilde{R}$ against $P_\perp^{jet}$ for JADE and CELLO data

Question and Discussion

J. GUNION (UC Davis): You are approaching a level of agreement between the QPM and data at $p_T \gtrsim 5$ at which you should include the $\gamma\gamma \to$ box $\to$ gg 2 jet $1/p_T^4$ contributions ($\sim$ 15% of $\gamma\gamma \to q\bar{q}$).

F. FOSTER: I agree with this comment.

H. KOLANOSKI (DESY): Is the excess of jet production above the QPM in the CELLO and Jade case not more significant than you quoted?

F. FOSTER: In the CELLO case the Table gives:

$$\frac{\text{Events with 2 clusters } (p_t \geq 3 \text{ GeV})}{\text{Born Expectation}} = 2.6 \pm 1.3$$

but the figure suggests a smaller uncertainty, unless some uncertainty in the Born diagram prediction (by Monte Carlo) is also included. In the JADE case there is no disagreement between the jet p_T^2 distributions and the $\tilde{R}_{\gamma\gamma}$ values.

S. BRODSKY (SLAC): It seems remarkable that one is analyzing jets at $p_T \geq 2$ GeV considering that jets are just discernible at SPEAR in e^+e^- annihilation and very high p_T is needed to see p_T^{-4} behavior in hadron-hadron collisions. Could you comment?

F. FOSTER: I agree it is remarkable that $P_\perp^{-4}$ behavior is seen when $p_\perp^{jet}$ is between 2 and 5 GeV, but there may be a straightforward explanation: the total cross section appears to be dominated by the incoherent sum of the "low $P_\perp$", 2-jet V.D.M. and the "high $P_\perp$" 2-jet Q.P.M. Unlike hadron-hadron collisions the high $P_\perp$ two photon events do not simultaneously have the dominant beam line jets. The Q.P.M. cross section falls like $1/W^2$ compared with the constant V.D.M. cross section so at the present relatively low values of W(4-15 GeV) the hard scattering beats the exponential V.D.M. at $P_\perp^{jet} \gtrsim 2.5$ GeV. The PLUTO data also shows that the effect becomes clearer as Q^2 increases since the hadronic behavior is further suppressed.

S. BRODSKY: How are $\gamma\gamma \to c\bar{c}$ jet events handled, how are phase-space and fragmentation corrections handled?

F. FOSTER: The quark pairs are generated using the Vermaseren four vector event generator and then allowed to fragment using the Field Feynman prescription. Usually the charm fragmentation has a harder fragmentation function than the u,d,s. The PLUTO group comment that the procedure gives a good representation of the annihilation process of comparable energies.

RESONANCES IN PHOTON PHOTON SCATTERING

Michael S. Chanowitz
Lawrence Berkeley Laboratory
University of California
Berkeley, CA 94720

ABSTRACT

A quantity called "stickiness" is introduced which should be largest for $J \neq 0$ glueballs and can be measured in two photon scattering and radiative J/ψ decay. An argument is reviewed suggesting that light $J = 0$ glueballs may have large couplings to two photons. The analysis of radiative decays of η and η' is reviewed and a plea made to desist from false claims that they are related to $\Gamma(\pi^0 \to \gamma\gamma)$ by SU(3) symmetry. It is shown that two photon studies can refute the difficult-to-refute hypothesis that $\xi(2220)$ or $\zeta(8320)$ are Higgs bosons. A gallery of rogue resonances and resonance candidates is presented which could usefully be studied in $\gamma\gamma$ scattering, including especially the low mass dipion.

1. Introduction

It is gratifying to see the many new experimental results on two photon widths of resonances obtained in the last year. Clearly the strength of the $\gamma\gamma$ coupling is a fundamental parameter of any resonance. We can use this information to test our understanding of the $\bar{q}q$ meson spectrum and to aid in the identification of new states, for instance, states with gluonic constituents. Naively we expect unmixed glueball states to have small $\gamma\gamma$ couplings, but I will discuss a remarkable low energy theorem, due to Novikov, Shifman, Vainshtein, and Zakharov, which implies that light $J = 0$ glueballs may have large $\gamma\gamma$ couplings, at least as large as $\bar{q}q$ mesons.[1]

My talk is in five parts. In Section 2 I will discuss the $\gamma\gamma$ coupling of glueballs, beginning with the naive expectation that pure glueballs have small $\gamma\gamma$ widths. I propose a quantitative measure, stickiness, which for a given state X is the ratio $\Gamma(\psi \rightarrow \gamma X)/$ $\Gamma(X \rightarrow \gamma\gamma)$ with phase space factors removed. Stickiness is a measure of the color charge of the constituents relative to their electric charge, so that glueballs should be much stickier than $\bar{q}q$ states. The available experimental data in the pseudoscalar and tensor channels indicates that $\iota(1440)$ and $\Theta(1700)$ are both rather sticky objects.

In Section 3 I will present the low energy theorem of Novikov et al., which is a consequence of the trace[2] and chiral anomalies[3] in the flavor singlet channel. Here my discussion largely follows Sharpe's talk at the Spring, 1984 Vanderbilt conference.[4] The conclusion is that the lightest particle in the scalar or pseudoscalar channel with large couplings to two gluons must also have large couplings to two photons. In the pseudoscalar channel this is probably $\eta'(958)$ (whose large coupling to two gluons can be understood without invoking a large glueball admixture).[5] We might also anticipate some enhancement for heavier states such as $\iota(1440)$ but it is very difficult to make even a semi-quantitative estimate. A light glueball might still be the lightest particle with large gluonic couplings in the scalar channel, in which case it could be copiously produced in $\gamma\gamma$ collisions. In his talk at this conference[6] Kück has reported on the PLUTO measurement of $\gamma\gamma \rightarrow \pi^+\pi^-$ which shows an apparent dip at $\sim$ 600 MeV. This might be the manifestation of an enhancement at $\leq$ 500 MeV (where the experimental efficiency dies) or it could be the effect of interference effects, as discussed by Sharpe, Jaffe, and Pennington.[7] Despite the apparent lack of structure in the I=0 s-wave $\pi\pi$ phase shift, it is still important to look carefully at the low mass dipion in $\gamma\gamma \rightarrow \pi\pi$ and $\psi \rightarrow \gamma\pi\pi$.

Section 4 concerns the $\gamma\gamma$ decays of η and η'. Frequently, at this conference and elsewhere, we are shown so-called SU(3) predictions for these decays, which are used to determine the η-η' mixing angle. In fact these predictions do <u>not</u> follow from SU(3) but are based on a dynamical assumption, tantamount to ideal mixing, which is very badly violated in the η-η' system (this violation of ideal mixing is the essence of the U(1) problem). In Section 4 I present an analysis which does not require this problematical assumption.[8] In this analysis the decays $\eta \to \gamma\gamma$, $\eta' \to \gamma\gamma$ and $\eta \to \pi+\pi-\gamma$ are all deduced from the chiral anomaly. This allows comparison with the data without the need for an ideal-mixing or so-called "nonet symmetry" assumption. A second analysis applies vector dominance to $\eta \to \gamma\gamma$, $\rho \to \eta\gamma$, $\eta' \to \gamma\gamma$ and $\eta' \to \rho\gamma$ to test the quark charge assignments of QCD.[9] While this second analysis depends largely on the well known ratio $\Gamma(\eta' \to \gamma\gamma)/\Gamma(\eta' \to \rho\gamma)$, the first analysis requires clarification of the presently confused experimental results for $\eta \to \gamma\gamma$ and $\eta' \to \gamma\gamma$.

In Section 5 I describe an important test that can be made in $\gamma\gamma$ physics of the hypotheses that $\xi(2220)$ or $\zeta(8320)$ (but not both) are Higgs bosons in two doublet models with enhanced couplings to $Q = +2/3$ or $Q = -1/3$ charge quarks respectively. These are difficult hypotheses to test experimentally.   If either particle can be detected in $\gamma\gamma$ scattering it would imply that it is <u>not</u> such a Higgs boson.

In Section 6 I conclude with a Rogues Gallery of particles and particle-candidates which could profitably be studied in $\gamma\gamma$ collisions. We would like to either measure or bound the $\gamma\gamma$ widths of all of these rogues.

2. Stickiness

Naively, using perturbation theory as a guide, the decay of a hypothetical pure glueball G to two photons proceeds by a qq loop. Therefore comparing its $\gamma\gamma$ width to the $\gamma\gamma$ width of a meson M of the same spin-parity and comparable mass, we expect a suppression

$$\frac{\Gamma(G \to \gamma\gamma)}{\Gamma(M \to \gamma\gamma)} \sim \left(\frac{\alpha_s}{\pi}\right)^2 \tag{1}$$

Similarly in perturbation theory the ratio of radiative decay widths from the ψ is enhanced by

$$\frac{\Gamma(\psi \to \gamma G)}{\Gamma(\psi \to \gamma M)} \sim \left(\frac{1}{\alpha_s}\right)^2 \tag{2}$$

This suggests that we consider the stickines S_X of the state X, defined as the ratio

$$S_X \equiv \frac{\Gamma(\psi \to \gamma X)}{\Gamma(X \to \gamma\gamma)} \; \frac{LIPS(X \to \gamma\gamma)}{LIPS(\psi \to \gamma X)} \tag{3}$$

where LIPS means Lorentz Invariant Phase Space. We do not attribute any significance to the absolute normalization of S_X. Now from Eqs. (1) and (2) we expect a glueball G to be much more sticky than a meson M of the same J^{PC},

$$S_G \gg S_M \tag{4}$$

Since S_X is proportional to the ratio $|<X|gg>|^2/|<X|\gamma\gamma>|^2$ it probes the relative color and electric charges of the constituents of X.

Stickiness has two advantages, one experimental, the other theoretical. The experimental advantage is that we do not need to know the branching ratio of the state X, $B(X \to$ final state), in order to measure S_X. For example, we can determine S_θ from measurements of $\Gamma(\psi \to \gamma\theta \to \gamma\bar{K}K)$ and $\sigma(\gamma\gamma \to \theta \to \bar{K}K)$ without knowing the branching ratio $B(\theta \to \bar{K}K)$. The theoretical advantage is that dynamical factors tend to cancel in the stickiness ratio, though the cancellation is not perfect since in $\psi \to \gamma X$ the two gluons which couple to X may be off-mass-shell whereas the two photons in $\gamma\gamma \to X$ are essentially on-shell. The imaginary part of $M(\psi \to \gamma X)$ corresponds to on-shell gluons by the Cutkosky-Landau rules, and the off-shell effects are only in the real part.

Consider the pseudoscalar and tensor channels, which are of interest in the glueball hunt. For $J^{PC}(X) = 0^{-+}$ both $\psi \to \gamma X$ and $X \to \gamma\gamma$ have p-wave phase space, so with an arbitrary normalization N I define

$$S_X(0^{-+}) = N \left(\frac{m_X}{k_{\psi\gamma X}} \right)^3 \frac{\Gamma(\psi \to \gamma X)}{\Gamma(X \to \gamma\gamma)}$$

$$(5)$$

where $k_{\psi\gamma X}$ is the energy of the energy of the photon in $\psi \to \gamma X$ in the ψ rest frame. I choose N so that $S_\eta = 1$. Then using the experimental data I find

$$S_\iota : S_{\eta'} : S_\eta : S_{\pi^0} \cong (>10) : 2\tfrac{1}{2} : 1 : .02$$

$$(6)$$

with experimental errors omitted. The small value of S_{π^0} is the expected consequence of the pion isospin. The lower limit for S_ι is based on the experimental upper limit $\Gamma(\iota \to \gamma\gamma) \cdot B(\iota \to \bar{K}K\pi) \leq$ 7 KeV. Notice that $\iota(1440)$ is apppreciably more sticky than $\eta'(958)$:

$$S_\iota / S_{\eta'} \gtrsim 4$$

$$(7)$$

For the tensors, $J^{PC}(X) = 2^{++}$, $X \to \gamma\gamma$ and $\psi \to \gamma X$ can proceed by s-wave or d-wave. I will assume the s-wave dominates and define

$$S_X(2^{++}) = N \frac{m_X}{k_{\psi\gamma X}} \cdot \frac{\Gamma(\psi \to \gamma X)}{\Gamma(X \to \gamma\gamma)}$$

$$(8)$$

with N chosen to give $S_f = 1$. Then the experimental data implies

$$S_\xi : S_\theta : S_{f'} : S_f \cong (>1) : (>20) : 14 : 1$$

$$(9)$$

where again I have omitted the experimental errors, which are in some cases considerable. For the sake of comparison I have assumed $J(\xi) = 2$. The lower limits for S_ξ and S_θ are based on new preliminary 95% CL upper limits from TASSO first presented at this conference,[10] $\Gamma(\xi \to \gamma\gamma) \cdot B(\xi \to \bar{K}K) < 0.5$ KeV and $\Gamma(\theta \to \gamma\gamma) \cdot B(\theta \to \bar{K}K) < 0.14$ KeV. We see that θ is very sticky compared to f, though f' is also rather sticky.

The large value of $S_{f'}/S_f$ is not unexpected. If we assume ideal mixing, $f= 1/\sqrt{2}(\bar{u}u + \bar{d}d)$ and $f' = \bar{s}s$, then a naive calculation gives

$$S_{f'}/S_f = \frac{1}{2} \Bigg/ \left[\frac{\frac{1}{9}}{\frac{1}{\sqrt{2}}\left(\frac{4}{9} + \frac{1}{9}\right)} \right]^2 \tag{10}$$

The factor 1/2 is the expected ratio $\Gamma(\psi \to \gamma f')/\Gamma(\psi \to \gamma f)$ with phase space removed. The remaining factor 25/2 is the square of the weighted sum of the square of the constituent quark charges, the naive expectation for $\Gamma(f \to \gamma\gamma)/\Gamma(f' \to \gamma\gamma)$ with phase space removed. Including the experimental uncertainties which were omitted in (9), I find combining errors in quadrature that

$$S_{f'}/S_f = 14 \pm 8 \tag{11}$$

This is consistent with the naive estimate (10) at the 1σ level.

As the upper limit on $\Gamma(\theta \to \gamma\gamma)$ improves, the stickiness of θ becomes increasingly evident. It suggests that θ is not a $\bar{q}q$ meson, since of the $\bar{q}q$ mesons only $\bar{s}s$ would be very sticky, while θ is much too light to be the $\bar{s}s$ radial excitation of f'.

3. Low Energy Theorems for J = 0 States.

Contrary to the previous section, light $J = 0$ glueballs may have $\gamma\gamma$ couplings as large as $\bar{q}q$ mesons. This conclusion applies to the lightest scalar and pseudoscalar particle which couples appreciably to two gluons. For pseudoscalars this is probably $\eta'(958)$, whose large coupling to two gluons can be understood without assuming a large glueball admixture (see below and Ref. 5). The implications for a heavier pseudoscalar glueball - such as $\iota(1440)$ may be - are less clear, though some enhancement above the naive expectation of Section 2 seems likely. In the scalar channel, many theoretical estimates have suggested a very light glueball, ≤ 1 GeV, and there are amusing hints in $\psi \to \omega\pi\pi$[11] and $\Upsilon(3S) \to \Upsilon(1S)\pi\pi$[12] for structure around $\sim 500-600$ MeV. The low energy theorem suggests that $\gamma\gamma \to \pi\pi$ is well worth studying down to the lowest possible dipion masses. The report[6] of a rise toward threshold below 600 MeV certainly deserves further investigation. We need to measure the spectrum as close to threshold as possible, in both $\gamma\gamma \to \pi\pi$ and $\psi \to \gamma\pi\pi$.

The remarkable low energy theorems[1] follow from the trace[2] and chiral anomalies.[3] Let $\theta^{\mu\nu}$ be the QCD stress tensor for just the light matter fields u, d, s, and let A_1^μ be the SU(3) flavor singlet axial current. We then consider the matrix elements between the vacuum and a two photon state of the trace of the stress tensor $\theta \equiv \theta_\mu^{\ \mu}$ and the divergence of the axial current, ∂A_1. We neglect the light quark masses, $m_{u,d,s} \to 0$, but include anomalies of both QCD and QED. The result is

$$\langle 0| \theta |\gamma_1\gamma_2\rangle = \langle 0| \frac{\beta}{2g} G\cdot G + \frac{\alpha R}{6\pi} F\cdot F |\gamma_1\gamma_2\rangle$$

$$\tag{12a}$$

$$\langle 0| \partial A_1 |\gamma_1\gamma_2\rangle = \langle 0| \frac{\sqrt{3}\alpha_s}{2\sqrt{2}\pi} G\cdot\tilde{G} + \frac{\alpha}{\sqrt{6}\pi} F\cdot\tilde{F} |\gamma_1\gamma_2\rangle$$

$$\tag{12b}$$

Here $\beta = \beta(g)$ is the QCD β function for three flavors, $R = 2$ is the R of e^+e^- annihilation for the three light flavors, $G\cdot G = G_{\mu\nu}G^{\mu\nu}$ is the square of the gluon field strength tensor, $G\cdot\tilde{G} = \frac{1}{2}\epsilon_{\mu\nu\alpha\beta} G^{\mu\nu}G^{\alpha\beta}$, and $F^{\mu\nu}$ is the photon field strength tensor. Because of the Adler-Bardeen theorem, Eq. (12b) is exact to any finite order in α and α_s. Equation (12a) does have $O(\alpha)$ corrections but the important and remarkable feature is that there are no $O(\alpha_s)$ QCD corrections to the $O(\alpha)$ QED anomaly.[13] This means that the low energy theorem derived from Eq. (12a) does not have uncalculable QCD corrections.

The crucial observation of Ref. (1) is that the left sides of Eqs. (12) are $O(k^4)$, where k_i are the photon momenta, so that the $O(k^2)$ terms on the right side must cancel. If there are no singularities, then Lorentz invariance, gauge invariance, and Bose statistics imply that the left sides are $O(k^4)$. In the massless quark limit the triangle diagrams do contribute $1/k^2$ singularities to the left sides,[14] but these are not really present in a confining theory. Goldstone bosons could also contribute $1/k^2$ poles to the left side, but these are not expected in QCD because of nonperturbative effects – the $U(1)$ effects for the pseudoscalars[15] and analogous effects for the scalars.[16]

The cancellation of the QED and QCD anomaly terms then gives the following remarkable results:

$$\langle 0 | \frac{\beta}{2g} G \cdot G | \gamma_1 \gamma_2 \rangle = \frac{\alpha R}{6\pi} F_1 \cdot F_2 + O(k^4)$$

(13a)

$$\langle 0 | \frac{\sqrt{3}\,\alpha_s}{2\sqrt{2}\,\pi} G \cdot \tilde{G} | \gamma_1 \gamma_2 \rangle = \frac{\alpha}{\sqrt{6}\,\pi} F_1 \cdot \tilde{F}_2 + O(k^4)$$

(13b)

where $F_i^{\mu\nu} = k_i^{\mu}\epsilon_i^{\nu} - k_i^{\nu}\epsilon_i^{\mu}$, k_i and ϵ_i being the momentum and polarization tensor of the photon γ_i. Equation (13) is remarkable because there are no powers of α_s on the right side!

We can easily use Eq. (13) to obtain low energy theorems for the $\gamma\gamma$ widths of the lightest $J = 0$ particles which dominate the two gluon channel. Consider first the scalar, $J^{PC} = 0^{++}$. Suppose there is a very light scalar glueball S, which couples with strength F_S (analogous to F_π) to the $G \cdot G$ operator:

$$\langle 0 | \frac{\beta}{2g} G \cdot G | S \rangle = F_S m_S^2$$

(14)

We assume that the left side of Eq. (13a) is dominated by the S pole,

$$\langle 0 | \frac{\beta}{2g} G \cdot G | \gamma_1 \gamma_2 \rangle = \frac{F_S m_S^2 \cdot \langle S | \gamma_1 \gamma_2 \rangle}{m_S^2 - (k_1 + k_2)^2}$$

(15)

Then from Eq. (13)

$$\langle S | \gamma_1, \gamma_2 \rangle = \frac{1}{F_s} \frac{\alpha R}{6\pi} F_1 \cdot F_2 + O(k^4)$$

$$(16)$$

If S is light enough we can neglect the $O(k^4)$ terms in (16) and compute the two photon width

$$\Gamma(S \to \gamma\gamma) \cong \frac{\alpha^2 R^2 m_s^3}{144 \pi^3 F_s^2}$$

$$(17)$$

With $S \to \sigma$ this is just the old POT/PCDC low energy theorem that Crewther, Ellis and I found for the would-be "dilaton", – the would-be Goldstone bosons of scale invariance.[2] Under the stated assumptions, Eqs. (16) and (17) should be approximately valid even though there is no Goldstone limit of scale invariance in QCD.

If F_S is of the order of the PCAC constant F_π, then $\Gamma(S \to \gamma\gamma)$ would be of the same order as the $\gamma\gamma$ widths of $\overline{q}q$ mesons. The actual value of F_S depends on dynamics which is difficult to estimate reliably. The ITEP sum rules imply[17] $F_S \simeq 300$ MeV which for $m_S = 500$ MeV gives $\Gamma(S \to \gamma\gamma) \simeq 70$ eV. Another estimate,[18] based on a different approach to the ITEP sum rules, finds a value of F_S smaller by one order of magnitude, implying a value for $\Gamma(S \to \gamma\gamma)$ larger by two orders of magnitude.

The result in the pseudoscalar channel is analogous. We assume that the left side of Eq. (13b) is dominated by the pole of a single light pseudoscalar P. Defining F_P by

$$\langle 0 | \frac{\sqrt{3}\alpha_s}{2\sqrt{2}\pi} G \cdot \tilde{G} | P \rangle = F_P m_P^2$$

$$(18)$$

we find the low energy theorem

$$\langle P | \gamma_1, \gamma_2 \rangle = \frac{1}{F_P} \frac{\alpha}{\sqrt{6}\pi} F_1 \cdot \tilde{F}_2 + O(k^4)$$

$$(19)$$

and the width

$$\Gamma(P \to \gamma\gamma) \cong \frac{\alpha^2 m_P^3}{24\pi^3 F_P^2} \tag{20}$$

If the lightest pseudoscalar P happens to be the flavor singlet meson

$$\eta_1 = \frac{1}{\sqrt{3}}(\bar{u}u + \bar{d}d + \bar{s}s) \tag{21}$$

then Eq. (19). is just the familiar low energy theorem for $\eta_1 \to \gamma\gamma$. We define the singlet PCAC constant F_1 by

$$\langle 0| A_1^\mu | \eta_1(k)\rangle = i k^\mu F_1 \tag{22}$$

F_1 is analogous to F_π but they are <u>not</u> related by any symmetry. Now in the chiral limit and neglecting electroweak corrections, ∂A_1 is just given by the QCD anomaly, so that

$$\langle 0| \partial A_1 | \eta_1\rangle = \langle 0| \frac{\sqrt{3}\,\alpha_s}{2\sqrt{2}\,\pi} G \cdot \tilde{G} | \eta_1\rangle \tag{23}$$

or, comparing Eqs. (18) and (22),

$$F_1 = F_P \tag{24}$$

With $F_P = F_1$ and $P = \eta_1$, Eq. (19) is the usual low energy theorem, which can be obtained more directly using broken chiral symmetry and η_1 pole dominance.

Incidentally, Eq. (24) shows that the large coupling of $\eta'(958)$ to gluons, which is indicated by the large value of $\Gamma(\psi \to \gamma\eta')$, can be understood as a consequence of approximate chiral SU(3) and the chiral anomaly, with no need to assume a large glueball admixtude in η'. That is, we see from Eqs. (23) and (24) that as $m_{u,d,s} \to 0$, the QCD chiral anomaly requires the $\bar{q}q$ state η_1 to have the same coupling to the gluon bilinear operator $(\sqrt{3}\,\alpha_s/2\,\sqrt{2}\,\pi)\,G\cdot G$ as it has to the $\bar{q}q$ bilinear ∂A_1! The relevant masses $m_{u,d,s}$ are the <u>current</u> quark masses, which are indeed small. The qualitative conclusion which follows from approximate chiral symmetry is that the large coupling of η' to gluons is expected in the conventional picture, which identifies η' predominantly with η_1.

It is straightforward to include the effect of $\eta_1 - \eta_8$ mixing on the low energy theorems for $\eta \to \gamma\gamma$ and $\eta' \to \gamma\gamma$. We consider broken chiral symmetry, $m_{u,d,s} \neq 0$, and allow an arbitrary $\eta - \eta'$ mixing angle θ, defined conventionally as

$$\eta = \cos\theta\ \eta_8 - \sin\theta\ \eta_1$$
$$\eta' = \sin\theta\ \eta_8 + \cos\theta\ \eta_1 \tag{25}$$

The PCAC constants are defined by

$$\langle 0 |\, A_8^\mu \,|\, \eta_8 \rangle = i k^\mu F_8$$
$$\langle 0 |\, A_1^\mu \,|\, \eta_1 \rangle = i k^\mu F_1 \tag{26}$$

We assume that SU(3) symmetry holds for the current matrix elements,

$$\langle 0 |\, A_8^\mu \,|\, \eta_1 \rangle = 0$$
$$\langle 0 |\, A_1^\mu \,|\, \eta_8 \rangle = 0 \tag{27}$$

so that for instance

$$\langle 0 |\, A_1^\mu \,|\, \eta' \rangle = i k^\mu \cos\theta\ F_1 \tag{28}$$

The constant F_8 is related to F_π by SU(3) symmetry

$$F_8 \cong F_\pi \tag{29}$$

but there is no symmetry relating F_1 to F_8 or F_π. Because of the presence of the anomaly and nonperturbative gluonic effects in the flavor singlet channel, there is no simple dynamical reason to expect F_1 to equal F_π. They are equal to lending order in the $1/N_{color}$ expansion, but in the same order the η' would be lighter than $\sqrt{3}\, m_\pi$ – a rather unsuccessful prediction. The value of F_1/F_π depends on complicated dynamics. The experimental data should be analyzed without assuming $F_1 = F_\pi$ or any other value. Instead we should try to determine F_1 from the data, as discussed in the next Section.

4. Radiative Decays of η and η'

At this conference and others, the widths for $\eta \to \gamma\gamma$ and $\eta' \to \gamma\gamma$ are used to determine the $\eta-\eta'$ mixng angle defined by means of a so-called "SU(3)" comparison with $\pi^0 \to \gamma\gamma$. This is an incorrect argument. SU(3) relates $\eta_8 \to \gamma\gamma$ to $\pi^0 \to \gamma\gamma$ but says nothing at all about $\eta_1 \to \gamma\gamma$. What is really assumed is not SU(3) symmetry but the OIZ rule. More precisely, in the context of a $\bar{q}q$ bound state, the assumption is that the wave function at the origin is the same for η_1 and η_8. The assumption that octet and singlet wave functions are approximately equal implies approximately equal binding energies, and this in turn implies ideal mixing. This is a good assumption for the vector mesons, as shown by the success of the ideal mixing sum rule,

$$m_\omega^2 + m_\phi^2 = 2 m_{K^*}^2 \tag{30}$$

It is however a terrible assumption for the pseudoscalars, for which the corresponding sum rule is badly violated. This failure is the "U(1) problem". It means that η_1 and η_8 have very different binding energies. It is therefore very dangerous to assume they have equal wave functions. In this section I present an analysis which avoids this assumption.

The low energy theorems which follow from pole dominance and the chiral anomaly are

$$\mathcal{F}(\eta \to \gamma\gamma) = -\frac{\gamma}{\sqrt{3}\,\pi}\left(\frac{\cos\theta}{F_8} - 2\sqrt{2}\,\xi\,\frac{\sin\theta}{F_1}\right)$$

$$\mathcal{F}(\eta' \to \gamma\gamma) = -\frac{\gamma}{\sqrt{3}\,\pi}\left(\frac{\sin\theta}{F_8} + 2\sqrt{2}\,\xi\,\frac{\cos\theta}{F_1}\right)$$

$$(31)$$

where F_1, F_8, and θ are defined in Eqs. (25) and (26). The parameter ξ is defined as $\xi = 1$ in QCD and $\xi = 2$ in the Han Nambu model.[19] (In the chrial limit the amplitude for $\eta_1 \to \gamma\gamma$ was derived in the previous section. For broken chiral symmetry it can be derived in an analogous way to the familiar derivation of $\pi^0 \to \gamma\gamma$).[20] SU(3) symmetry implies $F_8 \simeq F_\pi$ but tells us nothing abut F_1. At this point we have two unknown parameters, θ and F_1, which we might fit with the two experimental quantities $\Gamma(\eta \to \gamma\gamma)$ and $\Gamma(\eta' \to \gamma\gamma)$. We can do better than this by considering $\Gamma(\eta \to \pi^+\pi^-\gamma)$, which is also fixed by pole dominance and the chiral anomaly.

Before discussing what is learned from $\eta \to \pi\pi\gamma$, I want to add a few words about the most remarkable aspect of the chiral (and trace) anomaly. The anomalies are especially fascinating because they connect low and high energy phenomena. The anomaly in the VVA Ward identity that determines $\Gamma(\pi^0 \to \gamma\gamma)$ arises because of the behavior <u>at infinite momentum</u> of the fermion loop in the VVA triangle graph.[20] So by measuring a low energy quantity, $\Gamma(\pi^0 \to \gamma\gamma)$, we are learning about properties of the theory at very high energy. This is shown explicitly by Crewther[2] who related the $\pi^0 \to \gamma\gamma$ amplitude to KR, where R is the familiar R of e^+e^- annihilation and K is a quantity measured in deep inelastic election scattering. Similarly, as shown in Eqs. (13a) and (17) above, the trace anomaly is fixed by R alone.[2]

It is this remarkable dependence on the high energy behavior which makes $\eta \to \gamma\gamma$ sensitive to the color degrees of freedom in the Han Nambu model that give $\xi = 2$ in Eq. (31). (See however footnote (19).) This is in contrast to $\sigma(e^+e^- \to \text{hadrons})$ and $\sigma(\gamma\gamma \to \text{hadrons})$ which are not sensitive to the color degrees of freedom at energies far below color threshold.

108

There are chiral anomalies not only in the VVA and AAA triangles but also in box and pentagon diagrams with odd numbers of axial currents. In particular, there are low energy theorems for $\eta \to \pi\pi\gamma$ and $\eta' \to \pi\pi\gamma$ which follow from the VAAA and VVA anomalies. The low energy theorems are[8]

$$\mathcal{F}(\eta \to \pi^+\pi^-\gamma) = \frac{-e}{4\sqrt{3}\,\pi^2}\,\frac{1}{F_\pi^2}\left(\frac{\cos\theta}{F_8} - \sqrt{2}\,\frac{\sin\theta}{F_1}\right)$$

$$\mathcal{F}(\eta' \to \pi^+\pi^-\gamma) = \frac{-e}{4\sqrt{3}\,\pi^2}\,\frac{1}{F_\pi^2}\left(\frac{\sin\theta}{F_8} + \sqrt{2}\,\frac{\cos\theta}{F_1}\right) \tag{32}$$

The low energy theorems (31) and (32) apply at the unphysical low energy point where all four-momenta vanish. For $\eta \to \gamma\gamma$ and $\eta' \to \gamma\gamma$ we assume η and η' pole dominance which implies approximate equality of the physical amplitude to the amplitude at the unphysical low energy point. But for $\eta \to \pi^+\pi-\gamma$ and $\eta' \to \pi^+\pi-\gamma$ we must also include the effect of the ρ meson on the extrapolation. The Dalitz plot for $\eta \to \pi^+\pi-\gamma$ shows that the dipion is dominated by the ρ, even though we are well below the ρ threshold. We incorporate this effect by adding a Breit-Wigner factor to the amplitude which is normalized to unity at zero dipion mass as dictated by the low energy theorem. So we replace Eq. (32) by

$$\mathcal{F}(\eta \to \pi^+\pi^-\gamma) \longrightarrow \mathcal{F}(\eta \to \pi^+\pi^-\gamma) \cdot \frac{-m_\rho^2 + im_\rho\Gamma_\rho}{p_{\pi\pi}^2 - m_\rho^2 + im_\rho\Gamma_\rho} \tag{33}$$

where $p_{\pi\pi}$ is the four-momentum of the dipion.

For $\eta' \to \pi\pi\gamma$ we are above the ρ threshold and in fact the decay is completely dominated by $\eta' \to \rho\gamma$. In this case the relevance of the low energy theorem Eq. (32) is unclear, since we must extrapolate across the ρ pole. In addition, there may be two separate components: an elementary ρ amplitude, $\eta' \to \rho\gamma$, which has nothing to do with Eq. (32) as well as $\eta' \to \pi\pi\gamma$ component given by Eq. (32). Because of these uncertainties I will not include $\eta' \to \pi^+\pi^-\gamma$ in the analysis.

So we have finally three experimental quantities to determine the two parameters F_1 and θ for the cases $\xi = 1$ and $\xi = 2$. The amplitudes in Eqs. (31) and (32) are related to the widths by

$$\Gamma(X \to \gamma\gamma) = \frac{m_X^3}{64\pi} \left| \mathcal{F}(X \to \gamma\gamma) \right|^2 \tag{34}$$

for $X = \eta, \eta'$ and

$$\Gamma(\eta \to \pi^+\pi^-\gamma) = \frac{3 I_\eta}{2 m_\eta (2\pi)^5} \left| \mathcal{F}(\eta \to \pi^+\pi^-\gamma) \right|^2 \tag{35}$$

where $I_\eta = 4.75 \cdot 10^{-4} \, m_\eta^8$ is the result for the three body phase space including the ρ final state interaction of Eq. (33) (which contributes an enhancement of 1.85 to I_η).

While there is now a large spread in experimental values for the three widths $\Gamma(\eta \to \gamma\gamma)$, $\Gamma(\eta' \to \gamma\gamma)$, and $\Gamma(\eta \to \pi\pi\gamma)$, the ratio $\Gamma(\eta \to \gamma\gamma)/\Gamma(\eta \to \pi\pi\gamma)$ seems well determined. We may use the experimental value[22]

$$\frac{\Gamma(\eta \to \gamma\gamma)}{\Gamma(\eta \to \pi^+\pi^-\gamma)} = 7.94 \pm .027 \tag{36}$$

and Eqs. (32)–(35) to find an interesting constraint on F_1 and θ:

$$\frac{F_8}{F_1} \tan\theta = \begin{cases} -.21 & \xi = 1 \\ -.06 & \xi = 2 \end{cases} \tag{37}$$

This should be counted as at least a qualitative success, since it implies that if F_8/F_1 is positive and of order one then θ is small and negative – as we would expect from the naive quark model calculation of $\eta - \eta'$ mixing which gives $\theta = -11^0$.

The next step is to use the constraint of Eq. (37) and $F_8 \simeq F_\pi$ to determine $\Gamma(\eta \to \gamma\gamma)$ in terms of θ alone:

$$\Gamma(\eta \to \gamma\gamma) = \begin{cases} 432 \text{ eV } \cos^2\theta & \xi = 1 \\ 304 \text{ eV } \cos^2\theta & \xi = 2 \end{cases} \tag{38}$$

Experimentally, the old Cornell value[23] (from Primakoff photoproduction) is 324 ± 46 eV while from $\gamma\gamma$ scattering the Crystal Ball has 560 ± 120 ± 100 eV and JADE has 560 ± 50 ± 80.[24] For small θ the prediction for $\xi = 1$ is strategically located between the experimental values while the prediction for $\xi = 2$ would not survive confirmation of the higher value.

Finally we include $\Gamma(\eta' \to \gamma\gamma)$ in the analysis. We impose the constraint Eq. (37) and then compute $\Gamma(\eta \to \gamma\gamma)$ and $\Gamma(\eta' \to \gamma\gamma)$ for a range of values of F_8/F_1. For $F_8/F_1 = 1/2, 1, 2$ the results are shown in Table 1.

Table 1: Predicted values for θ, $\Gamma(\eta \to \gamma\gamma)$ and $\Gamma(\eta' \to \gamma\gamma)$ as a function of $F_8/F_1 = 1/2, 1, 2$. Predictions are given for fractional ($\xi = 1$) and integral ($\xi = 2$) charge quarks.

$F_8/F_1 =$		$\tfrac{1}{2}$	1	2
$\xi = 1$	$\theta =$	-23^0	-12^0	-6^0
	$\dfrac{\Gamma_{\eta \to \gamma\gamma}}{\Gamma_{\eta' \to \gamma\gamma}} =$	366 0.75	413 6	427 eV 27 KeV
$\xi = 2$	$\theta =$	-7^0	$-3\tfrac{1}{2}^0$	$-1\tfrac{1}{2}^0$
	$\dfrac{\Gamma_{\eta \to \gamma\gamma}}{\Gamma_{\eta' \to \gamma\gamma}} =$	300 $6\tfrac{1}{2}$	300 28	300 eV 115 KeV

The experimental situation for $\Gamma(\eta' \to \gamma\gamma)$ is also very unclear,[24] with results quoted between ~3 and ~6 KeV. We see from Table 1 that the Han Nambu model, $\xi = 2$, could fit the data for $F_8/F_1 \leq \tfrac{1}{2}$ if $\Gamma(\eta \to \gamma\gamma)$

$\leq$ 300 eV. The QCD predictions could be consistent with $\Gamma(\eta \to \gamma\gamma)$ ~400 eV and $\Gamma(\eta' \to \gamma\gamma)$ ~$4\frac{1}{2}$ KeV for F_8/F_1 a little smaller than 1 and θ a little smaller than -11°. It will be interesting to see the final experimental determinations of $\Gamma(\eta \to \gamma\gamma)$ and $\Gamma(\eta' \to \gamma\gamma)$.

I conclude this section by briefly reviewing a second analysis of the radiative η and η' decays.[9] This analysis uses vector meson dominance to get at the parameter ξ and is independent of the values of F_1, F_8, and θ. The main point is that vector vector meson dominance with ρ, ω, ϕ only applies to the color singlet part of the photon. Therefore the vector dominance relation is

$$\mathcal{F}(\eta_1 \to \gamma\gamma) = \xi \sum_{V = \rho, \omega, \phi} \frac{e}{f_V} \mathcal{M}(\eta_1 \to V\gamma)$$

(39)

For QCD, $\xi = 1$, and this is just the usual statement of vector dominance. For the Han-Nambu model, $\xi = 2$, $(\eta_1 \to \gamma\gamma)$ has two contributions,

$$\langle \eta_1 | J_{EM}^{HN} J_{EM}^{HN} | 0 \rangle = \langle \eta_1 | J_{EM}^{(1)} J_{EM}^{(1)} | 0 \rangle$$
$$+ \langle \eta_1 | J_{EM}^{(8)} J_{EM}^{(8)} | 0 \rangle$$

(40)

Here J_{EM}^{HN} is the electromagnetic current in the Han Nambu model,

$$J_{EM}^{HN} = J_{EM}^{(1)} + J_{EM}^{(8)}$$

(41)

$J_{EM}^{(1)}$ is the color singlet current which is equal to the usual electromagnetic current of QCD, and $J_{EM}^{(8)}$ is a color octet current that gives the quarks their integer charge assignments. From the anomaly at the low energy point we know that

$$\langle \eta_1 | J_{EM}^{(1)} J_{EM}^{(1)} | 0 \rangle = \langle \eta_1 | J_{EM}^{(8)} J_{EM}^{(8)} | 0 \rangle$$

(42)

and I assume (42) is still approximately correct on the mass shell.
Vector dominance with ρ, ω, ϕ applies only to $J_{EM}^{(1)}$, so the right
side of Eq. (39) is multiplied by $\xi = 2$ to include the contribution of
$< \eta_1 | J_{EM}^{(8)} J_{EM}^{(8)} | 0 >$.

It is an easy exercise in SU(3) flavor symmetry to show that Eq.
(39) implies

$$\mathcal{F}(\eta_1 \to \gamma\gamma) = \frac{4}{3} \frac{e}{f_\rho} \mathcal{M}(\eta_1 \to \rho\gamma) \tag{43}$$

Next we use the theorem

$$\sin^2\theta + \cos^2\theta = 1 \tag{44}$$

to write

$$|\mathcal{F}(\eta_1 \to \gamma\gamma)|^2 = |\mathcal{F}(\eta \to \gamma\gamma)|^2 + |\mathcal{F}(\eta' \to \gamma\gamma)|^2 - |\mathcal{F}(\eta_8 \to \gamma\gamma)|^2 \tag{45}$$

$$|\mathcal{M}(\eta_1 \to \rho\gamma)|^2 = |\mathcal{M}(\rho \to \eta\gamma)|^2 + |\mathcal{M}(\eta' \to \rho\gamma)|^2 - |\mathcal{M}(\eta_8 \to \rho\gamma)|^2 \tag{46}$$

By SU(3) flavor symmetry,

$$|\mathcal{F}(\eta_8 \to \gamma\gamma)|^2 = \frac{1}{3} |\mathcal{F}(\pi^0 \to \gamma\gamma)|^2 \tag{47}$$

and

$$|\mathcal{M}(\eta_8 \to \rho\gamma)|^2 = \frac{1}{3} |\mathcal{M}(\omega \to \pi^0\gamma)|^2 \tag{48}$$

Finally we can solve for ξ in terms of six experimentally measurable
quantities:

$$\xi^2 = \left(\frac{3f_\rho}{4e}\right)^2 \frac{|\mathcal{F}_{\eta'\to\gamma\gamma}|^2 + |\mathcal{F}_{\eta\to\gamma\gamma}|^2 - \frac{1}{3}|\mathcal{F}_{\pi^0\to\gamma\gamma}|^2}{|\mathcal{M}_{\eta'\to\rho\gamma}|^2 + |\mathcal{M}_{\rho\to\eta\gamma}|^2 - \frac{1}{3}|\mathcal{M}_{\omega\to\pi^0\gamma}|^2} \tag{49}$$

Fortunately the present uncertainties in the total η' and η widths are correlated in the numerator and denominator of (49) and therefore tend to cancel. The principal uncertainty in ξ^2 is due to $B(\eta \to \rho\gamma)$ which figures only in the denominator.[28] The extraction of the $\mathcal{F}$ and $\mathcal{M}$ amplitudes from the experimental widths is straightforward except for $\mathcal{M}(\eta' \to \rho\gamma)$, for which it is essential to include the effect of the large ρ total width, $\Gamma_\rho = 154 \pm 5$ MeV. This is done by evaluating the three body phase space, $\eta' \to \pi^+\pi^-\gamma$, using a Breit-Wigner rho pole amplitude[21] as in Eq. (35).

The 1980 evaluation of the data gave[9] $\xi^2 = 1.15 \pm .25$. With the present experimental situation I find

$$\xi^2 = .87 \, {}^{+.50}_{-.22} \tag{50}$$

in good agreement with QCD, $\xi^2 = 1$, and in sharp disagreement with the naive expectation for integrally charged quarks, $\xi^2 = 4$. To arrive at this value of ξ^2, I crudely summarized the experimental situation for η and η' by $\Gamma(\eta \to \gamma\gamma) = 440 \pm 120$ eV and $\Gamma(\eta' \to \gamma\gamma) = 4\frac{1}{2} \pm 1\frac{1}{2}$ KeV and took $B(\rho \to \eta\gamma)$ from Ref. (25). As in Ref. (9) I followed the Yennie prescription[26] for the extrapolation of f_ρ from $q^2 = m_\rho^2$ to $q^2 = 0$, according to which self energy effects cause $f_\rho^2/4\pi = 1.93$ to be replaced by $f_{\rho\pi\pi}^2/4 = 2.96$. Had I used the unrenormalized $f_\rho^2/4 = 1.93$, I would have found $\xi^2 = .51^{+.32}_{-.22}$ still consistent with QCD at the one "sigma" level.

For QCD Eq. (49) is just the vector meson dominance relationship between $\eta \to \gamma\gamma$ and $\eta_1 \to (\rho, \omega, \phi)\gamma$. It is therefore just the analogue of the Gell Mann-Sharpe-Wagner relation between $\pi^0 \to \gamma\gamma$ and $\omega \to \pi^0\gamma$, which in my notation is

$$|\mathcal{F}(\pi^0 \to \gamma\gamma)|^2 = \frac{4}{9}\frac{e^2}{f_\rho^2}|\mathcal{M}(\omega \to \pi^0\gamma)|^2 \tag{51}$$

How well does this work? Using the prescription of Ref. (24), $f_\rho \rightarrow f_{\rho\pi\pi}$, it works embarassingly well, the left side being $6.42\cdot10^{-10}$ MeV^{-2} and the right side $6.49\cdot10^{-10}\mathrm{MeV}^{-2}$. If instead we use $f_\rho^2/4\pi = 1.93$, the right side overestimates the left by ~55%, just as ξ^2 drop from .87 to .57.

5. ξ/ζ and the Higgs Hypothesis

$\xi(2220)$ and $\zeta(8320)$ are fascinating creatures, which both stand in need of experimental confirmation. If either (or both) is a real effect and if their widths turn out to be too small to be hadronic, then it is natural to consider the possibility that they might be Higgs bosons. Neither can be the Higgs boson of the standard model, since $\Gamma(\psi \rightarrow \gamma\xi)$ and $\Gamma(T \rightarrow \gamma\zeta)$ are respectively ~10 and ~50 times too large. The simplest-variation is to consider two doublet models[27,28,29] in which one double Φ_1 couples to the weak isospin $+1/2$ fermions and Φ_2 couples to $-1/2$. Then for instance ξ could be dominantly from Φ_1 and the factor ~10 enhancement in $\psi \rightarrow \gamma\xi$ can be accomodated by having the vacuum expectation $v_1 \simeq v/3$. Here v is the vacuum expectation value of the standard one doublet model, and to get M_W right we need

$$v_1^2 + v_2^2 = v^2$$

(52)

Similarly ζ could be from Φ_2 with $v_2 \simeq v/7$. Because of (52) these two hypotheses are not simultaneously tenable.

It turns out that this explanation of ξ or ζ is extremely difficult to test in a conclusive way. For instance, to measure the spin of ξ would require $\geq 20,000,000$ ψ decays, which is $\geq 2\ 1/2$ times the present largest sample. To improve the limits on the widths is also very difficult. For ξ, the best test would be to look for toponium $\rightarrow \gamma\xi$.

Study of the two photon channel can make an important contribution to this problem. If ξ or ζ are the Φ_1 or Φ_2 of a two doublet model, then their $\gamma\gamma$ couplings must be far too small to be observable in $\gamma\gamma$

collisions. Therefore the detection of either in $\gamma\gamma$ scattering would rule out this interpretation.

The decay of a Higgs boson to two photons occurs by intermediate W-boson and fermion loop diagrams.[30] For fermions these are the same diagrams that give rise to the trace anomaly, and as discussed in preceding sections they have a remarkable sensitivity to the high energy structure of the theory. In particular _all_ fermions heavier than the Higgs can contribute to the $\gamma\gamma$ width. In the standard one doublet model, the width is

$$\Gamma(H \to \gamma\gamma) = \frac{\alpha^2 G_F m_H^3}{8\sqrt{2}\,\pi^3} \left| I_W + \sum_f I_f \right|^2 \tag{53}$$

where $I_W = -7/4$ and I_f goes from zero at $m_f \ll m_H$ to $R_f/3$ for $m_f \gg m_H$, R_f being the contribution of $\bar{f}f$ to the R of e^+e^- annihilation.

In two doublet models the fermion contribution to the $\gamma\gamma$ width of a boson predominantly from doublet Φ_i will be enhanced by (v/v_i) in amplitude. Therefore, assuming three generations of fermions, we have in the two doublet interpretation of ξ or ζ

$$\Gamma(\xi \to \gamma\gamma) \cong \frac{\alpha^2 G_F m_\xi^3}{8\sqrt{2}\,\pi^3} \left| I_W + 3(I_c + I_t) \right|^2$$
$$\cong .02 \text{ eV} \tag{54}$$

and

$$\Gamma(\zeta \to \gamma\gamma) \cong \frac{\alpha^2 G_F m_\zeta^3}{8\sqrt{2}\,\pi^3} \left| I_W + 7 I_b \right|^2$$
$$\cong 1 \text{ eV}. \tag{55}$$

Both are unobservably small. If ξ or ζ can be observed in $\gamma\gamma$ scattering, it will mean they cannot be the Higgs bosons of two doublet models. It would in fact make any Higgs interpretation seem very unlikely, since to make the $\gamma\gamma$ widths observably large would require a dubious tuning of vacuum expectation values in more complicated models.

116

6. Rogues Gallery

The main points in this talk are:

(1) By measuring the stickiness, _i.e._, comparing $\Gamma(X \to \gamma\gamma)$ and $\Gamma(\psi \to \gamma X)$, we probe the quark and glue content for all mesons X except possibly the J=0 states.

(2) Contrary to intuition, light J=0 glueballs may have substantial $\gamma\gamma$ widths.

(3) $\Gamma(\eta \to \gamma\gamma)$ and $\Gamma(\eta' \to \gamma\gamma)$ seem well described by QCD and not by integral charge quark models, though the experimental situation needs clarification. Everyone is to desist from claims that $\eta \to \gamma\gamma$ and $\eta' \to \gamma\gamma$ are related to $\pi^0 \to \gamma\gamma$ by SU(3) symmetry.

(4) Observation of $\xi(2220)$ or $\zeta(8320)$ in $\gamma\gamma$ scattering would contradict the difficult-to-contradict hypothesis that ξ or ζ are Higgs bosons in two doublet models.

Table 2 is a rogues gallery of interesting objects which can profitably be studied in $\gamma\gamma$ scattering. It is very important to put bounds on $\iota(1440)$ and $\theta(1700)$ in order to evaluate their status as possible glueballs. $\zeta(1270)$ is a possible $J^{PC} = 0^{-+}$ $\eta\pi\pi$ resonance[31] which is important in understanding whether $\iota(1440)$ is a glueball or a radially excited $\bar{q}q$ state.[32] It has the same mass and $\eta\pi\pi$ decay mode as the $J^{PC} = 1^{++}$ D(1270). $\gamma\gamma$ collisions are a good place to look for $\zeta(1270)$, since the D(1270) will not couple to $\gamma\gamma$ because of the Landau-Yang theorem. G(1590) is an interesting object seen at Serphukov,[32] for which a glueball interpretation has been advanced.[34] And in view of Section III and the report of Kück,[6] it is important to study $\gamma\gamma \to \pi\pi$ and $\psi \to \gamma\pi\pi$ at the lowest possible dipion masses.

The Rogues Gallery

Table 2: Resonances and resonance candidates

	Γ(MeV)	J^{PC}	Production/Decay	$\gamma\gamma$ Limit (KeV)
ι(1440)	76 ± 10	0^{-+}	$\psi\to\gamma\iota$, pp/$KK\pi$,$\rho\gamma$(?)	< 7-8/B($KK\pi$)
θ(1720)	130 ± 25	2^{++}	$\psi\to\gamma\theta$/KK,$\eta\eta$	< 0.14/B(KK)
ζ(1270)	70 ± 25	0^{-+}	$\pi^-p\to\zeta n/\delta\pi\to\eta\pi\pi$	
ξ(2320)	< 40		$\psi\to\gamma\xi$/KK	< .5/B($K_S K_S$)
G(1590)	210 ± 40	0^{++}	$\pi^-p\to GX/\eta\eta$,$\eta\eta$	
?(2.1)			$\psi\to\gamma?/\pi^+\pi^-$	
$\phi\phi$(2-2.3)		2^{++}	$\pi p\to\phi\phi n$	
$\rho\rho$		$0^{++}/2^{++}$	$\gamma\gamma$	seen
$\rho\rho/\omega\omega$ 1.8-1.9		0^{-+}	$\psi\to\gamma\rho\rho,\gamma\omega\omega$	
ζ(8320)	< 80		$\Upsilon\to\gamma\zeta$/multihadron	
Low mass dipion		0^{++}	$\gamma\gamma\to\pi\pi$ & $\psi\to\gamma\pi\pi$?	

Acknowledgments: I thank Stephen Sharpe for numerous discussions of the low energy theorems in Section 3.

References

1. V. Novikov, M. Shifman, A. Vainshtein, and V. Zakharov, Nucl. Phys. B191, 301 (1981).

2. R. Crewther, Phys. Rev. Lett. 28, 1421 (1972);
    M. Chanowitz and J. Ellis, Phys. Lett. 40B, 397 (1972) and Phys. Rev. D7, 2490 (1973). Higher order corrections are computed by J. Collins, A. Duncan and S. Juglekar, Phys. Rev. D16, 438 (1977);
    N. K. Nielsen, Nucl. Phys. B120, 212 (1977);
    P. Minkowski, U. Berne preprint-76-0813 (1976).

3. S. Adler, Phys. Rev. 117, 2426 (1969);
    J. S. Bell and R. Jackiw, Nuovo. Cim. 60A, 47 (1969).

4. S. Sharpe, Talk presented at the Symposium on High Energy e^+e^- Interactions, Vanderbilt Univ., 1984 (to be published in the proceedings).

5. V. Novikov, M. Shifman, A. Vainshtein, and V. Zakharov, Nucl. Phys. B165, 55 (1980).

6. H. Kück, these proceedings.

7. S. Sharpe, R. Jaffe, and M. Pennington, H.U.T.P.-84/A017.

8. M. Chanowitz, Phys. Rev. Lett. 35, 977 (1975).

9. M. Chanowitz, Phys. Rev. Lett. 44, 59 (1980).

10. Presented in the talk of F. Erné, these proceedings.

11. G. Gidal, presented at the 1981 N.Y. A.P.S. meeting (Jan., 1981);
     D. Scharre, Orbis Scientiae 1982 (Ed. B. Kursonogolu).

12. J. Green et al., Phys. Rev. Lett. 49, 617 (1982);
     G. Mageras et al., Phys. Lett. 118B, 453 (1982).

13. S. Collins, A. Duncan, and S. Joglekar, op. cit.;
     K. Nielsen, op. cit.

14. A. Dolgov and V. Zakharov, Nucl. Phys. B27, 925 (1971).

15. G. 't Hooft, Phys. Rev. Lett. 37, 8 (1976).

16. Scale invariance is broken perturbatively and by instanton effects like those discussed in Ref. (15) – see M. Chanowitz, Phys. Lett. 68B, 180 (1977).

17. The estimate is due to S. Sharpe, Ref. (5), based on the sum rules of Ref. (1).

18. S. Narison, CERN-TH-3796 (1984).

19. In at least some realizations of the Han-Nambu model (e.g., A. de Rujula, R. Giles, and R. Jaffe, Phys. Rev. $\underline{D17}$, 285 (1978)) dynamical effects can conspire to make $\xi = 1$ rather than the naive prediction $\xi = 2$. I thank R. Jaffe for a discussion.

20. S. Adler, 1970 Brandeis Summer Inst. (eds. S. Deser et al.) p. 5 (M.I.T. Press, Cambridge, MA).

21. To be precise, the prescription used is that in Eq. (11) of J. D. Jackson, Nuovo. Cim. XXXIV 1644 (1964).

22. Particle Data Group, Rev. Mod. Phys. $\underline{56}$, 51 (1984).

23. A. Browman et al., Phys. Rev. Lett. $\underline{32}$, 1067 (1974).

24. See the summary of the data by A. Cordier, these proceedings.

25. D. Andrews et al., Phys. Rev. Lett. $\underline{38}$, 198 (1977).

26. T. Bauer, R. Spital, D. Yennie, and F. Pipkin, Rev. Mod. Phys. $\underline{50}$, 261 (1978).

27. H. Haber and G. Kane, Pys. Lett. $\underline{135B}$, 196 (1984).

28. R. Willey, Phys. Rev. Lett. $\underline{52}$, 585 (1984).

29. M. Barnett, G. Senjanovici, L. Wolfenstein, and D. Wyler, Phys. Lett. $\underline{136B}$, 191 (1984);
M. Barnett, G. Senjanovici, and D. Wyler, NSF-ITP-84-45 (1984).

30. J. Ellis, M. Gaillard, and D. Nanopoulas Nucl. Phys. $\underline{B106}$, 292 (1976).

31. N. Stanton et al., Phys. Rev. Lett. $\underline{42}$, 346 (1979).

32. M. Chanowitz, Multiparticle Dynanics 1983 (eds. P. Yager and J. Gunion), p. 716 (1983).

33. F. Binon et al., Yad. Fiz. $\underline{38}$, 934 (1983).

34. S. Gerstein et al., (HEP 83-148 (1983).

120

Questions and Discussion

G. FARRAR (Rutgers): How do the predictions for $\Gamma(\eta' \to \gamma\gamma)$ change when the effect of glue is included? If we assume QCD is the correct theory (so that we put aside the integral charge quark question) can we deduce the presence or absence of new states?

M. CHANOWITZ: I think the ITEP low energy theorem implies that $\gamma\gamma$ widths are not going to help us discriminate between $J = 0$ glueballs and qq mesons. We will have to use the other kinds of evidence which help us to recognize the nonets and the states which do not fit neatly into nonets.

J.H. FIELD (L.P.N.H.E.): In the PCAC approach the radiative widths of the pseudoscalars are essentially determined by the Adler anomalies i.e., by the short distance behavior of the constituent quarks, where, from QCD, point like γ-quark coupling is expected. Is it not rather inconsistent to use VDM, characteristic of larger distance scales, in such situations? I should also like to remark that VDM calculations of the η,η 12γ widths typically overestimate the experimental values by a factor of 2-3. Taking this into account your ξ parameter then becomes consistent with the integer charge quark model value.

M. CHANOWITZ: There is no inconsistency between the use of the anomaly to compute $\Gamma(\pi^\circ \to \gamma\gamma)$ and the use of vector dominance to relate $\Gamma(\pi^\circ \to \gamma\gamma)$ with $\Gamma(\omega \to \pi^\circ\gamma)$. $\pi^\circ \to \gamma\gamma$ is certainly a low energy process; it happens to be sensitive to the high energy structure of the theory because it proceeds by a fermion loop which is dominated by the large loop momenta. But the <u>external</u> momenta are small and it is entirely appropriate to apply vector dominance.

As to your second point, my own calculations show good results in applying vector dominance to $\pi^\circ \to \gamma\gamma$ and $\eta/\eta' \to \gamma\gamma$. I do not find the factor 2-3 discrepancy which you cite.

Note added: While this talk was being prepared for publication, Field informed me that he had found a sign error in the VDM calculations of Dalitz and Sutherland which were the basis for his remark. With the sign error corrected he also finds good agreement with vector dominance.

PSEUDOSCALAR AND TENSOR MESON
PRODUCTION IN $\gamma\gamma$ REACTION

A. *CORDIER*

Laboratoire de l'Accélérateur Linéaire
91405 ORSAY CEDEX France

ABSTRACT

This revue covers the experimental situation concerning
the pseudoscalar and tensor mesons produced in $\gamma\gamma$ interactions. Results on other exclusive $\gamma\gamma$ processes are
presented by Fritz ERNE in the following report. A comparison of the results with SU(3) is made.

INTRODUCTION

Since F. Low's[1] suggestion in 1960 to measure the π° radiative
width at e^+e^- storage rings great progress have been done in the study
of two photon resonances. Since 1979 almost three dozens of experimental results have been published concerning in particular the "classical"
resonances which belong to the pseudoscalar meson multiplet = π°, η, η'
or to the tensor meson multiplet = f°, A_2, f'. Moreover for the first
time, at this Conference, final results on the Q^2 dependence of the
meson radiative widths have been reported.

Before considering the reasons of this interest let us recall
briefly the general features of this resonance production :

- the only resonances accessible via real photon interactions have
C-parity $C = +1$ and spin $J = 0$ or 2. For tensor mesons the helicities
$\lambda = 0$ and $|\lambda| = 2$ have to be considered

- the cross section of the $e^+e^- \rightarrow e^+e^- R$ reaction is given by :

$$\sigma(e^+e^- \to e^+e^- R) = \int \sigma_{\gamma\gamma \to R}(s) \; L_{\gamma\gamma}(z) \; dz$$

where s is the c.m. energy squared and $z = \sqrt{s}/2E_b$ (E_b = electron beam energy). $L_{\gamma\gamma}$ is the $\gamma\gamma$ luminosity function[2], $\sigma_{\gamma\gamma \to R}$ is the cross section for production of a resonance R with spin J. For two real photons :

$$\sigma_{\gamma\gamma \to R} = 8\pi(2J+1) \; \frac{\Gamma_R \; \Gamma_{R\gamma\gamma}}{(S - M_R^2)^2 + M_R^2 \; \Gamma_R^2}$$

Using the equivalent photon approximation and assuming that R is a narrow resonance ($\Gamma_R/M_R \ll 1$) :

$$\sigma(ee \to eeR) = \frac{16\alpha^2}{M_R^3} \; \ln^2\left(\frac{E_b}{m_e}\right) f(z). (2J+1). \Gamma_{R\gamma\gamma}$$

where the Low function $f(z) = (2 + z^2)^2 \, \ln(1/z) - (1-z^2)(3+z^2)$ is a slowly varying function of M_R and E_b.

Three remarks can be made :

* σ increases slowly with E_b, like $\ln E_b^2$
* σ decreases fastly with M_R, like M_R^3, meaning that there is a severe limitation on high mass resonance production
* $\Gamma_{R\gamma\gamma}$ wich is the interesting variable to determine is directly proportional to σ the measured cross section.

We will come back later on the case of non real photons.

Why measured $\Gamma_{R\gamma\gamma}$?

The interest attached to the measurement of the meson radiative widths comes from the fact that they are directly correlated to the quark content of these mesons (see e.g. ref. 3,4).

Indeed in the quark model mesons are described as coherent mixture of $q\bar{q}$ pairs with different flavours :

$$| R > = \sum_q C_q \, |q\bar{q} >$$

The coupling of two photons (with a given $\gamma\gamma$ helicity) to a quark pair is proportional to the square of the quark changes :

$$< q\bar{q}/\gamma\gamma > \ \alpha \ e_q^2 \ \psi_q(o) \qquad (\text{S wave})$$

$$< q\bar{q}/\gamma\gamma > \ \alpha \ e_q^2 \ \psi_q'(o) \qquad (\text{P wave})$$

where $\psi_q(o)$ is the radial quark wave function at the origin and $\psi_q'(o)$ is derivative.

Consequently, assuming $\psi_q(o)$ independant of the quark flavour, the $\gamma\gamma$ coupling constant and then the radiative width are directly correlated to the quark charges, therefore to the quark content of the mesons :

$$\Gamma_{R\gamma\gamma} \ \alpha \ m_R^3 \ g_{R\gamma\gamma}^2 \ \alpha \ < R|\gamma\gamma >^2 \ \alpha \ \left(\sum_q c_q \ e_q^2 \right)^2 \ = \ < e_q^2 >^2$$

Then, by measuring meson radiative widths we can understand how they are made. This understanding is somewhat complicated by the fact that while the coefficient c_q are well defined in the SU(3) framework, this SU(3) symmetry is broken by the larger mass of the s quark : the observable isoscalar $\eta - \eta'$ and $f - f'$ are mixtures of the SU(3) singlet and octet states expressed in terms of mixing angles θ that $\Gamma_{R\gamma\gamma}$ measurements have to determine. From the previous relations and with some weak assumptions linear relations can be written which connect the radiative widths of the three flavour neutral members of the pseudo-scalar or tensor nonet. First experimental results will be presented then we will use some of these relations to determine the mixing angles.

Let us start with some experimental remarks before presenting these results :
i) all experiments used the same criteria to measure the meson radiative widths :
   - low total energy ($< 0.4 \ E_b$) and low multiplicity to select $\gamma\gamma$ events

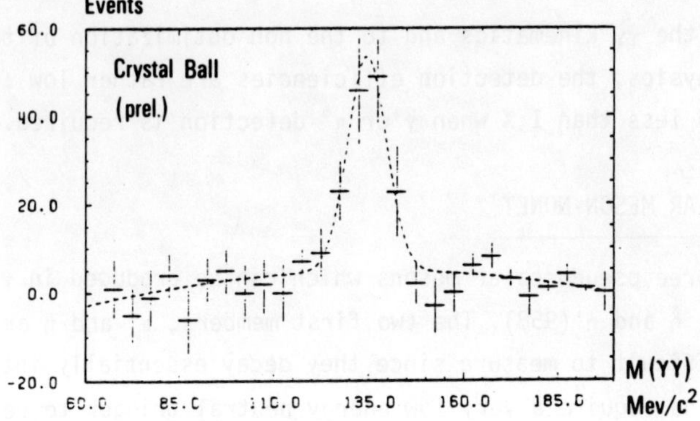

Fig. 1 : π° measurement from Crystal Ball :
Subtracted γγ mass spectrum

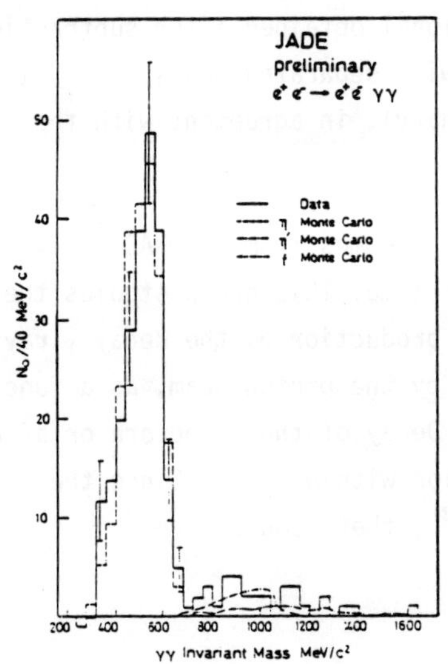

Fig. 2 : η measurement from JADE :
γγ mass spectrum

    - transverse momentum balance compatible with the apparatus
      resolution to select exclusive final state.

ii) due to the $\gamma\gamma$ kinematics and to the non optimization of the detector for this physics, the detection efficiencies are rather low : ~ 15 % for $\pi^+\pi^-$ to less than 1 % when γ or π° detection is required.

PSEUDO-SCALAR MESON NONET

The three pseudoscalar mesons which can be produced in $\gamma\gamma$ reaction are the π°, η and $\eta'(958)$. The two first members, π° and η are particularly difficult to measure since they decay essentially into neutral particles and require a very low energy neutral trigger to be detected.

1. $\pi^\circ \rightarrow \gamma\gamma$

In $\gamma\gamma$ interactions the only result came from the CRYSTAL BALL[5] at DORIS. They have collected 6.7 pb^{-1} at the T4S with a special trigger requiring more than 40 MeV in each hemisphere and a total energy greater than 90 MeV. Fig. 1 shows the nice π° signal obtained after subtraction of the beam gas background as measured with separated beams. $\Gamma_{\pi^\circ\gamma\gamma}$ is found to be $7.9 \pm 1.4 \pm 1.6$ eV (Preliminary), in agreement with the old Cornell[6] result.

The most precise measurement of $\Gamma_{\pi^\circ\gamma\gamma}$ comes from the NA30[7] experiment at CERN which measures the π° lifetime. This group studies the positron yield originating in e^+e^- pair production by the decay γ rays in a target of two gold foils traversed by the proton beam, as a function of the distance between the foils. Decay of the π° before or after the second foil allows a τ_{π° determination within 1.2 %. Since the branching ratio $B(\pi^\circ \rightarrow \gamma\gamma)$ is well known[8], they deduce

$$\Gamma_{\pi^\circ\gamma\gamma} = 7.59 \pm 0.09 \text{ eV}$$

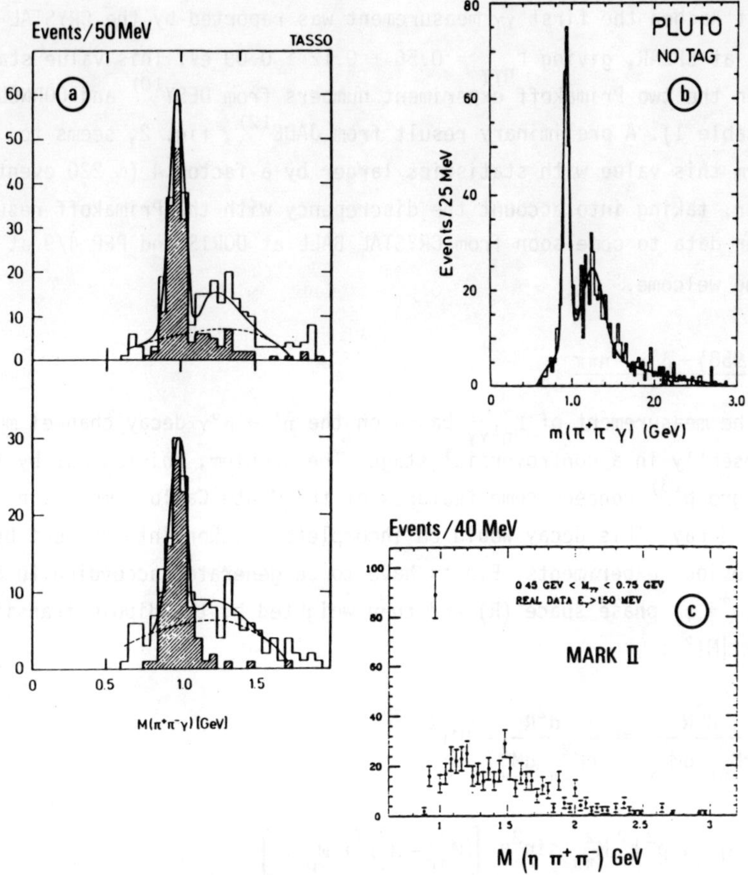

<u>Fig. 3</u> : η' measurement :
   a) TASSO : π⁺π⁻γ mass spectrum in the LABC (top)
      and in the HASH. The shaded areas correspond
      to a ρ° interval of $0.60 \leq M_{\pi^+\pi^-} \leq 0.85$ GeV

   b) PLUTO : π⁺π⁻γ mass spectrum

   c) MARK II : ηπ⁺π⁻ mass spectrum

2. $\eta \to \gamma\gamma$

At Aachen the first $\gamma\gamma$ measurement was reported by the CRYSTAL BALL[9] at SPEAR, giving $\Gamma_{\eta\gamma\gamma} = 0.56 \pm 0.12 \pm 0.09$ eV. This value stands between the two Primakoff experiment numbers from DESY[10] and CORNELL[11] (see table 1). A preliminary result from JADE[12], fig. 2, seems to confirm this value with statistics larger by a factor 4 (~ 220 events). However, taking into account the discrepency with the Primakoff results, further data to come soon from CRYSTAL BALL at DORIS and PEP 4/9 at PEP will be welcome.

3. $\eta'(958) \to \rho^\circ\gamma, \eta\pi\pi$

The measurement of $\Gamma_{\eta'\gamma\gamma}$ based on the $\eta' \to \rho^\circ\gamma$ decay channel mode is presently in a controversial stage. The problem, pointed out by the PLUTO group[13] concern some features of the Monte Carlo simulation of the η' decay. This decay would be incompletely taken into account by the previous experiments. Events have to be generated according to 3 body ($\pi^+\pi^-\gamma$) phase space (R) and then weighted by the dipole transition matrix $|M|^2$:

$$\frac{d^2N}{dM^2_{\pi\pi} dM^2_{\pi\gamma}} = \frac{d^2R}{dM^2_{\pi\pi} dM^2_{\pi\gamma}} |M|^2$$

$$|M|^2 \alpha \; p^2 k^2 M^2_{\pi\pi} \sin^2\theta \left[(M^2_{\pi\pi} - M^2_\rho)^2 + M^2_\rho \Gamma^2\right]^{-1}$$

where $p(k)$ is the pion (photon) momentum and θ the angle between the pion and the photon direction (all given in the ρ center of mass system). Moreover the energy dependence of the ρ width Γ has to be taken into account

$$\Gamma = \Gamma_0 (p/p_0)^3 M_\rho/M_{\pi\pi}$$

This method gives photon energies larger on average than expected from a pure phase space description of the $\rho\gamma$ final state and conse-

Table 1 : Radiative widths of pseudoscalar mesons

Experiment	Mode	$\Gamma_{\pi^\circ\gamma\gamma}$ (eV)	Ref.
Browman et al	Primakof	7.85 ± 0.54	6
CRYSTAL BALL	$\gamma\gamma$	7.9 ± 1.4 ± 1.6 (Prel.)	5
NA30	τ_{π°	7.59 ± 0.09 (Prel.)	7

Experiment	Mode	$\Gamma_{\eta\gamma\gamma}$ (keV)	Ref.
DESY	Primakof	1.00 ± 0.22	8-10
Browman et al	"	0.324 ± 0.046	11
CRYSTAL BALL	$\gamma\gamma$	0.56 ± 0.12 ± 0.09	9
JADE	$\gamma\gamma$	0.56 ± 0.05 ± 0.08 (Prel.)	

Experiment	Mode	$\Gamma_{\eta'\gamma\gamma}$ (keV)	Ref.
Binnie et al	Missing mass	5.4 ± 2.1	47
MARK II	$\rho\gamma$	5.8 ± 1.1 1.2	48
CELLO	$\rho\gamma$	6.2 (6.0)* ± 1.1 ± 0.8	15
JADE	$\rho\gamma$	5.0 ± 0.5 ± 0.9	49
TASSO	$\rho\gamma$	5.1* ± 0.4 ± 0.7	14
PLUTO	$\rho\gamma$	3.8* ± 0.26 ± 0.43	13
MARK II	$\eta\pi\pi$	3.7 ± 1.0 ± (Prel.)	16
Weighted mean		4.42 ± 0.34	

*Results using "full" $\eta' \to \rho\gamma$ generator

130

quently it leads to a higher detection efficiency. The PLUTO[13] value
for $\Gamma_{\eta'\gamma\gamma}$, fig. 3, is lower (3.8 ± 0.7 keV) than the previous measurements which are around 5 keV, (see table 1). The PLUTO group suggest it
would be due to this improved simulation. However with the same simulation TASSO[14] reports 5.1 ± 0.4 keV, fig. 3, and CELLO[15] confirmed its
old result 6.0 ± 1.1 keV within a minor change. Clarification may come
from the MARK II group[16] who has analyzed the $\eta' \to \eta\pi\pi$ decay avoiding
the ambiguities of the Monte Carlo simulation. They report, fig. 3, a
preliminary value of 3.7 ± 1.0 keV. Other measurements, particularly in
the $\eta\pi\pi$ decay mode seem necessary.

4. <u>Other pseudoscalars</u>

The only other candidate is the i(1440). The controversial nature
of this meson (radial excitation of the η', gluonium,...) is discussed
in F. ERNE's report.

5. <u>Conclusion on pseudoscalars mesons</u>

Using the relations between pseudoscalars meson radiative widths
describe above, we can determine the SU(3) mixing angle θ between the
η and the η' :

$$\eta = \cos\theta\, \eta_8 - \sin\theta\, \eta_1 \qquad\qquad \eta' = \sin\theta\, \eta_8 + \cos\theta\, \eta_1$$

The most precise measurement is the $\pi°$ radiative width obtained
by NA30. Therefore we use it as an input in the following equations :

$$\Gamma_{\eta\gamma\gamma} = \Gamma_{\pi°\gamma\gamma} \left(\frac{m_\eta}{m_{\pi°}}\right)^3 \cdot \frac{1}{3} (\sqrt{8}\,\sin\theta - \cos\theta)^2$$

$$\Gamma_{\eta'\gamma\gamma} = \Gamma_{\pi°\gamma\gamma} \left(\frac{m_{\eta'}}{m_{\pi°}}\right)^3 \cdot \frac{1}{3} (\sin\theta + \sqrt{8}\,\cos\theta)^2$$

to obtain the curves of fig. 4. The angle θ can be obtained from the
measured values of $\Gamma_{\eta\gamma\gamma}$ and $\Gamma_{\eta'\gamma\gamma}$.

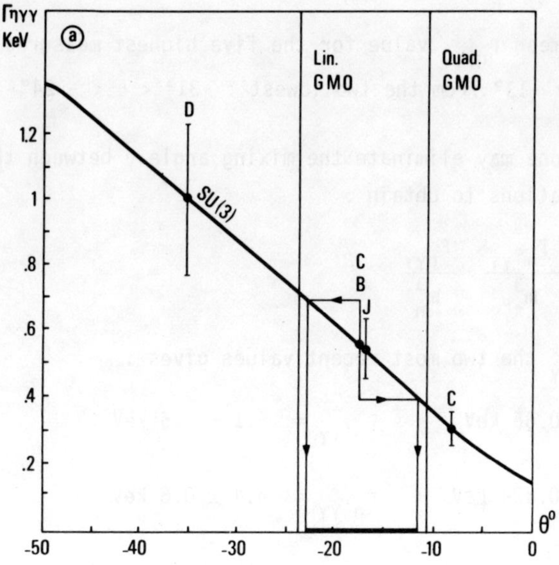

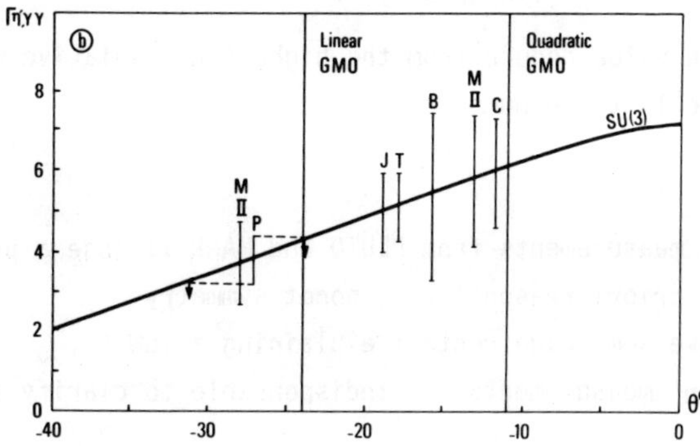

<u>Fig. 4</u> : $\Gamma_{\eta\gamma\gamma}$ and $\Gamma_{\eta'\gamma\gamma}$ as function of the mixing angle θ.

Imput : $\Gamma_{\pi^\circ\gamma\gamma}$ = 7.56 eV.

The experimental values are shown as data points.

From the mean $\Gamma_{\eta'\gamma\gamma}$ value for the five highest measures we deduce : $-20° < \theta < -13°$ from the two lowest : $-31° < \theta < -24°$

Moreover one may eliminate the mixing angle θ between the radiative width relations to obtain :

$$\frac{\Gamma_{\eta'\gamma\gamma}}{m_{\eta'}^3} = \frac{3\,\Gamma_{\pi°\gamma\gamma}}{m_{\pi°}^3} - \frac{\Gamma_{\eta\gamma\gamma}}{m_{\eta}^3}$$

Taking for $\Gamma_{\eta\gamma\gamma}$ the two most recent values gives :

$$\Gamma_{\eta\gamma\gamma} = 0.56 \text{ keV} \rightarrow \Gamma_{\eta'\gamma\gamma} = 5.1 \pm 0.5 \text{ keV}$$

$$\Gamma_{\eta\gamma\gamma} = 0.324 \text{ keV} \rightarrow \Gamma_{\eta'\gamma\gamma} = 6.4 \pm 0.6 \text{ keV}$$

These values are in better agreement with the highest η' measurements.

Notice also that the ideal mixing corresponds to $\theta = -35.3°$ and that the Gell-Mann-Okubo mass formulae give :

$$\theta = -11.1° \pm 0.2° \text{ (quadratic GMO)}$$
$$\theta = -24° \pm 0.5° \text{ (linear GMO)}$$

The mixing angle value deduce from the highest η' radiative widths fall in between these last two numbers.

To conclude :

- the recent η' measurements from PLUTO and MARK II pose a problem
- there is no a priori reason for 0^- nonet symmetry
- the η' may have some glue content explaining a low $\Gamma_{\eta'\gamma\gamma}$
- better η and η' measurements are indispensable to clarify the 0^{-+} nonet.

TENSOR MESON NONET

We will only consider the flavour neutral states of the 2^{++} nonet, i.e. the isovector $A_2(1230)$ and the isoscalars $f(1270)$, $f'(1525)$ abondantly produced in $\gamma\gamma$ interactions. Since the Aachen Conference results

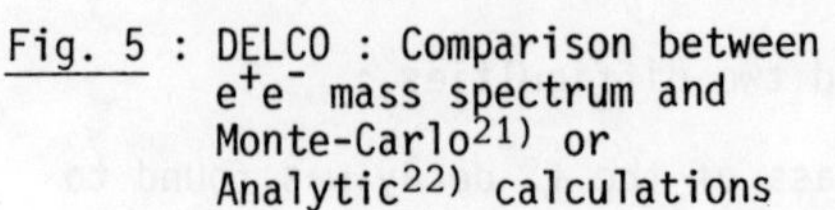

Fig. 5 : DELCO : Comparison between
e^+e^- mass spectrum and
Monte-Carlo[21] or
Analytic[22] calculations

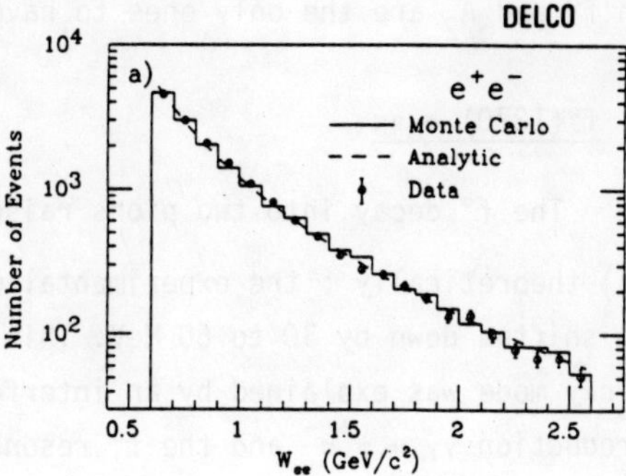

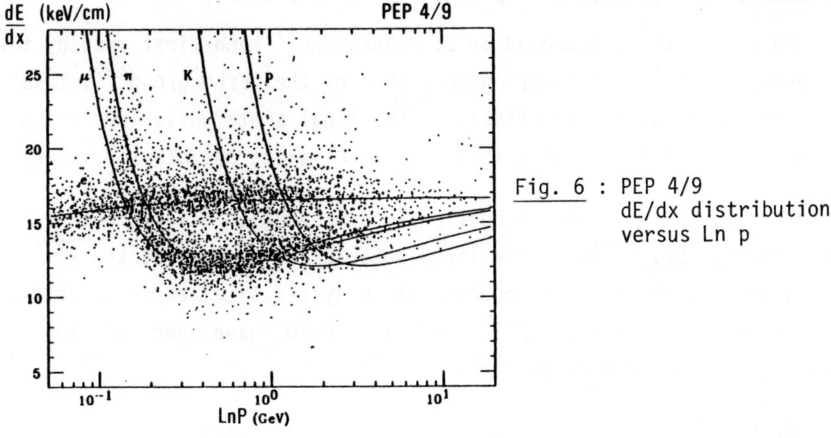

Fig. 6 : PEP 4/9
dE/dx distribution
versus Ln p

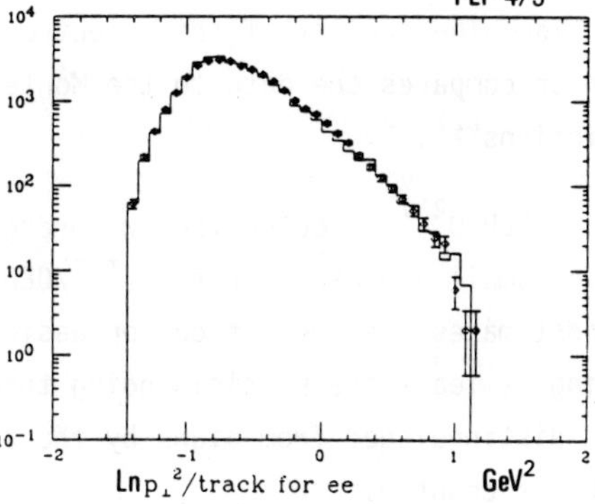

Fig. 7 : PEP 4/9
Ln(Impulsion transverse)^2
per track for $e^+e^- \rightarrow e^+e^-e^+e^-$
events. Comparison to
Monte-Carlo[25]

on f° and A_2 are the only ones to have been published.

1. $\underline{f^\circ(1270) \to \pi\pi}$

The f° decay into two pions raised two difficulties :

i) theoretically : the experimental mass of the f° decay was found to be shifted down by 30 to 60 MeV. This shift observed in the charged decay mode was explained by an interference between the direct Born production $\gamma\gamma \to \pi^+\pi^-$ and the f° resonant production. However a shift observed in the neutral decay $f^\circ \to \pi^\circ\pi^\circ$ could not be explained by such an interference (direct $\pi^\circ\pi^\circ$ production is forbidden).

An explanation, proposed by G. MENNESSIER[17] and first used by the DM1 group at DCI[18] at lower energy, then by the CELLO group[19], takes into account rescattering effects in the final state while preserving unitarity. We will see below that this model gives satisfactory results.

ii) experimentally : the two body charged production below 2 GeV is dominated by QED, leading to a large subtraction to extract the $\pi^+\pi^-$ signal and consequently introducing large systematical errors. For the first time experiments (PEP4/9, DELCO and PLUTO) have been able to identify QED and/or $\pi^+\pi^-$ particles.

DELCO[20] identifies electrons in Cerenkov counters covering $53^\circ < \theta < 127^\circ$. Then they deduce the $\gamma\gamma$ luminosity and the $\mu^+\mu^-$ contribution. The quality of this electron identification is shown in fig. 5 which compares the data to the Monte Carlo[21] and to the analytic predictions[22].

PLUTO[23] detected the two prongs in their forward spectrometer. At such small angles the ratio $\pi^+\pi^-$/QED is more favorable and the Lorentz boost makes the identification easier. Mesons are identified by requiring at least one particle going through an iron filter. Electrons can be distinguished from pions by the amount of energy deposited in a shower counter.

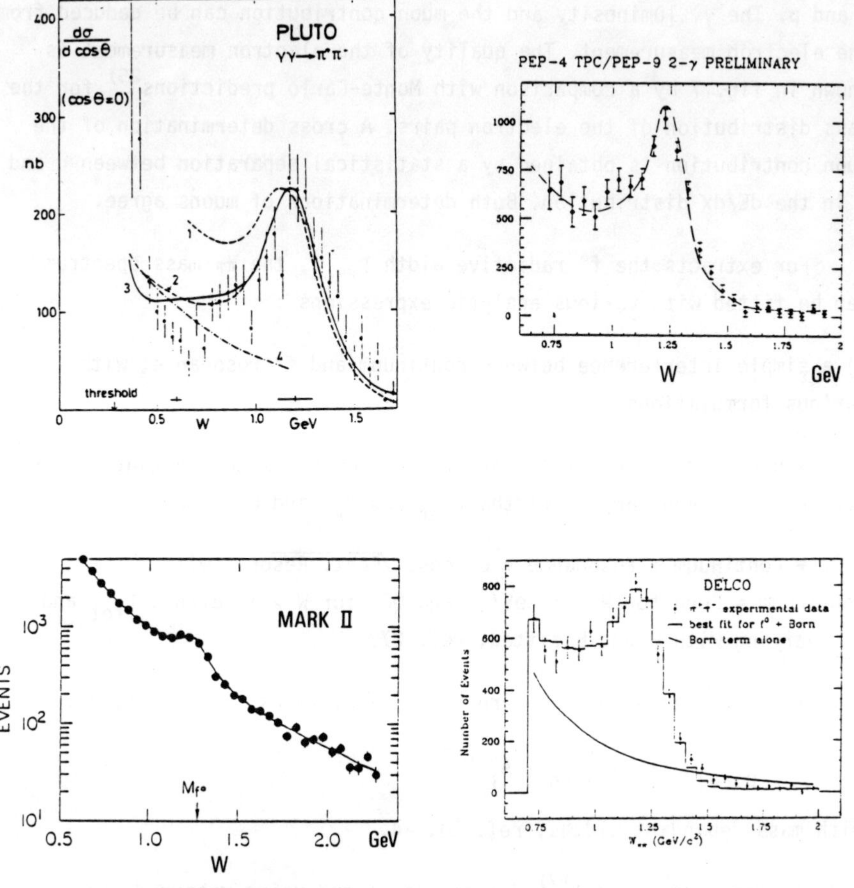

<u>Fig. 8</u> : f° (1270) measurements : π⁺π⁻ invariant mass distributions.

136

PEP 4/9[24] used the Time Projection Chamber which provides dE/dx information to identify the different particles. The dE/dx versus momentum distribution, fig. 6, shows a clear separation between e, (μ, π), K and p. The $\gamma\gamma$ luminosity and the muon contribution can be deduced from the electron measurement. The quality of the electron measurement is shown in fig. 7 by a comparison with Monte-Carlo predictions[25] for the mass distribution of the electron pairs. A cross determination of the muon contribution is obtained by a statistical separation between μ and π in the dE/dx distribution. Both determinations of muons agree.

For extracts the $f°$ radiative width $\Gamma_{f°\gamma\gamma}$, the $\pi\pi$ mass spectrum can be fitted with various analytic expressions :

i) a simple interference between continuum and $f°$ resonance, with various formulations :

* Born + resonance + B. $\cos\delta$ $\sqrt{\text{Born. Reson.}}$, with constant values for the total and partial widths Γ_{tot} and $\Gamma_{\pi\pi}$ and B fitted.

* Continuum + resonance + B. $\cos\delta$ $\sqrt{\text{Cont. Reson.}}$, with Continuum = Born for $W < 1$ GeV/c^2 and QCD for $W > 1$ GeV/c^2, Γ_{tot} and $\Gamma_{\pi\pi}$ varying with W and B fitted, ref. 27.

* Interference between Born and $f°$ amplitudes in helicity 2 :

$$\left| A_{Born}^{\lambda=0} \right|^2 \; + \; \left| A_{Born}^{\lambda=2} + A_{f°}^{\lambda=2} \right|^2$$

with mass dependant widths, ref. 20, 24.

ii) the Mennessier model[17] defines above and using measured phase shifts in the coupled channels $\pi^+\pi^-$, $\pi°\pi°$ and K$\bar{\text{K}}$. This model fits the $f°$ mass shift in the charged decay mode and also predicts a shift in the neutral decay mode by about 30 MeV explaining the observations made by CRYSTAL BALL[28] and JADE[29].

Fig. 8, shows the $\pi\pi$ mass distributions and fits to the most recent experiments.

Table 2 : f° radiative widths (assuming $|\lambda| = 2$)

experiment	Mode	$\Gamma_{\gamma\gamma f°}$ (keV)		Ref.
PLUTO	$\pi^+\pi^-$	$2.3 \pm 0.5 \pm 0.35$		50
MARK II	$\pi^+\pi^-$	$3.6 \pm 0.3 \pm 0.5$		26
TASSO	$\pi^+\pi^-$	$3.2 \pm 0.2 \pm 0.6$		51
CRYSTAL BALL	$\pi°\pi°$	$2.7 \pm 0.2 \pm 0.6$	$\lambda = 2$	
		$2.9\,^{+0.6}_{-0.4} \pm 0.6$	λ Fit	28
CELLO	$\pi^+\pi^-$	$2.5 \pm 0.1 \pm 0.5$		19
JADE	$\pi°\pi°$	$2.3 \pm 0.2 \pm 0.5$		29
MARK II	$\pi^+\pi^-$	$2.52 \pm 0.13 \pm 0.38$		27
PLUTO	$\pi^+\pi^-$	$2.85 \pm 0.25 \pm 0.5$		23
DELCO	$\pi^+\pi^-$	2.7 ± 0.18		20
PEP 4/9	$\pi^+\pi^-$	$2.39 \pm 0.06 \pm 0.30$		24
Weighted mean		2.65 ± 0.12		

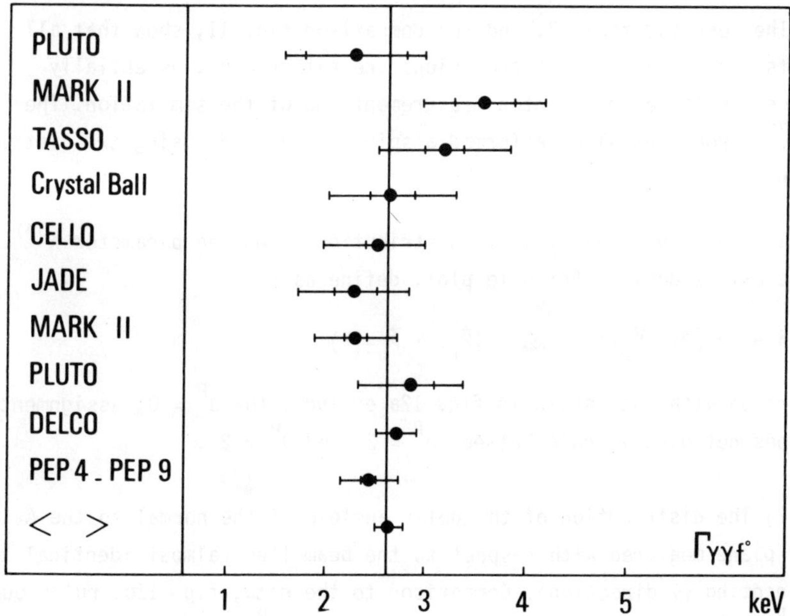

Fig. 9 : f° radiative widths. Comparison between experiments

The $f°$ radiative widths $\Gamma_{f°\gamma\gamma}$ determined by these different fits in the $f°$ region are quite similar. Table 2 and fig. 9 show that the results of the different experiments are compatible. In the last experiments, errors are dominated by systematics. All these measurements assumed a $|\lambda| = 2$ helicity for the $f°$. This has only been checked by the CRYSTALL BALL Collaboration[28]. Such an assumption affects dramatically the systematic errors since acceptances can vary by a factor greater than 2 when the helicity is assumed to be 0 rather than 2.

2. $\underline{A_2(1320) \rightarrow \pi^+\pi^-\pi°, \eta\pi°}$

This isovector partner of the $f°$ in the 2^{++} multiplet is not easy to measure because of the presence of neutral components in its decay products, mainly $\pi^+\pi^-\pi°$ and $\eta\pi°$. Consequently there were just three measurements before the Aachen Conference : CRYSTAL BALL[30] at SPEAR, CELLO[31] and JADE[29] and only two new ones : PLUTO[32] and CRYSTAL BALL at DORIS[33], fig. 10.

The results, table 3, and the comparison fig. 11, show that all results are compatible but precisions are rather poor, essentially because the low energy photon measurement and of the simulation. The PLUTO[32] group has also performed a spin parity study using two distributions :

1) The three pion c. of m. distribution using the parameter Λ[34] of the Dalitz density triangle plot, define as :

$$\Lambda = |\vec{P}_{\pi^+} \times \vec{P}_{\pi^-}|^2 \, / \, \max \, (|\vec{P}_{\pi^+} \times \vec{P}_{\pi^-}|^2).$$

Comparison with data shown in fig. 12a excludes the $J^P = 0^-$ assignment but does not discriminate between $J^P = 2^+$ and $J^P = 2^-$.

2) The distribution of the polar angle α of the normal to the A_2 decay plane measured with respect to the beam line (almost identical to the incoming $\gamma\gamma$ direction). Comparison to the data, fig. 12b, rules out the $J^P = 2^-$ assignment but is consistent with $J^P = 2^+$ with helicity

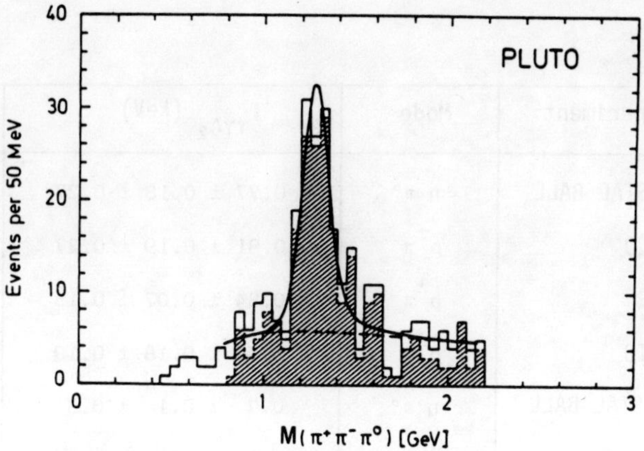

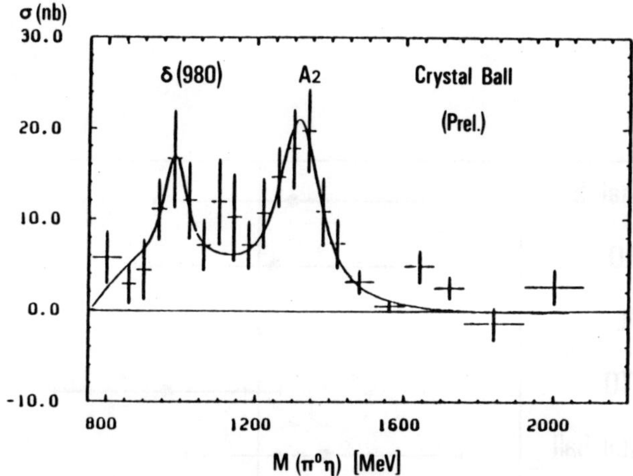

Fig. 10 : A_2 (1320) measurements :

a) PLUTO ($\pi^+\pi^-\pi^°$) invariant mass.
Shaded area corresponds to events with a $\pi^+\pi^°$ or $\pi^-\pi^°$ mass combination in the ρ mass band.

b) CRYSTAL BALL ($\pi^°\eta$) mass spectrum

Table 3 : A_2 radiative widths

experiment	Mode	$\Gamma_{\gamma\gamma A_2}$ (keV)	Ref.
CRYSTAL BALL	$\eta\ \pi^\circ$	$0.77 \pm 0.18 \pm 0.27$	30
CELLO	$\rho^\pm \pi$	$0.81 \pm 0.19 \pm 0.27$	31
JADE	$\rho^\pm \pi$	$0.84 \pm 0.07 \pm 0.15$	29
PLUTO	$\rho^\pm \pi$	$1.06 \pm 0.18 \pm 0.19$	32
CRYSTAL BALL	$\eta\ \pi^\circ$	$0.7\ \pm 0.1\ \pm 0.3$	33
Wreighted mean		$0.85 \pm .11$	

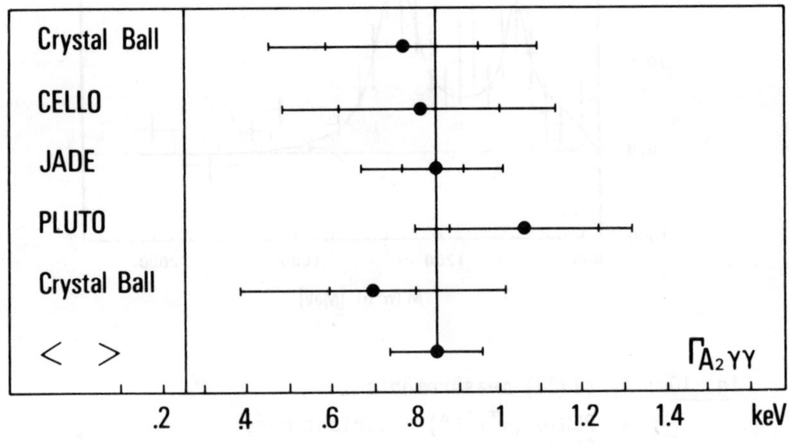

Fig. 11 : A_2 radiative widths. Comparison between experiments

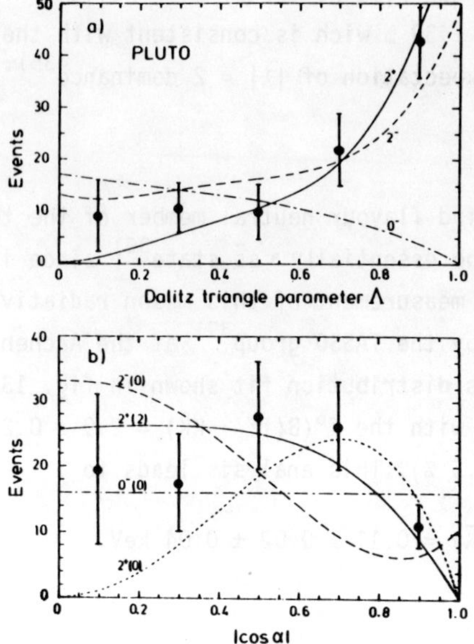

Fig. 12 : PLUTO. A_2 events - Distributions of :
  a) Dalitz triangle parameter (see text)

  b) $|\cos \alpha|$, where α is the polar angle of the
     normal to the decay plane.

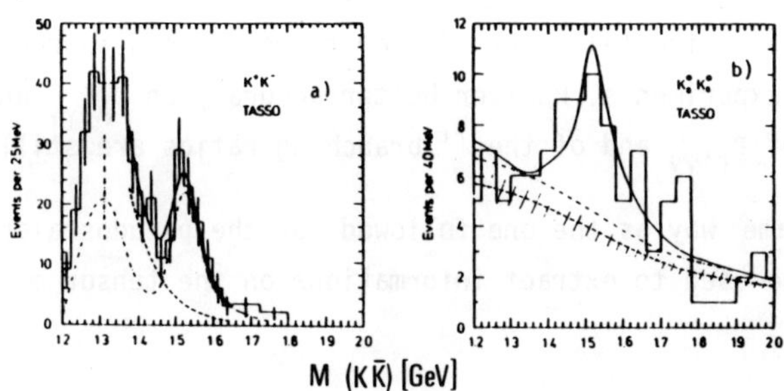

Fig. 13 : TASSO. f'(1525) measurement
  a) Mass spectrum of charged Kaon pairs
  b) Mass spectrum of neutral Kaon pairs

0 or 2. A fit to a linear combination of both helicity states yields a $|\lambda|$ = 2 fraction of 62 ± 39 % wich is consistent with the JADE[29] results and the theoretical expectation of $|\lambda|$ = 2 dominance[35].

3. $f'(1525) \rightarrow K\bar{K}$

The f'(1525), third flavour neutral member of the tensor meson nonet, is assumed to be essentially a $s\bar{s}$ state[36] since its main decay mode is $K\bar{K}$. The only measurement of this meson radiative width has already been presented by the TASSO group[37] at the Aachen Conference. The K^+K^- and $K_S^0 K_S^0$ mass distribution fit shown in fig. 13 takes into account interferences with the $f°(B(f° \rightarrow K\bar{K})$ = 2.9 ± 0.2 %) and the A_2 $(B(A_2 \rightarrow K\bar{K})$ = 4.8 ± 0.5 %). This analysis leads to :

$$\Gamma_{f'\gamma\gamma} \cdot B(f' \rightarrow K\bar{K}) = 0.11 \pm 0.02 \pm 0.04 \text{ keV}.$$

4. Other tensor mesons ?

The only candidate, the $\theta(1640)$ discovered in the radiative J/ψ decays[38] has not been seen in $\gamma\gamma$ interaction. Limits are given in F. ERNE'S revue.

5. Conclusions on the Tensor nonet

From the experimental point of view there is no desagreement between the different experiments. However better accuracy on $\Gamma_{A_2\gamma\gamma}$ and more measurements of $\Gamma_{f'\gamma\gamma}$ and of the f' branching ratios are still necessary.

In the same way as the one followed for the pseudoscalars mesons, the data can be used to extract informations on the tensor meson quarks content.

Fig. 14 shows the ratio of the A_2 to f° and f' to f° radiative widths as a function of the 2^{++} mixing angle θ, given by the SU(3) relations :

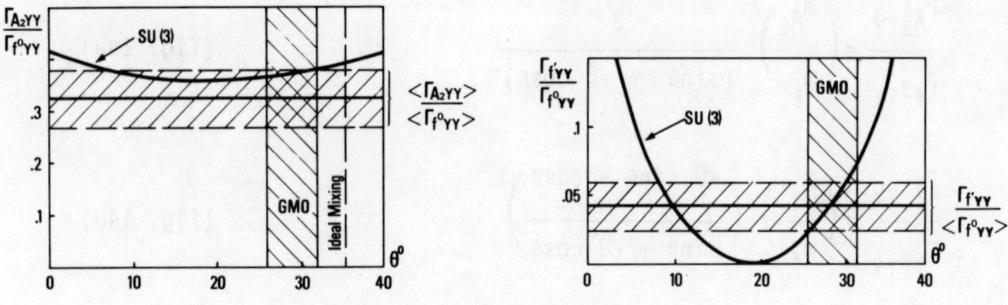

Fig. 14 : Ratios $\Gamma_{A_2\gamma\gamma}/\Gamma_{f^\circ\gamma\gamma}$ and $\Gamma_{f'\gamma\gamma}/\Gamma_{f^\circ\gamma\gamma}/$(assuming $B(f' \to K\bar{K})=1$) as function of mixing angle θ

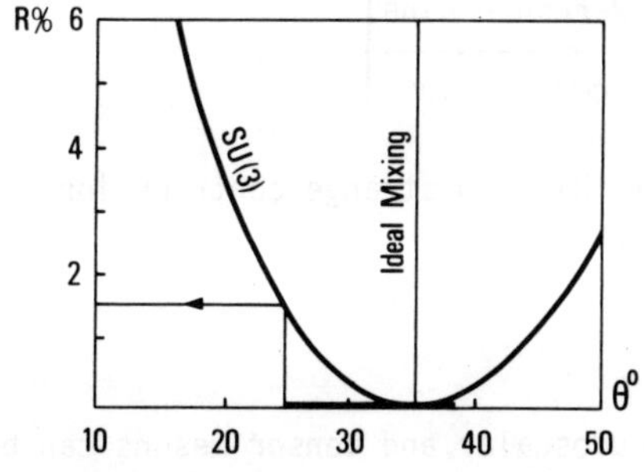

Fig. 15 : Percentage R of non-strange quarks in f' as function of the mixing angle θ

$$R = \left| \frac{u\bar{u} + d\bar{d}}{s\bar{s}} \right|^2 = \left[\frac{1/\sqrt{2}\,\cos\theta - \sin\theta}{\sqrt{2}\,\cos\theta + \sin\theta} \right]^2$$

$$\frac{\Gamma_{A_2\gamma\gamma}}{\Gamma_{f^\circ\gamma\gamma}} = \left(\frac{m_{A_2}}{m_{f^\circ}}\right)^3 \frac{3}{(\sin\theta + \sqrt{8}\,\cos\theta)^2} \qquad \text{(fig. 14a)}$$

$$\frac{\Gamma_{f'\gamma\gamma}}{\Gamma_{f^\circ\gamma\gamma}} = \left(\frac{m_{f'}}{m_{f^\circ}}\right)^3 \left(\frac{\sqrt{8}\,\sin\theta - \cos\theta}{\sin\theta + \sqrt{8}\,\cos\theta}\right)^2 \qquad \text{(fig. 14b)}$$

Comparison with the experimental ratios (assuming $Br(F' \to K\bar{K}) \gtrsim 50\,\%$)

- show a consistency with nonet symmetry and mass formula within the errors,

- leads to a mixing angle value $25° < \theta < 39°$

From this θ determination, one can deduce the content of non-strange quarks in f'. In order to do so we consider the intercept of the θ limits with the curve representing the ratio of non strange to strange quarks given by :

$$R = \left|\frac{u\bar{u} + d\bar{d}}{s\bar{s}}\right|^2 = \left|\frac{1/\sqrt{2}\,\cos\theta - \sin\theta}{\sqrt{2}\,\cos\theta + \sin\theta}\right|^2$$

Fig. 15 shows that this non strange contribution is less than 2 %.

SUM RULE TESTING

The results on pseudoscalar and tensor mesons can be used to test a sum rule[39] which relates the cross sections for $\gamma\gamma$ helicities 0 and 2 :

$$\int_0^\infty \frac{dW_{\gamma\gamma}^2}{W_{\gamma\gamma}^2} \left(\sigma^{\lambda=0}(w_{\gamma\gamma}) - \sigma^{\lambda=2}(w_{\gamma\gamma})\right) = 0$$

Saturating this sum rule with the low lying scalar, pseudoscalar and tensor states gives, assuming narrow resonances :

$$\sum_s \frac{\Gamma_{s\gamma\gamma}}{m_s^3} + 5 \sum_T \frac{\Gamma_{T\gamma\gamma}(\lambda=0)}{m_T^3} \approx 5 \sum_T \frac{\Gamma_{T\gamma\gamma}(\lambda=2)}{m_T^3} - \sum_p \frac{\Gamma_{p\gamma\gamma}}{m_p^3}$$

Notice than the narrow resonance assumption used above is question-able for the f° resonance interfering with the $\pi^+\pi^-$ continuum. Taking into account this interference effect and using the radiative widths presented above the right hand side (RHS) gives :

RHS = 0.2 ± 1.0 keV/GeV^3.

This result which is close from zero seems to indicate a suppression of the scalar mesons at the $\Gamma_{\gamma\gamma} < 1\text{-}2$ keV level.

Experimentally, the DM1 experiment[18] at DCI suggests an $\varepsilon(600)$ contribution to explain an excess of $\pi^+\pi^-$ events at low mass, but this result is not confirmed by PLUTO[23] nor by DELCO[20].

Upper limits have also been published (at 95 % CL) :

- TASSO[40] : $\Gamma_{\varepsilon\gamma\gamma}$. $B(\varepsilon \to \pi^+\pi^-) < 1.5$ keV for $1.3 < M_\varepsilon < 1.5$ GeV

- JADE[26] : $\Gamma_{S^*\gamma\gamma} < 0.8$ keV

- CRYSTAL BALL[33] : $\Gamma_{\delta\gamma\gamma}$. $B(\delta \to \pi\,\eta) = 0.10 \pm 0.04 \pm 0.06$ keV

$$\text{(fig.10)}$$

TEST OF THE QUARK CHARGES

Tests of the quark charges using the meson radiative widths have been presented at this conference[41] in particular by J.H. FIELD. These tests exclude gauge integer charge quark models and predict significant gluonium amplitudes in the η' and η'.

Q^2 DEPENDENCE OF SINGLE MESON PRODUCTION

Novel and important contributions to this conference have been made which concern the Q^2 dependence of the η' and f° production.

This Q^2 dependence has previously been described in the framework of the Vector Dominance Model (VDM) by using a ρ° form factor.

Additional contributions coming from the ω and ϕ mesons, the higher resonances, the continuum and the longitudinal-transverse photon combinations are taken into account in the Generalized Vector Dominance Model (GVDM)[42].

The Q^2 dependence has also been studied in the parton model[43], in the non relativistic[44] or relativistic[45] quark model as well as the framework of perturbative QCD[46].

We, now have experimental results :

- on the η', the PLUTO group[13] published a signal of 35 ± 9 events in four Q^2 bins with $< Q^2 > = 0.4$ GeV^2/c^2, fig. 16 ;

- on the f° :
  * the MARK II group[27] published the f° radiative widths for $Q^2 < 0.26$ GeV/c^2 and $0.26 < Q^2 < 1.4$ GeV^2/c^2, fig. 17
  * the PEP 4/9 group[24] reported f° radiative widths in four Q^2 bins Q^2 bins up to a $Q^2 = 3$ GeV^2/c^2.

However these nice measurements suffer from lack of statistics and the Q^2 dependences can be simply described by VDM or GVDM without allowing a test of more sophisticated models.

CONCLUSION

In the part few years and in particular since the last, Aachen Conference, many radiative width measurements have been published. Yet it would be useful to have more data for the understanding of the η'

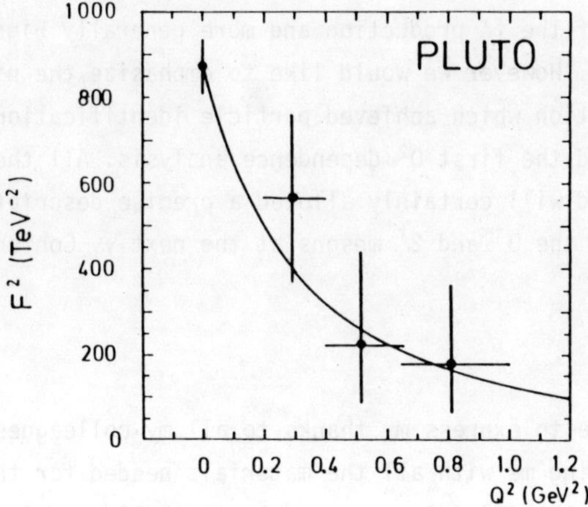

Fig. 16 : PLUTO. The $\eta' \to \gamma\gamma$ transition form factor.
The line is a simple ρ° pole normalised to
the no tag result.

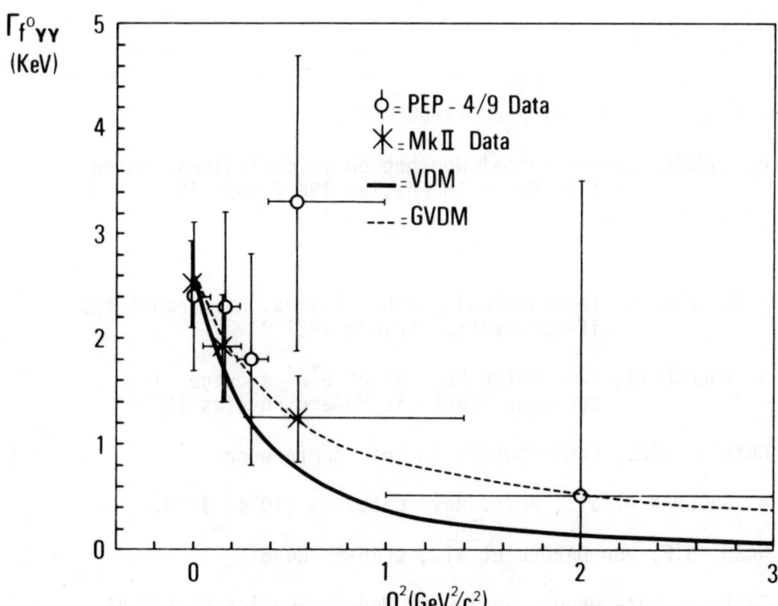

Fig. 17 : PEP 4/9 - MARK II

Q^2 dependence of the f$^\circ$ radiative width

production and of the f' production and more generally higher statistics are still needed. However we would like to emphasize the nice results on the f° production which achieved particle identification, the first results on $\pi°$ and the first Q^2 dependence analysis. All these results are promising and will certainly allowed a precise description of the quark content of the 0^- and 2^+ mesons at the next $\gamma\gamma$ Conference.

ACKNOWLEDGMENTS

I would like to express my thanks to all my colleagues at PEP and PETRA for providing me with all the materials needed for this review. I am greatful to the PEP 4/9 group and to A. COURAU and J. HAISSINSKI for their friendly and efficient help. I would like also to thank the organizers, especially R. LANDER, for inviting me to such interesting and fruitful conference.

REFERENCES

1.   F.E. LOW, Phys. Rev. 120 (1960) 582.

2.   A. COURAU, International Worshop on $\gamma\gamma$ collisions, Amiens, Lecture Notes in Physics 134 (1980) 19.

     J. FIELD, Nucl. Phys. B168 (1980) 477, Nucl. Phys. B176 (1980) 545.

3.   F.M. RENARD, Electron-Positron Collisions, Ed. Frontières Gif-sur-Yvette, France 1981.

4.   H. KOLANOSKI, Two Photon Physics at e^+e^- storage ring, Springer Tracts in Modern Physics 1984.

5.   CRYSTAL BALL, Contribution to this conference.

6.   A. BROWMAN et al., Phys. Rev. Lett. 33 (1974) 1400.

7.   NA30, G.F. von DARDEL et al., CERN-EP 84-67.

8.   Particle Data group, Review of Modern Physics 56 (1984).

9.   CRYSTAL BALL, A. WEINSTEIN et al., Phys. Rev. D28 (1983) 2896.

10.   BEMPORAD et al., Phys. Lett. 25B (1967) 380.

11.   A. BROWMAN et al., Phys. Rev. Lett. 32 (1974) 1067.

12.   JADE, W. BARTEL et al., Contribution to the XXII International
        Conference on High Energy Physics, Leipzig (1984).

13.   PLUTO, Ch. BERGER et al., Phys. Lett. 142B (1984) 125.

14.   TASSO, M. ALTHOFF et al., DESY 84-069, Submitted to Phys. Lett.

15.   CELLO, H.J. BEHREND et al., Phys. Lett. 113B (1982) 378 (E)
                                    Phys. Lett. 125B (1983) 518.

16.   MARK II, G. GIDAL et al., Contribution to this conference.

17.   G. MENNESSIER, Z. Phys., C16 (1983) 241.

18.   DM1, A. COURAU et al., Phys. Lett. 96B (1980) 402.

19.   CELLO, H.J. BEHREND et al., Z. Phys. C23 (1984) 223.

20.   DELCO, A. COURAU et al., SLAC Pub 3362,
                                 Phys. Lett. 174B (1984) 227.

21.   A. COURAU, SLAC Pub 3363, to be submitted to Phys. Rev.

22.   A. COURAU, Phys. Rev. Lett. 49 (1982) 963
                   Phys. Rev. D29 (1984) 26.

23.   PLUTO, Ch. BERGER et al., DESY 84-074,
                Submitted to Phys. Lett.

24.   PEP 4/9, H. AIHARA et al., Contribution to the XXII International
        Conference on High Energy Physics, Leipzig (1984).

25.   J.A.M. VERMASEREN, J. SMITH and G. GRAMMER, Phys. Rev. D15
                           Phys. Rev. D15 (1977) 3280.

26.   MARK II, A. ROUSSARIE et al., Phys. Lett. 105B (1981) 304.

27.   MARK II, J.R. SMITH, Phys. Rev. D30 (1984) 851.

28.   CRYSTAL BALL, C. EDWARDS et al., Phys. Lett. 110B (1982) 82.

29.   J.E. OLSSON, Proceedings of the 5th International Colloquium on
        $\gamma\gamma$ interaction, Aachen (1983).

30.   CRYSTALL BALL, C. EDWARDS et al., Phys. Lett. 110B (1982) 82.

31.   CELLO, H.J. BEHREND et al., Phys. Lett. 114B (1982) 378 (E)
                                    Phys. Lett. 125B (1983) 518.

32. PLUTO, Ch. BERGER et al., DESY 84-084, Submitted to Phys. Lett.

33. CRYSTAL BALL, Contribution to this conference.

34. G. GOLDHABER et al., Phys. Rev. Lett. 15(1965) 118.

35. See e.g. P. GRASSBERGER and R. KOGERLER, Nucl. Phys. $\underline{B106}$ (1976) 451.

36. L. MONTANET, Rep. Prog. Phys. $\underline{46}$ (1983) 337.

37. TASSO, M. ALTHOFF et al., Phys. Lett. $\underline{121B}$ (1983) 216.

38. CRYSTAL BALL, C. EDWARDS et al., Phys. Rev. Lett. $\underline{48}$ (1982) 458.

39. P. ROY, Phys. Rev. 9 (1974) 2631.

40. TASSO, R. BRANDELIK et al., Z. Phys. C10 (1981) 117.

41. J.H. FIELD, see contribution to this conference.

42. I.F. GINSBURG and V.G. SERBO, Phys. Lett. $\underline{109B}$ (1982) 231.

43. G. KOPP, T.F. WALSH and P. ZERWAS, Nucl. Phys. $\underline{B184}$ (1981) 269.

44. H. KRASEMANN and J.A.M. VERMASEREN, Nucl. Phys. $\underline{B184}$ (1981) 269.

45. L. BERGSTROM, G. HULTH and H. SNELLMAN, ref. TH.3381-CERN (1982).

46. S.J. BRODSKY and J.P. LEPAGE, Phys. Rev. $\underline{D24}$ (1981) 1801.

47. D.M. BINNIE et al., Phys. Lett. $\underline{83B}$ (1979) 141.

48. MARK II, G.S. ABRAMS et al., Phys. Rev. Lett. 43 (1979) 477.

49. JADE, W. BARTEL et al., Phys. Lett. $\underline{113B}$ (1982) 190.

50. PLUTO, Ch. BERGER et al., Phys. Lett. $\underline{94B}$ (1980) 254.

51. TASSO, R. BRANDELIK et al., Z. Phys. C.10 (1981) 117.

Resonance production in $\gamma\gamma$ collisions, II.

F.C. Erné.

National Institute for Nuclear and High Energy Physics, Amsterdam, The Netherlands.

I. Introduction.

The assumption of vector dominance for both photons in $\gamma\gamma$-collisions leads us to the expectation that elastic scattering and charge-exchange reactions will occur from a variety of incoming vector mesons, with the attendant resonance formation and t-channel exchanges. We will examine data on vector-meson production and its Q^2-dependence to test this picture. In doing so we encounter hints of a variety of intriguing objects permitted by QCD, such as four-quark states and glueballs. We finally give a status report on $\gamma\gamma$-excitation of Charmonium states, a subject still in its infancy.

II. Vector meson production.

The reaction $\gamma\gamma \rightarrow 2\pi^+2\pi^-$ with prominent $\rho^0\rho^0$ production, peaking at $\approx$ 1.4 GeV, was found in 1980/81 by TASSO[1] and MarkII,[2] has now been studied further by TASSO,[3] CELLO[4] and PEP4/PEP9.[5] There is reasonable agreement on the magnitude of the 4π as well as the $\rho^0\rho^0$ cross sections, as is evident in Fig. 1. The share of $\rho^0\pi^+\pi^-$ and phase space production increases smoothly with increasing mass (W); this is shown in Fig. 2. It may be of interest to note that the $\rho^0\pi^+\pi^-$ contribution was not correctly parametrized in these studies, as was pointed out recently by Achasov et al..[6] The even charge conjugation of the $\gamma\gamma$ system imposes P,F,.. waves on the $\pi^+\pi^-$ phase space part, which results in similar threshold behaviour for ρ and $\pi^+\pi^-$. The $\rho^0\pi^+\pi^-$ final state could look very similar to $\rho^0\rho^0$ at low masses. A PEP4/PEP9 analysis with a tagged photon shows that the shape of the cross section persists approximately as it is at $Q^2 = 0$, but damped by a ρ-type formfactor, out to $Q^2 \approx 2$ GeV^2; beyond this value some flattening in Q^2 occurs. This is shown in Fig. 3.

A spin-parity analysis on the $\rho^0\rho^0$ signal was carried out by the TASSO group.[3] They find a large $J^P = 0^+$ component at lower masses and $J^P = 2^+$ dominance above 1.7 GeV in mass. The PEP4/PEP9 analysis[5] favors also $J^P = 2^+$

152

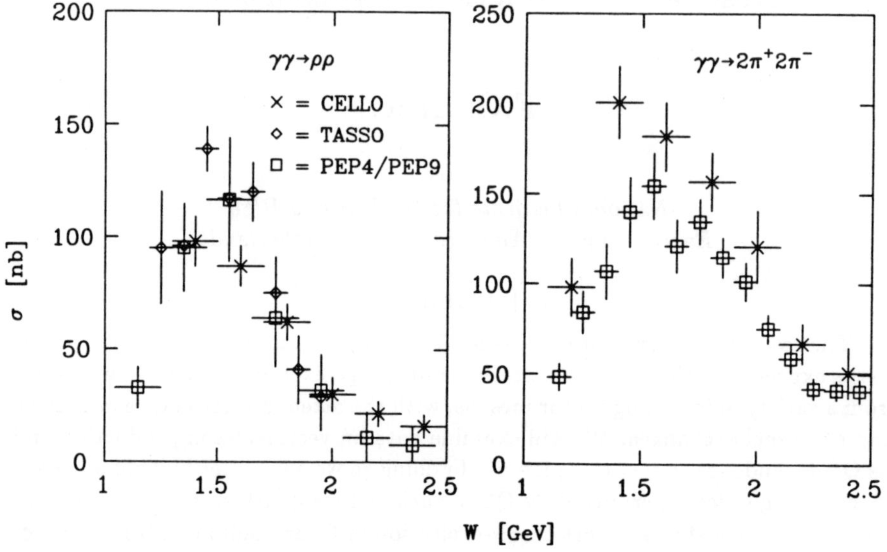

Fig. 1 *The Cross Sections for $\gamma\gamma \to \rho^0\rho^0$ and $\gamma\gamma \to 2\pi^+2\pi^-$ vs W.*

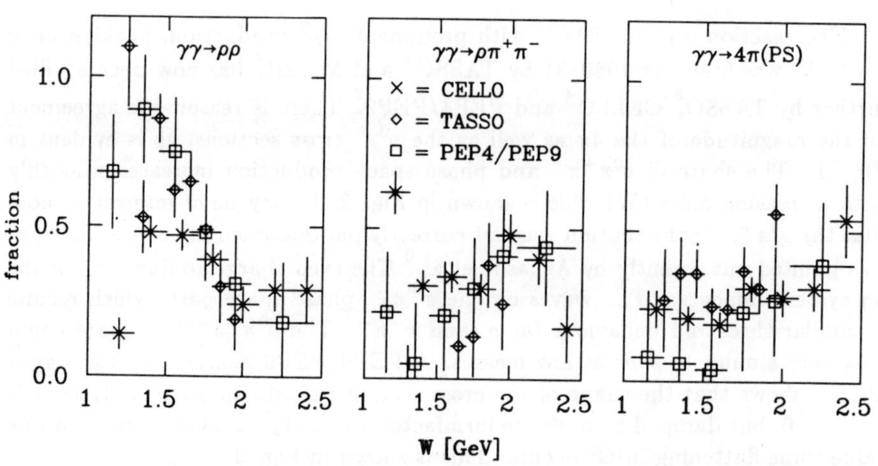

Fig. 2 *The $\rho^0\rho^0$, $\rho^0\pi^+\pi^-$ and 4π (phase space) fractions vs W.*

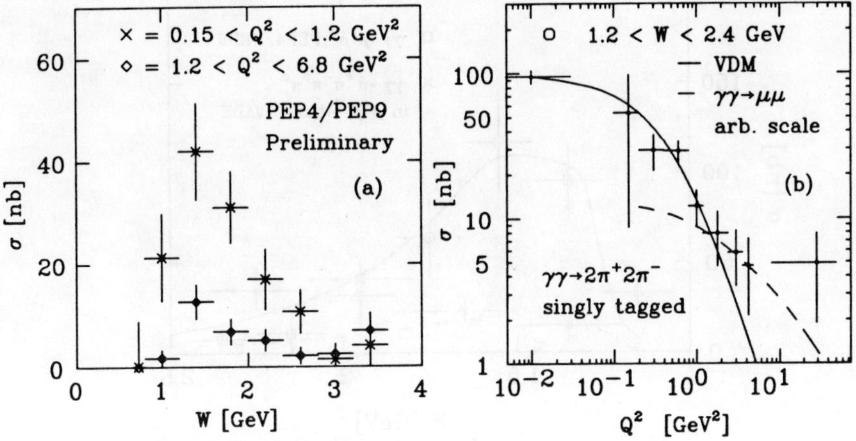

Fig. 3 *The Cross Section for* $\gamma\gamma \rightarrow 2\pi^+2\pi^-$ *at non-zero* Q^2
   a. The Cross Section vs W for two intervals of Q^2.
   b. The Q^2-*dependence for* $1.2 < W < 2.4$ *GeV. Full curve:* ρ-*form*
   factor, dashed curve: $\gamma\gamma \rightarrow \mu^+\mu^-$ *with arbitrary normalization.*

in the lower, $W = 1.3 - 1.6$ GeV, mass range. A CELLO analysis of ρ−decay angular distributions in the ρ−helicity frame showed an isotropic distribution, except for forward ρ's, $\cos\theta_\rho > 0.8$, where the distribution goes as $\sin^2\theta_H$, indicative for ρ−production with helicity ± 1. The TASSO and CELLO results were reported at the Aachen workshop last year.[7]

The JADE group[8] studied the $\gamma\gamma \rightarrow \pi^+\pi^-\pi^0\pi^0$ reaction. The cross section in the $\rho^+\rho^-$ band is generally lower than that of $\gamma\gamma \rightarrow \rho^0\rho^0$ and has quite a different behaviour as a function of $W_{\gamma\gamma}$, as is shown in Fig. 4.

The measurements on $\rho^0\rho^0$ and $\rho^+\rho^-$ indicate that the $\rho^0\rho^0$ enhancement cannot come from a single resonance with well-defined isospin.
Assuming ρ-dominance, the amplitudes for $\rho^0\rho^0$ production in $I = 0$ and $I = 2$ respectively can be written as:

$$A(\rho^0\rho^0 \rightarrow \rho^0\rho^0) = \frac{2}{3}A(2) + \frac{1}{3}A(0)$$

$$A(\rho^0\rho^0 \rightarrow \rho^+\rho^-) = \frac{\sqrt{2}}{3}A(2) - \frac{\sqrt{2}}{3}A(0)$$

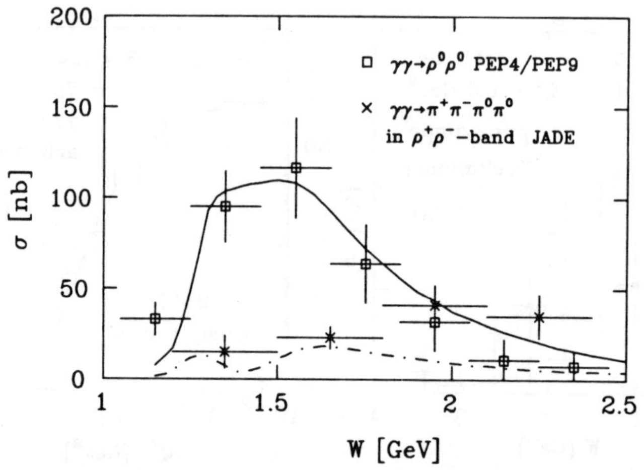

Fig. 4 *The Cross Sections for $\gamma\gamma \to \rho^0\rho^0$ from PEP4/PEP9 compared with the Cross Section for $\gamma\gamma \to \pi^+\pi^-\pi^0\pi^0$ in the $\rho^+\rho^-$ band from JADE. The curves indicate a description with two four-quark resonances by Achasov et al.*

where $A(0)$ and $A(2)$ are the production amplitudes of the states in I=0 and I=2 respectively. For pure I=0 one expects

$$\sigma(\rho^+\rho^-) = 2\sigma(\rho^0\rho^0)$$

and for $I = 2$

$$\sigma(\rho^+\rho^-) = \frac{1}{2}\,\sigma(\rho^0\rho^0)$$

Neither case is observed experimentally.

Early comparisons showed, moreover, that the $\rho^0\rho^0$ cross sections are about an order of magnitude larger than expected from Pomeron exchange.[1,2] Of the many interpretations advanced, I will discuss what I see as the remaining contenders (proposed $q\bar{q}$, gg, $q\bar{q}g$ interpretations have $I = 0$).

II.1 FOUR-QUARK STATES

An interpretation in terms of MIT-bag-model $I = 0$ and $I = 2$ $q\bar{q}q\bar{q}$ states was proposed by Li and Liu[9] and by Achasov, Devyanin and Shestakov.[10] Let me digress briefly into a description of these unconventional states. In 1976 Jaffe and Johnson[12] discussed several possibilities: hadrons made of glue alone, $q\bar{q}q\bar{q}$

exotics and $q\bar{q}g$ or $q\bar{q}$ states with unconventional quantum numbers. They remarked that up to now no dynamic ingredient is known that forbids the existence of these states or elevates them to a high mass.

The classification of $q\bar{q}q\bar{q}$ light-quark states was worked out by Jaffe[12] in 1977 in terms of the MIT-bag model. The basic assumption is that quarks interact relatively weakly. The mass of a hadron thus should increase roughly linearly with the number of quarks; with (non-strange) $q\bar{q}$ mesons at $\approx$ 700 MeV, qqq baryons at 1100 MeV one expects the $q\bar{q}q\bar{q}$ mesons at $\approx$ 1500 MeV in mass. Given the splittings within SU(6) multiplets, the lightest $q\bar{q}q\bar{q}$ mesons might be expected at $\approx$ 1 GeV in mass. The states with quantum numbers that cannot be obtained from $q\bar{q}$ are labelled exotics (E) and others Cryptoexotics (C). States with $J^P = 0^+$ end up in $\underline{9}, \underline{9}^*$ and $\underline{36}, \underline{36}^*$ multiplets, $J^P = 2^+$ states end up in $\underline{9}$ and $\underline{36}$ multiplets.

A $q\bar{q}q\bar{q}$ state can simply fall apart (fission) into $q\bar{q}$ mesons. In contrast, a $q\bar{q}$ meson (ρ-meson for example) must first polarize an additional $q\bar{q}$ pair from the vacuum before decay into two mesons. The $q\bar{q}q\bar{q}$ states should be narrow only if there is an angular momentum barrier or if its fission decay channels are closed. Achasov et al.[10] obtain several acceptable fits to the $\rho^0\rho^0$ and $\rho^+\rho^-$ mass spectra by taking into account an exotic, I=2, and a cryptoexotic, I=0, state from a $(\underline{9}, 2^+)$ multiplet, and a small contribution from the f(1270). The production amplitudes interfere constructively for $\rho^0\rho^0$ and destructively for $\rho^+\rho^-$ production. An example of these fits is shown in Fig. 4. The resonance parameters are

$$\mathrm{m_E} = 1.33 \text{ GeV}, \Gamma_\mathrm{E}^0 = 0.23 \text{ GeV}, \Gamma_\mathrm{E} \rightarrow \gamma\gamma = 4.2 \text{ keV}.$$
$$\mathrm{m_C} = 1.6 \text{ GeV}, \Gamma_\mathrm{C}^0 = 0.6 \text{ GeV}, \Gamma_\mathrm{C} \rightarrow \gamma\gamma = 3 \text{ keV}.$$

The bag-model estimates place the masses of such states at around 1.65 GeV. Li and Liu[11] obtain descriptions of similar quality by considering the contributions from the two $q\bar{q}q\bar{q}$ states above and in addition a low-mass I=0 state. Examples of $q\bar{q}q\bar{q}$-model predictions for $\gamma\gamma$-excitation of another vector-meson pair, $\rho^0\omega$, are shown in Fig. 5, together with upper limits by PLUTO and JADE.

A t-channel exchange interpretation of the $\rho\rho$ phenomena was first hinted at by Alexander et al.[13] on the grounds of factorization with photoproduction on nucleons and nucleon-nucleon interactions; the high energy relation

$$\sigma(\gamma\gamma \rightarrow \rho\rho) \approx [\sigma(\gamma p \rightarrow \rho p)]^2 / \sigma(pp \rightarrow pp)$$

was adapted by including a pion-exchange contribution at fixed p_{cm}. This approach leads to a correct order-of-magnitude estimate of the $\rho^0\rho^0$ production

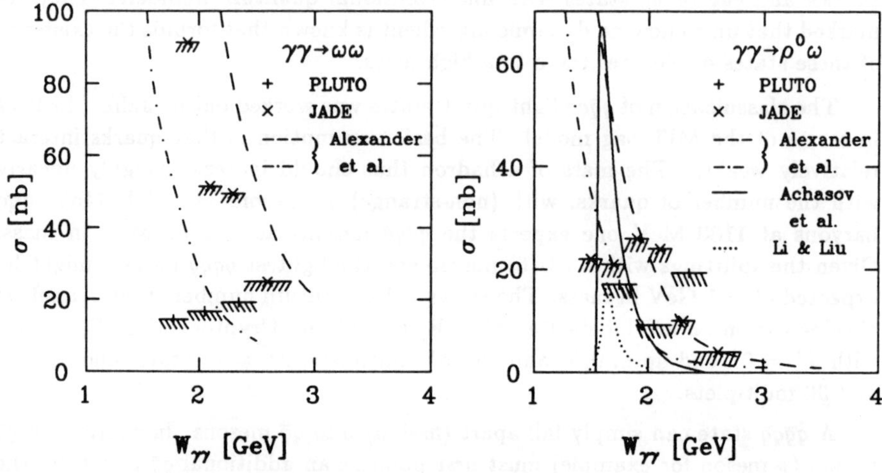

Fig. 5 *Upper limits on $\sigma(\gamma\gamma \to \rho^0\omega)$ and $\sigma(\gamma\gamma \to \omega\omega)$ derived by the PLUTO and JADE experiments. The curves represent predictions from 4-quark models by Achasov et al and Li and Liu, and a pion-exchange model by Alexander et al.*

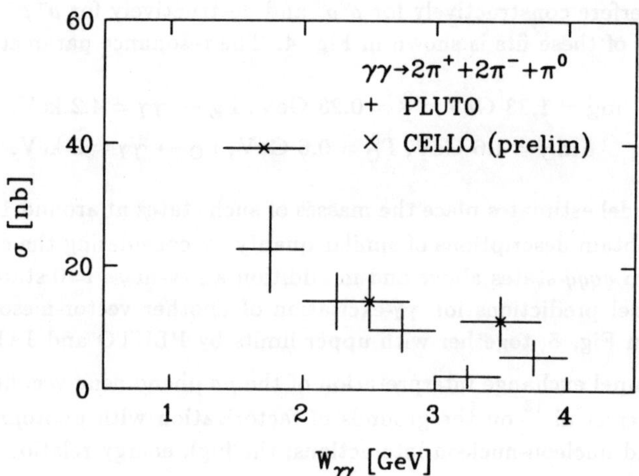

Fig. 6 *The Cross Section for $\gamma\gamma \to 2\pi^+ 2\pi^- \pi^0$ vs W measured by the PLUTO and CELLO experiments.*

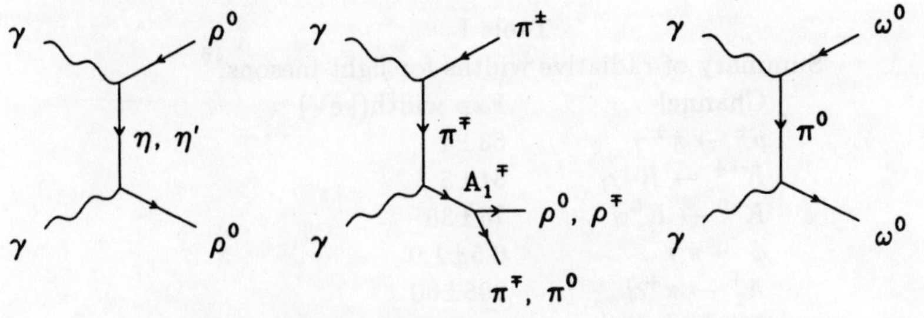

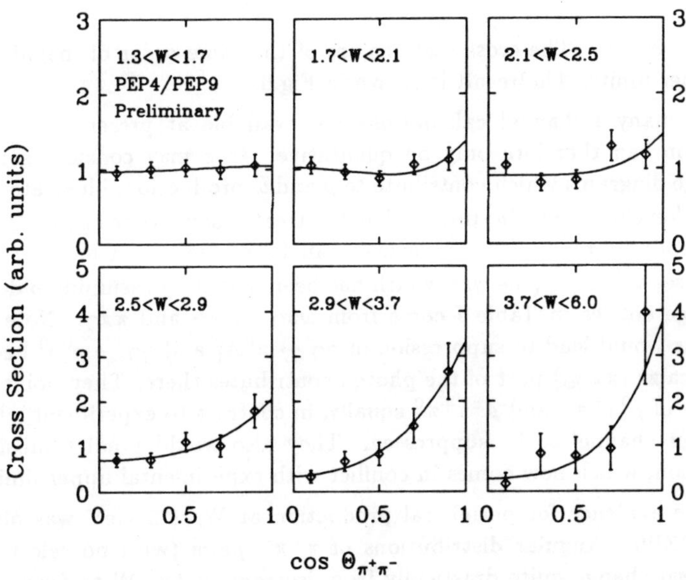

Fig. 7 *Meson-exchange diagrams.*

process and the high-mass end of its mass dependence. It also leads to the interesting prediction of substantial $\gamma\gamma \to \omega\omega$ production. Limits set by the JADE experiment[14] on this process last year still allowed a substantial amount of it.

This has changed since. Upper limits by PLUTO and JADE on $\gamma\gamma \to \rho^0\omega$ and $\gamma\gamma \to \omega\omega$ start to contradict the factorization assumption, as is shown in Fig. 5. The PLUTO and CELLO groups have made progress in identifying the final

Fig. 8 *Angular distributions of $\pi^+\pi^-$ pairs in the process $\gamma\gamma \to 2\pi^+2\pi^-$ at six intervals of W, measured by the PEP4/PEP9 experiment.*

Table I

Summary of radiative widths for light mesons.[16]

Channel	Exp width(keV)
$\rho^{\pm} \to \pi^{\pm}\gamma$	63±4
$K^{*+} \to K^{+}\gamma$	51±5
$K^{*0} \to K^{0}\gamma$	75±35
$\phi \to \pi\gamma$	6.5±1.9
$A_2^{+} \to \pi^{+}\gamma$	295±60
$K^{**+} \to K^{*+}\gamma$	240±45
$A_1^{+} \to \pi^{+}\gamma$	1000-1500
$B^{+} \to \pi^{+}\gamma$ [17]	230±60
$\omega \to \pi\gamma$	789±92
$\rho \to \eta\gamma$	72.5±14
$\eta' \to \rho\gamma$	93.1±25
$\phi \to \eta\gamma$	67.7±9
$\eta' \to \omega\gamma$	8.4±2.7
$\omega \to \eta\gamma$	3.2±2.6
$(\gamma \to \pi\pi$	$\approx 1100)$

state $2\pi^{+}2\pi^{-}\pi^{0}$. The cross sections are of the same order of magnitude as the $\rho^0\omega$ upper limits. The result is shown in Fig.6.

Not many t-channel calculations are available at present. The following discussion can therefore only be qualitative. One may consider some meson-exchange diagrams which contribute to ρ and ω production, illustrated in Fig. 7. The η, η'-exchange in $\rho^0\rho^0$ production was treated some time ago by C.Schmidt.[15] The coupling strength at the vertices can be derived from measured radiative widths, see table I[16] (the $\gamma\pi\pi$ width has been added by assuming ρ dominance). The large entries in Table I come from $\omega\pi\gamma$, $A_1\pi\gamma$ and $\pi\pi\gamma$. Note that such diagrams would lead to suppression of $\gamma\gamma \to \pi^0 A_1^0$ and $\gamma\gamma \to \rho^0\rho^0$ because only the isoscalar $(\pi\pi\gamma_\omega)$ part of the photon contributes there. They point to the occurrence of $\rho^0\pi^{+}\pi^{-}$ and $\rho^{\pm}\pi^{\mp}\pi^{0}$ equally, in contrast to experiment which shows the latter channel to be suppressed. They also would predict substantial $\omega\omega$ production, which now comes in conflict with experimental upper limits.

Some evidence for peripheral production at W > 2 GeV was obtained by PEP4/PEP9. Angular distributions of $\pi^{+}\pi^{-}$ pairs (with no selection on the pair mass) change quite drastically from isotropy at low W to forward peaking at higher W, as is shown in Fig. 8. The lines come from 3-term Legendre polynomial fits.

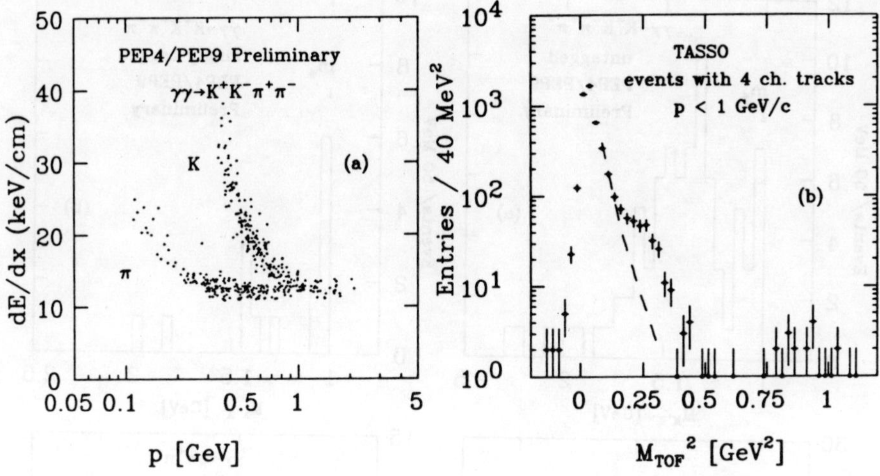

Fig. 9 *Kaon identification in 4-prong events in the PEP4/PEP9 experiment by dE/dx in the TPC and in the TASSO experiment by Time of Flight.*

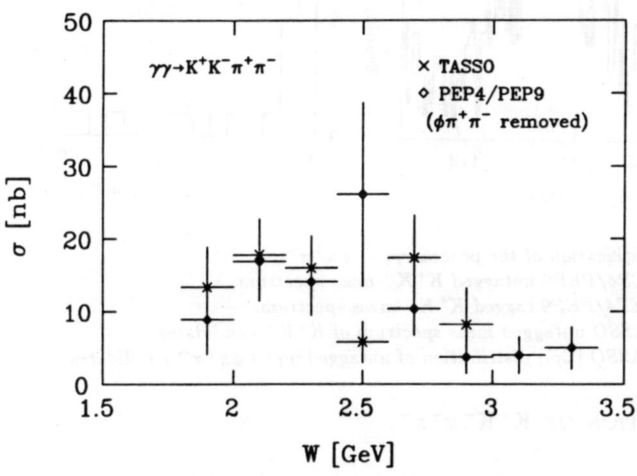

Fig. 10 *The Cross Section for $\gamma\gamma \to K^+K^-\pi^+\pi^-$ vs W.*

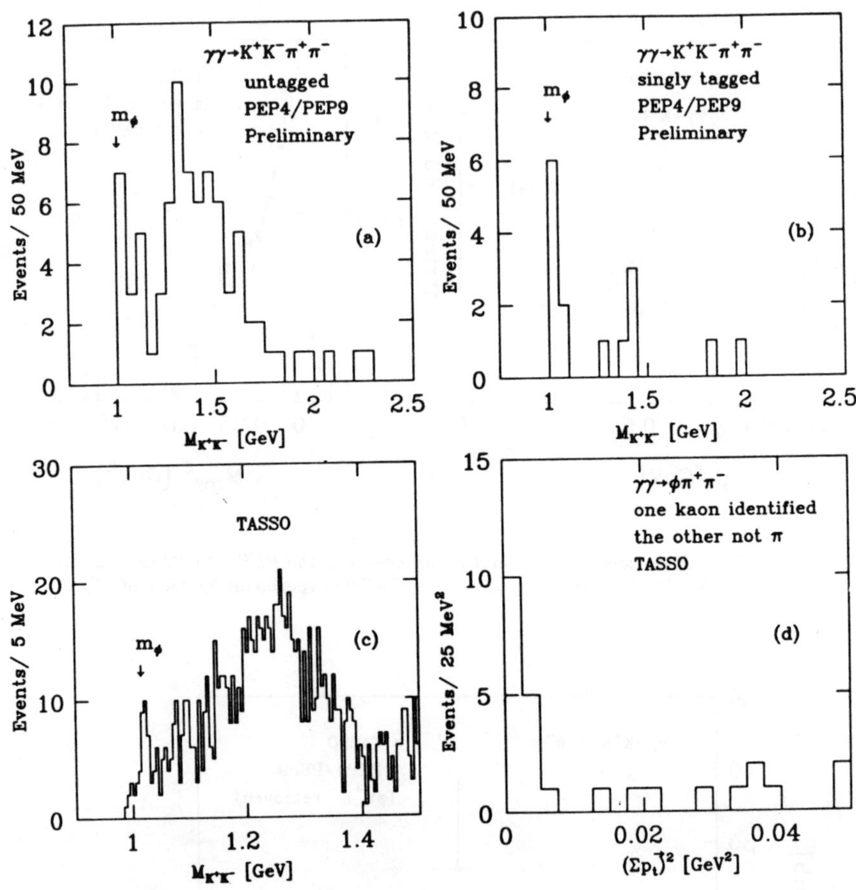

Fig. 11 *Identification of the process* $\gamma\gamma \to \phi\pi^+\pi^-$.
  a. *PEP4/PEP9 untagged* K^+K^- *mass spectrum.*
  b. *PEP4/PEP9 tagged* K^+K^- *mass spectrum.*
  c. *TASSO untagged mass spectrum of* K^+K^- *candidates.*
  d. *TASSO* $|\Sigma\vec{p}_t|$ *distribution of untagged* $\gamma\gamma \to \phi\pi^+\pi^-$ *candidates.*

II.2   PRODUCTION OF $K^+K^-\pi^+\pi^-$.

A preliminary report on the reaction $\gamma\gamma \to K^+K^-\pi^+\pi^-$ was given by the PEP4 collaboration at Brighton 1983.[14] In the meantime a systematic study was taken up by the TASSO and PEP4/PEP9 collaborations.

TASSO identifies particles by time of flight, and PEP4/PEP9's TPC by

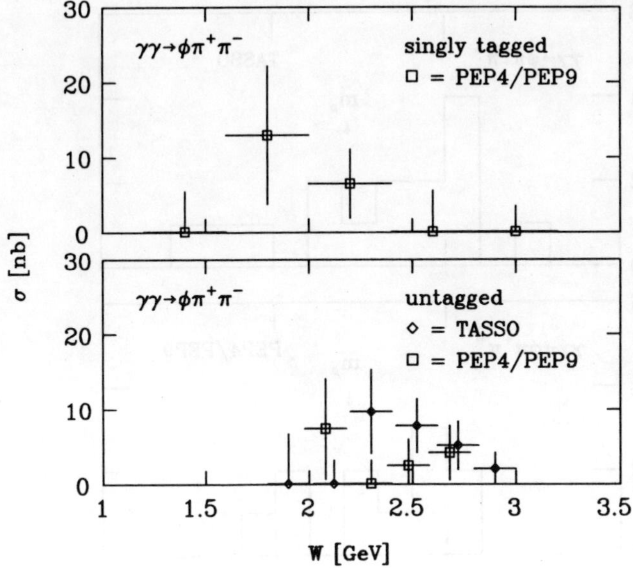

Fig. 12 *Untagged and singly tagged Cross Sections for* $\gamma\gamma \rightarrow \phi\pi^+\pi^-$.

dE/dx. Both methods give satisfactory results for low momentum kaons. This is shown in Fig. 9. There is agreement on a maximum cross section of $\approx$ 15 nb for the process $\gamma\gamma \rightarrow K^+K^-\pi^+\pi^-$, compared with $\approx$ 150 nb for $2\pi^+2\pi^-$ as is shown in Fig. 10. The K^+K^--mass spectrum, Fig.11a, 11b and 11c gives evidence for ϕ-production in both experiments. The PEP4/PEP9 tagged data show the ϕ more clearly than their untagged data; the observation of the ϕ signal is hindered by energy loss of low-momentum kaons in the material inside the detector. This hindrance is reduced at non-zero Q^2 and high W. The exclusive nature of the process is demonstrated clearly in Fig. 11d, which shows a peak near zero in the $|\Sigma p_t|$ spectrum.

The $\phi\pi^+\pi^-$ cross section is shown in Fig. 12. There is fair agreement between the two experiments in the untagged case. The singly tagged cross section is high (comparable to the untagged) and shifted towards lower W, illustrating the point made above. The statistical significance is low in all cases, and the acceptance calculations are perhaps not yet understood sufficiently. It is therefore premature to compare with model calculations. The $\pi^+\pi^-$ mass spectra from $\phi\pi^+\pi^-$ events are shown in Fig. 13. Both experiments agree that there is little evidence for ρ-excitation.

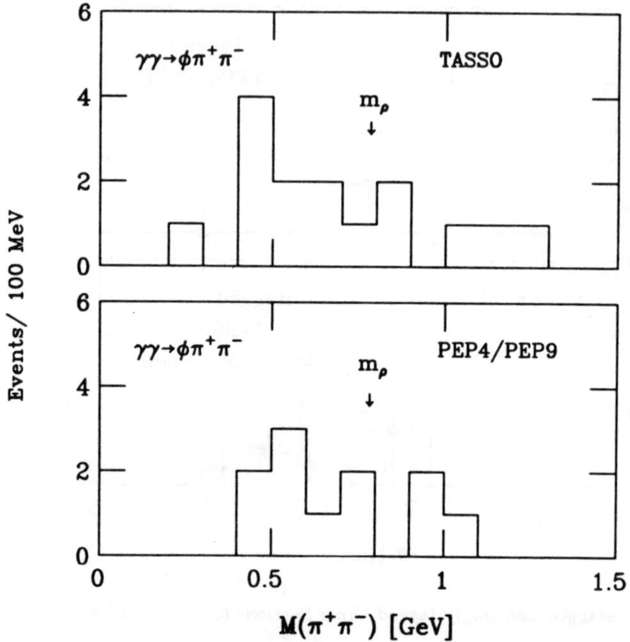

Fig. 13 *The $\pi^+\pi^-$ mass spectrum in the process $\gamma\gamma \rightarrow \phi\pi^+\pi^-$. No enhancement is seen at the ρ^0 mass.*

The nature of the remaining $K^+K^-\pi^+\pi^-$ events was examined by the PEP4-/PEP9 collaboration. Fig. 14 summarizes the mass structure in $K\pi$ pairs. There is good evidence for $K^{0*}K^{\pm}\pi^{\mp}$ production. A preliminary break-up of the number of events observed in various channels gives:

$\phi\pi^+\pi^-$	$7 \pm 3\%$
$K^{0*}K^{\pm}\pi^{\mp}$	$37 \pm 13\%$
$K^{0*}\overline{K}^{0*}$	0
$K^+K^-\pi^+\pi^-$	$56 \pm 18\%$.

Production angular distributions are compatible with isotropy. In summary, there is good evidence for $K^*K\pi$ and $\phi\pi\pi$ production, and none for $\phi\rho^0$ and $K^*\overline{K}^*$.

The observed production of $\phi\pi^+\pi^-$ is probably compatible with $q\bar{q}q\bar{q}$ models, which predict $\phi\rho^0$ production with a threshold enhancement of 10-70 nb[9,10] around 1.9 GeV in mass, where the present experiments have little acceptance,

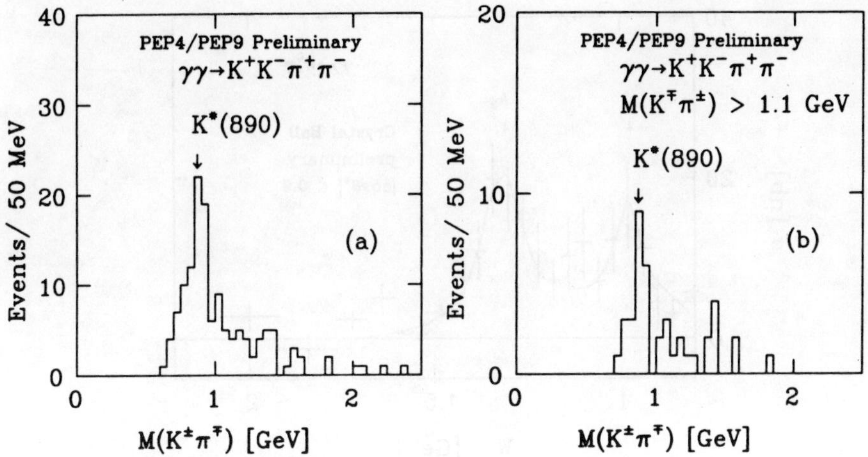

Fig. 14 *$K\pi$ spectra in the process $\gamma\gamma \to K^+K^-\pi^+\pi^-$.*
a. Projection of unlike sign $K\pi$ combinations, showing K^*890 excitation.
b. Unlike sign $K\pi$ combinations if the other combination has a mass greater than 1.1 GeV.

though a ρ^0 signal is conspicuously absent at higher mass. A description in terms of exchanges is more problematic, as only Pomeron exchange seems possible. It would be attractive to try K or K^* exchange on $K^*K\pi$ production.

II.3 PRODUCTION OF $\pi^0\eta$.

The Crystal Ball group measured the $\gamma\gamma$ cross section for $\pi^0\eta$ production at DORIS II. Their data constitute a substantial upgrade of an earlier report from SPEAR.[18] They find evidence for a δ and an A2 (see Fig 15). The excitation of the δ is of interest for $q\bar{q}q\bar{q}$ models. The state can be placed in the $(\underline{9},0^+)$ $q\bar{q}q\bar{q}$ nonet. With this assignment its coupling to vector mesons is very small, and Achasov et al[10] calculated its $\gamma\gamma$ width from 1-gluon exchange to be ≈ 0.27 keV. The Crystal Ball collaboration gives a preliminary value that is compatible with the $q\bar{q}q\bar{q}$ assignment:

$$\Gamma(\delta \to \gamma\gamma).\mathrm{Br}(\delta \to \eta\pi^0) = (0.10 \pm 0.04 \pm 0.06) \text{ keV},$$

The δ also has a place in a still controversial 0^+, L = 1, $q\bar{q}$ nonet. A calculation by Babcock and Rosner[19] gives

$$\Gamma(\delta \to \gamma\gamma) \approx 0 - 0.37 \text{ keV}.$$

The Crystal Ball value thus allows this assignment as well.

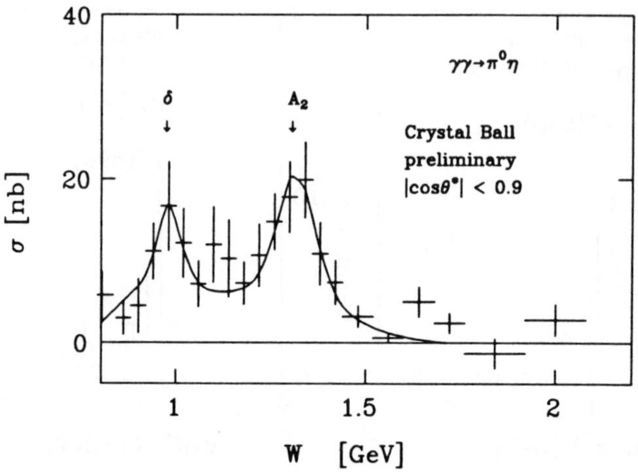

Fig. 15 *The Cross Section for the process $\gamma\gamma \to \pi^0\eta$, measured by the Crystal Ball Collaboration.*

III. Gluonium.

We now go to "upper limit land". If the ideas of Quantum Chromo Dynamics are correct, one can expect colorless, flavourless bound states of two or more gluons. Two-gluon states have charge conjugation C = +1, and radiative J/ψ decays which are thought to proceed via a photon and two gluons would produce them more easily than ordinary $q\bar{q}$-mesons.[20] A variety of well known $q\bar{q}$ states is excited in this way, like π^0, η, η', η_c, f, f', as well as the not yet classified $\iota(1440)\ 0^{-+}$, $\theta(1690)\ 2^{++}$ and $\xi(2200)\ 0^{++}, 2^{++}$ or 4^{++} states, and interesting structure is observed in $\rho\rho$ and $\omega\omega$ final states. Discussions on the gluonium content of these unclassified states abound in literature.[16,21,22,23] The $\gamma\gamma$-excitation of low-mass pseudo-scalar and tensor mesons, wich have possibly gluonium admixtures, is the subject of Cordier's talk at this conference.[24]

In lowest order $\gamma\gamma$ decays of gluonium states are suppressed, because the photons do not couple directly to gluons. However, this could be modified if a) the gluonium state has a $q\bar{q}$ admixture, or b) in next order one of the constituent gluons creates a $q\bar{q}$ pair to which the photon couples. As a consequence theorists are uncertain whether there is substantial suppression of $\gamma\gamma$ decay in these states.[21,22] The ι resonance is a case in point. Its possibly observed $\rho^0\gamma$ decay indicates at least a substantial $q\bar{q}$ admixture.

TASSO data on $\gamma\gamma \to K^0_s\overline{K^0_s}$ submitted to this conference (Fig. 16) allow

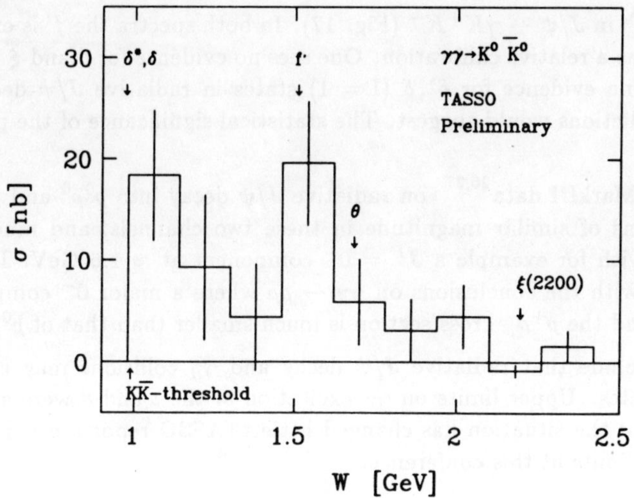

Fig. 16 *TASSO data on* $\gamma\gamma \to K^0\overline{K^0}$.

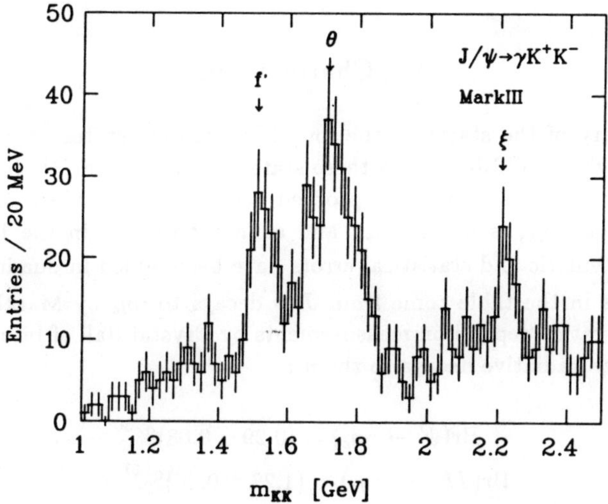

Fig. 17 *Mark III data on* $J/\psi \to \gamma\, K^+K^-$.

us to follow a suggestion by Chanowitz[21] and compare the same final state $K\overline{K}$ (MarkIII),[25] in $J/\psi \to \gamma K^+ K^-$ (Fig. 17). In both spectra the f' is excited, and that serves as a relative calibration. One sees no evidence for θ and ξ production in $\gamma\gamma$, and no evidence for δ^*, δ $(I = 1)$ states in radiative J/ψ decay, as the naivest predictions would suggest. The statistical significance of the present test is weak.

Recent MarkIII data[26,27] on radiative J/ψ decay into $\rho^0\rho^0$ and $\rho^+\rho^-$ show cross sections of similar magnitude in these two channels, and much detailed structure, with for example a $J^P = 0^-$ component at ≈ 1.6 GeV. This can be contrasted with the conclusions on $\gamma\gamma \to \rho\rho$ where a major 0^- component was excluded, and the $\rho^+\rho^-$ cross section is much smaller than that of $\rho^0\rho^0$.

We conclude that radiative J/ψ decay and $\gamma\gamma$ collisions may excite quite different states. Upper limits on $\gamma\gamma$ excitation of the ι and θ were summarized last year and the situation has changed little. TASSO reports preliminary 95% C.L. upper limits at this conference:

$$\Gamma(\gamma\gamma \to \xi) \cdot \mathrm{Br}(\xi \to K\overline{K}) < 0.5 \text{ keV},$$
$$\Gamma(\gamma\gamma \to \theta) \cdot \mathrm{Br}(\theta \to K\overline{K}) < 0.14 \text{ keV}.$$

IV. Charmonium.

A summary of the status of the four Charmonium states accessible to $\gamma\gamma$-excitation is given in Table II. As these states are below the $D\overline{D}$ (open Charm) threshold, the decay is thought to proceed via two gluons. Much of these data, especially on the η_c, is preliminary and cannot be found in the Partice Data Tables.[28] Systematic and statistical errors have been added in quadrature. Most recent entries in the table come from J/ψ decays to $\gamma\eta_c$ by MarkIII.[25,29] The η_c branching ratios depend on measurements by Crystal Ball of branching ratios for J/ψ and ψ' radiative decays to the η_c:

$$\mathrm{Br}(\psi' \to \gamma\eta_c) = (0.29 \pm 0.08)\%,[30]$$
$$\mathrm{Br}(J/\psi \to \gamma\eta_c) = (1.27 \pm 0.36)\%.[31]$$

The $\eta_c \to \gamma\gamma$ width was measured for the first time in the reaction $\overline{p}p \to \eta_c \to \gamma\gamma$ in the ISR jet experiment.[32] It depends on many factors, such as the knowledge of the ISR beam-momentum profile, the $\eta_c \to \overline{p}p$ BR, and the total η_c width.

Significantly, none of the entries come from $\gamma\gamma$- collision experiments, though the $\gamma\gamma$-widths are predicted to be of the order of a few keV, like for other $q\bar{q}$ resonances. The barriers are the low $\gamma\gamma$ luminosities at the realtively high mass, the small branching ratios to all-charged final states, the limited accuracy in mass reconstruction and the small solid-angular coverage of 4π detectors designed mainly for e^+e^- annihilation physics ($\approx 10\%$ for 4-prongs) and last but not least a lack of high machine luminosities needed to exploit this domain.

Up to now $\gamma\gamma$ excitation experiments derive only upper limits from inspection of the mass spectra in various channels. The PEP4/PEP9 experiment looked for various additional features which would enhance the $c\bar{c}$ component in the mass spectrum in $\gamma\gamma \to 2\pi^+2\pi^-$: the virtual photon Q^2-distribution, which should fall less steeply, and isotropic decay angular distributions. Their result represents an upper limit for the combined $2\pi^+\pi^-$ decays of the four resonances:

$$\Sigma_{\eta_c,\chi_0,\chi_2,\eta_c'}\Gamma(\gamma\gamma \to X) \cdot \mathrm{BR}(X \to 2\pi^+2\pi^-) < 3.5 \text{ keV } (95\%\text{c.l.}).$$

Upper limits on η_c excitation given last year[8] remain valid.

V. Conclusions.

i) Vector meson production.

From an analogy with hadron-hadron interactions and from vector dominance one would expect vector-meson pair production. Instead one observes the processes:

$$\gamma\gamma \to \rho^0\rho^0 \text{ or } \rho^0\pi^+\pi^-$$
$$\to K^{0*}K\pi$$
$$\to \phi\pi^+\pi^-$$

One does not observe:

$$\gamma\gamma \to \rho^0\omega$$
$$\to \omega\omega$$
$$\to \rho^0\phi$$
$$\to K^{0*}\overline{K}^{0*} \quad .$$

Empirically the data are compatible with the production of one vector meson instead of two.

ii) Gluonium states.

In the $K\overline{K}$ final state naive expectations that the $\gamma\gamma$-width of the gluonium candidates $\theta(1690)$ and $\xi(2200)$ are small are compatible with the present low statistics data.

iii) Charmonium states.

The $\eta_c \to \gamma\gamma$ width was measured for the first time in $\overline{p}p \to \gamma\gamma$. The value is close to what is expected.

Acknowledgements

I am indebted to many collegues for their cooperation in preparing this manuscript. In particular I want to mention J.G. Layter, W. Wagner, A. Buijs and W.G.J. Langeveld. I also like to thank Prof. R. Lander and his colleagues for hosting such an instructive and enjoyable conference.

Discussion.

H.Kolanoski (Bonn)

A comment on the spin-parity analysis of $\rho^0\rho^0$ by TASSO:

Our result of 0^+ dominance at low $W_{\gamma\gamma}$ has been obtained by a likelihood fit using the complete kinematics in the fitted matrix elements at each event. In particular the 0^+ dominance cannot easily be seen from the projected distributions, but is consistent with them. That means that the results have to rely on the chosen form of the matrix element, which is particularly sensitive below the nominal $\rho^0\rho^0$ threshold. Comparisons with the TASSO result should be made on the same basis.

Answer: For a more complete discussion of TASSO's conclusions I refer to your talk at the 1983 Aachen workshop.

Table II

Status on Widths and Branching Ratios (%) of Charmonium States accessible to excitation in $\gamma\gamma$ collisions.

	η_c	χ_0	χ_2	η_c'
J^{PC}	0^{-+} [36]	0^{++}	2^{++}	
M (GeV)	2.984	3.415	3.555	3.592
Γ_{tot} (MeV)	$11.5^{+4.5}_{-4.0}$ [31]	$16.1^{+1.5}_{-1.3}$ [33]	$2.1^{+1.0}_{-0.7}$ [33]	
$\Gamma_{\gamma\gamma}$ (keV)	4.5 ± 4.9 [32,23]		$1.3^{+0.9}_{-0.6}$ [33]	
$2\pi^+2\pi^-$	$2.0^{+1.5}_{-1.0}$ [34]	4.3 ± 0.9	2.3 ± 0.5	
$K^+K^-\pi^+\pi^-$	$1.4^{+2.1}_{-1.0}$ [34]	3.4 ± 0.9	2.0 ± 0.5	
$3\pi^+3\pi^-$		1.7 ± 0.6	1.2 ± 0.8	
$\pi^+\pi^-$		0.9 ± 0.2	0.20 ± 0.11	
K^+K^-		0.8 ± 0.2	0.16 ± 0.12	
$\bar{p}p\pi^+\pi^-$		0.6 ± 0.2	0.35 ± 0.14	
$K^\pm\pi^\mp K^0_s$	$5.2^{+3.1}_{-2.5}$ [34]			
$\eta\pi^+\pi^-$	2.4 ± 1.6 [35]			
	3.3 ± 1.1 [29]			
$\eta'\pi^+\pi^-$	2.8 ± 1.1 [29]			
$\bar{p}p$	0.11 ± 0.06 [29]		0.01 ± 0.005 [32]	
$\phi\phi$	0.8 ± 0.3 [29]			
$\gamma\gamma$	$(3.9 \pm 4.0)10^{-2}$ [32,23]		$(6.3 \pm 2)10^{-2}$ [33]	
ηK^+K^-	seen [29]			
$\pi^0 K^+K^-$	seen [29]			

REFERENCES

1. R. Brandelik et al., Phys. Lett. **97B**, 448 (1980).

2. D.L. Burke et al., Phys. Lett. **100B**, 153 (1981).

3. M. Althoff et al., Z. Phys. **C16**,13 (1982).

4. H.J. Behrend et al., Z. Physik **C21**, 205 (1984).

5. J.G. Layter et al., PEP4/PEP9 Contribution to Leipzig Conf. on High Energy Physics, July 1984.

6. N.N. Achasov, S.A. Devyanin, G.N. Shestakov, Preprint TPh-56(**137**), Dept of Theor. Physics 630090, Novosibirsk–90, March 1984.

7. H. Kolanoski, Proceedings of the 5th International Colloquium on $\gamma\gamma$ Interactions. Aachen. Ed. Ch. Berger, Lecture Notes in Physics, Vol. 191, Springer Verlag (1983).

8. J.E. Olsson, Proceedings of the 5th International Colloquium on $\gamma\gamma$ Interactions. Aachen. Ed. Ch. Berger, Lecture Notes in Physics, Vol. 191, Springer Verlag (1983).

9. B.A. Li and K.F. Liu, Phys. Rev. D **30**, 613 (1984), Phys. Letters **118B**, 435 (1982).

10. N.N. Achasov, S.A. Devyanin, and G.N. Shestakov, Phys. Lett. **108B** , 134(1982); Z.Phys.**C30**, 55 (1982).

11. B.A. Li and K.F. Liu, "Are $Q^2\overline{Q}^2$ states observable?", University of Kentucky Preprint, UK / 83-07.

12. R.L. Jaffe and K. Johnson, Phys. Letters **60B**, 201 (1976), R.L. Jaffe, Phys. Rev. **D15**, 267 (1977).

13. G. Alexander et al., Phys.Rev. **D26**, 1198(1982).

14. J.B. Dainton, Proceedings of the Int. Europhysics Conference on High Energy Physics, Brighton, 20-27 July, 1983, p. 652.

15. C. Schmidt, Priv. Comm.

16. D. Hitlin, "Radiative Decays and Glueball Searches", CALT-68-1071, 1983.

17. B. Collick et al, Phys. Rev. Letters **53** ,2374 (1984).

18. C. Edwards et al., Physics Letters **110B**, 82 (1982).

19. J. Babcock and J.L. Rosner, Phys Rev **D14**, 1286 (1976).

20. S.J. Brodsky et al, Phys Letters **73B**, 203 (1978).

21. M.S. Chanowitz, "Glueballs and Meiktons: a status report", invited talk at the XIV'th Int. Symposium on Multiparticle Dynamics at High Energies, Lake Tahoe, June 22–27, 1983; see also the lecture: "Resonances in photon photon scattering" at this conference.

22. P.M. Fishbane and S. Meshkov, "Glueballs", NBS 84-0567, submitted to Comments on Nuclear and Particle Physics.

23. J. Perrier, "Physics at the J/ψ and the status of glueballs",Invited talk presented at the Int. Conference "Physics in Collision IV", Santa Cruz, California, Aug. 22-24, 1984, SLAC-PUB-3436.

24. A.Cordier,"Resonances I", Invited talk at this conference.

25. W. Toki, SLAC-PUB-3262, Nov 1983, Presented at 11th SLAC Summer Institute on Particle Physics, Stanford, California, July 18-29, 1983.

26. N. Wermes, Invited talk at XIXth Rencontre de Moriond: New Particle production, La Plagne, France, March 4-10, 1984, SLAC-PUB-3312, April 1984.

27. J.D. Richman, Invited talk at the Symposium on High Energy e^+e^- Interactions, Vanderbilt University, April 5-7, 1984.

28. M. Roos et al., "Review of Particle Properties", Phys. Letters **111B**, i (1982).

29. MARK III Contribution to the Leipzig Conference, 1984, and Phys. Rev. Letters **52**, 2126 (1984).

30. F.C. Porter, "Recent Results form the Crystal Ball", Proc. of the SLAC Summer Institute on Particle Physics, July 27-August 7, 1981, SLAC-245, Jan 1982.

31. J.E. Gaiser, SLAC-0255, Aug 1982, Thesis.

32. C. Baglin et al., Contribution to the 4th Int. Conf. on Physics in Collision, Santa Cruz, Aug 22-24, 1984, CERN EP/EL-0027P.

33. E. Bloom, SLAC-PUB-2976, Sept 1982.

34. T.M. Himel et al., Phys. Rev. Letters **45**, 1146 (1980).

35. E. Partridge et al., Phys. Rev. Letters **45**, 1150 (1980).

36. R.M. Baltrusaitis et al., Phys. Rev. Letters **52**, 2126 (1984).

NEW PARTICLES AND TWO-PHOTON PHYSICS

Fridger Schrempp

II. Institut für Theoretische Physik
Universität Hamburg
D-2000 Hamburg 50
W.-GERMANY

ABSTRACT

In a first part, I review the general theoretical arguments
leading to new physics and new particles beyond the Standard
Model, either in terms of supersymmetry or compositeness.
Speculations about new particles expected within these schemes
are then discussed in the light of recent anomalous events
from the $p\bar{p}$ collider and from PETRA.
In a second part, I specifically try to evaluate the potential
of $\gamma\gamma$ and $e\gamma$ collisions at PETRA/PEP and LEP energies with
respect to new particle searches. Some interesting possibilities, including searches for spinless composite bosons, non-
standard enhanced Higgs particles, scalar electrons ($\tilde{e}$) and
$\gamma\gamma \to$ 'nothing' emerge.

1. New Physics beyond the Standard Model

The Standard Model of electroweak interactions [1]

$$\left[SU(2)_L \times U(1) \right]_{GSW} \tag{1}$$

is certainly in excellent health. It describes charged and neutral
current reactions with ease over a wide energy range and has predicted
the $W^{\pm}$ and Z^{o} vector bosons in the correct mass range. In addition to
the spectacular discovery of the weak bosons [2], evidence for the top-
quark has even been announced recently [3]. Thus, only the Higgs particle
is missing from the Standard Model point of view.

Let me motivate in this section why, nevertheless, a wealth of additional particles as manifestations of new physics may be expected.

The point is that besides providing a successful description of
electroweak phenomena, the Standard Model also leaves open a number of
important questions:
i) Which is the origin of the Fermi scale

$$\Lambda_F \equiv \left(\sqrt{2}\, G_F \right)^{-1/2} \sim 250 \text{ GeV } ? \tag{2}$$

On the one hand, Λ_F is one of the three fundamental and conspicuously
different scales we know of today

$$\Lambda_{color} \simeq 300\,MeV, \quad \Lambda_F \simeq 250\,GeV \quad and \quad \Lambda_{Planck} \simeq 10^{19}\,GeV. \quad (3)$$

The W,Z, Higgs, quark and lepton masses are all proportional to this single scale Λ_F of electroweak interactions. On the other hand, in the Standard Model, it arises very much <u>ad hoc</u>, as that value $\langle \phi \rangle$ of the scalar (Higgs) field ϕ for which the scalar potential $V(\phi)$ happens to have its absolute minimum.
ii) How can we hope to reduce the large number of free parameters (O(20)!) in the Standard Model? In a basic theory of electroweak interactions it should be possible to compute e.g. the quark and lepton masses in terms of Λ_F as well as the various KM-mixing angles.
iii) Which is the origin of the mysterious quark-lepton connection in form of families, with lepton e.m. charges being integer multiples of the quark charges?
iv) Why are there three (or more?) replicas of families, the generations?

Because of this inherent large degree of indetermination it is widely believed that the Standard Model only represents an effective theory in the wider sense; i.e. one expects that somewhere above the Fermi scale Λ_F there exists a new scale Λ_{new}, characterizing the onset of more fundamental physics. The general idea is then to relate the Fermi scale and the associated electroweak interactions to this new scale Λ_{new} and the new underlying physics.

Important clues for such a view come from the Higgs sector of the Standard Model.

Consider first the 'canonical' way of embedding electroweak interactions in schemes of grand unification (GUT), with

$$\Lambda_{new} = \Lambda_{GUT} \sim 10^{15}\,GeV. \quad (4)$$

As was first realized by Wilson [4] and subsequently emphasized by 't Hooft [5], such a large gap

$$\Lambda_{new}/\Lambda_F \gg 1, \quad (5)$$

combined with the requirement of perturbation theory being applicable all the way up, poses the problem of an unnatural finetuning of parameters in the scalar sector. The problem originates in the <u>quadratic</u> divergence of radiative corrections to the Higgs mass. With $\Lambda_{new} = \Lambda_{GUT}$ acting as a cut-off the loops involving the fermions and the bosons of the theory individualy give a large contribution (Fig. 1)

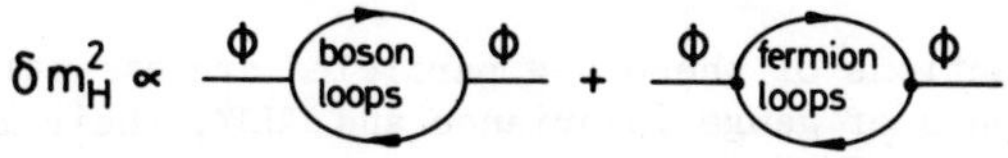

$$\delta m_H^2 \propto \frac{\phi \quad \overbrace{boson\ loops} \quad \phi}{} + \frac{\phi \quad \overbrace{fermion\ loops} \quad \phi}{}$$

Fig. 1: Quadratically divergent radiative corrections to the Higgs mass

$$\delta m_H^2 \propto e^2 \int dk^2 \,^{\Lambda_{new}^2} \propto e^2 \Lambda_{new}^2 \sim e^2 \Lambda_{GUT}^2 \qquad (6)$$

On the other hand the validity of perturbation theory requires

$$m_H^2 + \delta m_H^2 \lesssim \mathcal{O}(\Lambda_F^2) \qquad (7)$$

Thus, a tremendous finetuning of the input parameters is required at each level of perturbation theory, unless:

A) there is a symmetry, which enforces an almost perfect cancellation of the large loop corrections

or

B) the new scale Λ_{new} is actually close to Λ_F such that possible radiative corrections of type (6) are automatically small.

These two options have found their realization in form of two popular classes of schemes of very different underlying philosophy.

(A) <u>Supersymmetry</u>[6] (SUSY): As is well known SUSY relates fermions and bosons with $\Delta J = 1/2$. It turns out that by making the theory supersymmetric   one exactly effects the cancellation between boson and fermion loops [7], needed to avoid 'finetuning' if $\Lambda_{new}/\Lambda_F \gg 1$. The 'perturbative GUT spirit' can thus be consistently maintained. Models of supergravity [8] (SUGRA), involving SUSY in its local form, ambitiously hope to even relate the Fermi scale Λ_F directly to the 'ultimate' Planck scale

$$\Lambda_F \longleftrightarrow \Lambda_{Planck} \sim 10^{19} \text{ GeV.} \qquad (8)$$

An important implication of these ideas is, of course, that a <u>wealth of new particles</u>, the SUSY partners of q, l, W, Z, γ, g, ϕ, must exist (table 1).

<u>Table 1</u>: Particles and their SUSY partners

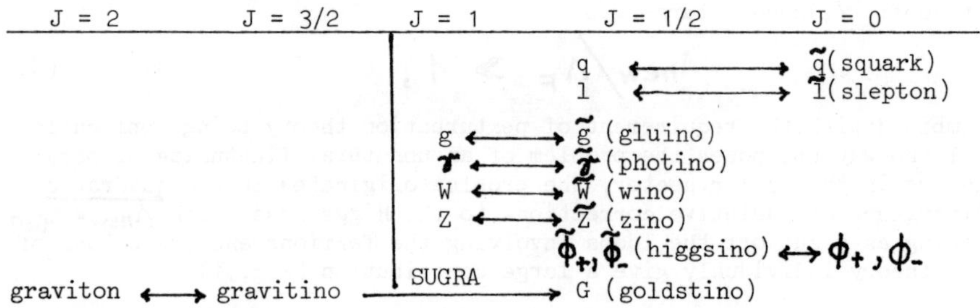

Whereas the interactions of these new particles are essentially fixed from the requirements of gauge invariance and SUSY, their masses are quite uncertain, since SUSY cannot be an exact symmetry at low energies. However, in order not to spoil the required cancellation of divergences, [c.f. eq. (6)] the mass splitting must not be too large

$$\left| m^2_{particle} - m^2_{sparticle} \right| \lesssim O(\Lambda^2_F).\qquad(9)$$

Some SUSY particles, such as the $\tilde{\gamma}$ are usually expected to be light, of the order of a few GEV, say. Intensive searches for SUSY particles are under way at PETRA/PEP and the $\bar{p}p$ collider. The familiar reaction

$$e\gamma \longrightarrow eX\qquad(10)$$

which traditionally serves to extract the photon-structure functions plays an important role in this context, as will be discussed in section 4.2.

Next, let me turn to the second possible solution of the 'fine-tuning' problem in the Higgs sector corresponding to $\Lambda_{new} \sim O(1 \div 5)\Lambda_F$.

(B) Compositeness:
The basic idea is that in some analogy to QCD, there exists a new gauge force ('technicolor', 'hypercolor', ...) which becomes strong in the vicinity of the Fermi scale Λ_F. Composites are formed from a set of new, strongly interacting constituents ('technifermions', 'preons', ...). Depending on how the new composite sector is thought to be interlocked with present physics, various possible scenarios emerge. Let me remind you of the essence behind the most popular schemes, proceeding in order of increasing 'radicalness'.

A crucial feature in all cases is that the Fermi scale Λ_F appears to be directly related to the confinement scale Λ_{TC} (HC) of the new underlying gauge theory.

i) Composite Higgs (Technicolor [9]): Only the Higgs scalar ϕ in the Standard Model is replaced by a boundstate

$$\phi \rightarrow \bar{F}\bar{F}\qquad(11)$$

of socalled technifermions F which, apart from the new technicolor forces, experience the same weak $\left[SU(2)_L \times U(1)\right]_{GSW}$ gauge interactions as the (elementary) quarks and leptons. The essential point is that the required spontaneous symmetry breaking in the Standard Model, is now effected dynamically through formation of technifermion condensates

$$\langle F\bar{F}\rangle \simeq \Lambda^3_{TC},\qquad(12)$$

in analogy to the quark condensates, $\langle q\bar{q}\rangle \simeq \Lambda^3_C$, being formed in QCD during the process of confinement. Simple rescaling from QCD gives

$$\Lambda_C/f_\pi \sim \Lambda_{TC}/\Lambda_F \quad or \quad \Lambda_{TC} \sim 4\Lambda_F \sim 1\,TeV.\qquad(13)$$

The Goldstone boson analogues of the pions are subsequently 'eaten' by the W,Z bosons, whereby the latter become massive as required.

Important possible signatures of such a technicolor mechanism at much lower energies, $E \ll \Lambda_{TC}$, are extra light (pseudo) Goldstone bosons [10]. Some may even carry color (e.g. leptoquarks!).

ii) <u>Composite quarks and leptons</u> [11]

Here, the idea is to consider the quarks and leptons as composites of a common set of constituents ('preons'), strongly bound by the new gauge force ('hypercolor') at $E \sim \Lambda_{HC} \gtrsim \Lambda_F$. By treating quarks and leptons as boundstates one hopes to eventually compute e.g. their masses as well as to understand the origin of families and generations.

There are two drastically different ways of how the weak interactions are incorporated. The first, more conservative possibility is to view the weak interactions as fundamental gauge interactions like in the Standard Model. Spontaneous symmetry breaking then has to be effected along the lines of technicolor via a dynamical Higgs scalar being also a composite of preons. Accordingly, the hypercolor scale is as in (13)

$$\Lambda_{HC} \sim 1 \, TeV. \tag{14}$$

The second possibility is much more radical [12]. The weak interactions are not associated with a fundamental gauge force and there is no Higgs mechanism at all. Instead, the weak interactions are viewed as residual hypercolor interactions among composite quarks and leptons just like the strong interactions among composite hadrons are known to be residual color interactions. The $W^{\pm}$ and Z^{o} bosons are, correspondingly, interpreted as prominent composite vector bosons not unlike the ρ mesons in strong interactions. Since in this case

$$\Lambda_{HC} \sim \Lambda_F \simeq 250 \, GeV, \tag{15}$$

this scheme is usually termed 'nearby' compositeness. As an important experimental signature of 'nearby' compositeness one expects a possibly <u>rich spectrum of new composite particles</u> with masses related to the Fermi scale. There should be further fermions, both colored and uncolored (q*, l*) as well as new colored and uncolored composite bosons of various spins. When discussing 'compositeness' in relation to experiments I shall mainly refer to this scenario, simply because the new particles are supposed to be relatively 'nearby'.

Finally, let me point out that despite intensive searches no satisfactory 'Standard Composite Model' has as yet emerged. This applies both for technicolor and for composite models of quarks and leptons. Hence, 'predictions' in compositeness are mostly of qualitative nature and heavily rely on the analogy to QCD and strong interactions.

2. Anomalous Events and New Particles?

A variety of socalled anomalous events has been reported from the UA1 and UA2 detectors [13,14] at the CERN $\bar{p}p$ collider. But even at the much lower energies of PETRA (DORIS) some surprises have been claimed to exist [15,16]. In view of the meagre statistics, some of these evidences presumably are fluctuations. On the other hand, the chances are not so small, that at least some of the unusual events will be confirmed and then might provide a first glimpse on the new physics expected in the vicinity of the Fermi scale.

There has been a flood of theoretical papers trying to interpret the anomalous events as evidence for new particles from either of the two competing schemes: supersymmetry and compositeness. Let me shortly summarize the most intriguing experimental findings (as of summer 1984) along with some representative theoretical speculations.

i) Radiative Z^0 decays (UA1/UA2)

Three out of 12 Z^0 events found by UA1/UA2 involve besides a lepton pair a hard photon in addition [2]. Taken at face value this would mean a radiative rate

$$R_{UA1/UA2} \sim 20 \div 25\% \, , \tag{16}$$

being in excess by about a factor 10 over the normal fraction for bremsstrahlung expected theoretically [17]. If to be interpreted in terms of new physics, these events are hard for SUSY. In contrast, there is a whole variety of suggestions from (brave) members of the compositeness camp like

(1) a new composite $J = 0$ partner [18] X^0 of the W,Z bosons of mass

$$m_X \simeq m(e^+e^-) \simeq 42 \div 50 \text{ GeV} \, , \tag{17}$$

causing

$$Z^0 \longrightarrow X^0 \, \gamma \, .$$
$$\qquad \hookrightarrow e^+e^-, \mu^+\mu^-, \dots \, . \tag{18}$$

I shall come back to this suggestion in sect. 3.2, since here is a case to illustrate how crucial, complementary information can come from the two-photon channel

$$e^+e^- \longrightarrow \gamma\gamma \tag{19}$$

at much lower energies [19] (PETRA).

(2) new (composite) excited leptons [20] of mass

$$m_{\ell^*} \simeq m(\ell\gamma) \simeq 80 \text{ GeV} \, , \tag{20}$$

causing

$$Z^\circ \longrightarrow \ell^* \bar{\ell} \; . $$
$$\hookrightarrow \ell \gamma \qquad\qquad (21)$$

(3) new (composite) states X degenerate [21] with the Z°
causing

$$X \longrightarrow Z^\circ \gamma \; . $$
$$\hookrightarrow \ell\bar{\ell} \qquad\qquad (22)$$

It should be noted, however, that none of these proposals seems entirely satisfactory. For instance, suggestions (1) and (2) have problems [22] to account for the 'distribution' of the $\ell^+\ell^-\gamma$ events in a Dalitz plot of $m^2(\ell^+\gamma)$, $m^2(\ell^-\gamma)$ and $m^2(\ell^+\ell^-)$.

ii) <u>'Monojets' and 'monophotons' (UA1)</u>
5 + (1) 'monojets' and 1 + (1) 'monophotons' have been reported [13] by UA1:

$$\bar{P}P \longrightarrow 1 \; hard \left\{ \begin{matrix} jet \\ '\gamma' \end{matrix} \right\}_{E_T \gtrsim 35\,GeV} + \; (jetlets) + \; E_T^{miss.} \gtrsim 35\,GeV \; . \quad (23)$$

This time the SUSY advocates are much more happy. 'Monojets' can arise in SUSY schemes as signals of the production and subsequent decay of SUSY-partners such as gluinos [23] ($\tilde{g}$) or – preferably-squarks [24] ($\tilde{q}$)

$$\bar{P}P \longrightarrow \tilde{q} \, \bar{\tilde{q}} + \ldots \; ; \quad \tilde{q} \longrightarrow q + \tilde{\gamma} \, (E_T^{miss}) \; . \quad (24)$$

The way how this process ends up fitting the experimental monojet distribution is somewhat tricky [24]. It happens through the combined effect of experimental cuts and 'jet merging' caused by the UA1 jet trigger and jet algorithms. As a result, the $\tilde{q}$ mass is claimed [24] to be in the range (Fig. 2)

$$m\tilde{q} \sim 25 \div 40 \; GeV, \qquad\qquad (25)$$

with some preference for $m_{\tilde{q}} \simeq 40$ GeV. A quite different possible SUSY-mechanism [25] for monojets and monophotons arises if SUSY happens to be spontaneously broken. In such case SUSY may be realized [26] (non-linearly) in terms of only one (massless) neutral fermion G , the goldstino being the Goldstone particle of broken SUSY *.

---

* note, however, that in the popular SUGRA schemes [8] the goldstino is 'eaten' by the gravitino (Super Higgs effect) which, in turn, gets a mass $m_{3/2}$.

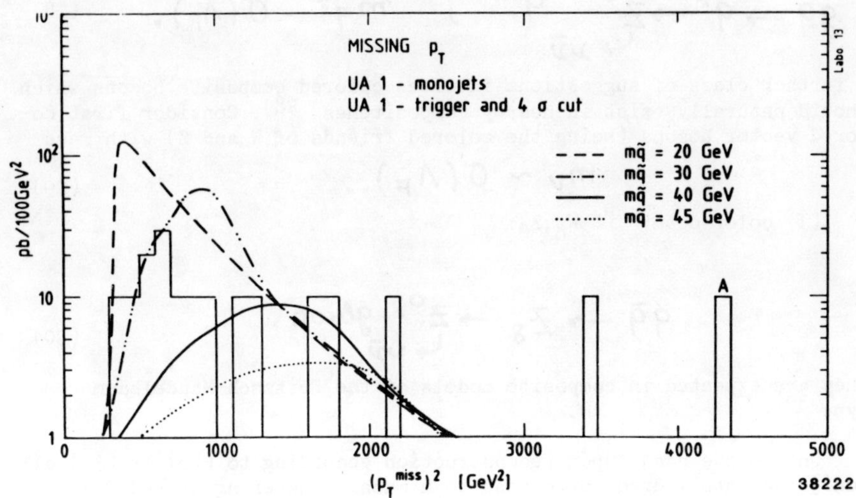

Fig. 2: Missing P_T distribution of UA1-monojets[13] compared to predictions from squark ($\tilde{q}$) production in SUSY. From J. Ellis et al. in ref. 24.

The goldstino $\mathbf{G}$ is pair produced [25] but not detected ($P_\perp^{miss}$)

$$\bar{P}P \longrightarrow jet + G + G + \cdots \qquad \text{(monojets)}$$

$$\bar{P}P \longrightarrow \gamma + G + G + \cdots \qquad \text{(monophotons)}. \tag{26}$$

The rates and distributions can be made to roughly agree in terms of one parameter, the SUSY breaking scale. Of interest is also the resulting parameter free prediction for e^+e^- collisions [25]

$$\sigma(e^+e^- \longrightarrow \gamma + G + G) \simeq (\sqrt{s}/35 GeV)^6 \cdot 10^{-2} pb \tag{27}$$

corresponding to ~ 1 event for a typical PETRA experiment.

In sect. 3.3 I shall speculate on possibly interesting implications of such a mechanism for two-photon physics.

Nearby compositeness (technicolor) schemes also like monojets!

One proposal is to consider, in correspondence to excited leptons in (21), heavy (composite) quarks [27] ('starks')

$$gg \rightarrow q^* \rightarrow Z^o \underset{\hookrightarrow \nu\bar{\nu}}{\quad} q \quad ; \quad m_{q^*} \sim \mathcal{O}(\Lambda_F). \qquad (28)$$

A further class of suggestions involves colored composite bosons which should naturally exist in nearby compositeness [28]. Consider first colored vector bosons (being the colored friends of W and Z) with mass

$$m_V \sim \mathcal{O}(\Lambda_F). \qquad (29)$$

(1) color octet [29] W_8, Z_8:

$$q\bar{q} \rightarrow Z_8 \rightarrow Z^o \underset{\hookrightarrow \nu\bar{\nu}}{\quad} gluon. \qquad (30)$$

They are expected in composite models of the Fritzsch-Mandelbaum type [30].

On the one hand, upon reconstruction according to (28) or (30) all monojet events indeed correspond to a mass, clustering around [31]

$$m_{Z_8(q^*)} \sim 160 \; GeV. \qquad (31)$$

On the other hand, the mechanisms (28), (30) predict a specific number of corresponding events with ν's replaced by charged leptons. Apparently, they are not seen, which seems to put these proposals on somewhat shaky ground.

(2) color triplet leptoquark bosons [28,32] V_3 with Q = 2/3 decaying as

$$V_3 \Rightarrow \begin{cases} q_{2/3} + \bar{\nu} \\ q_{-1/3} + \ell^+ . \end{cases} \qquad (32)$$

Monojet signals arise from the subprocess [28]

$$g\, q_{2/3} \longrightarrow V_3 \underset{\hookrightarrow q_{2/3}+\bar{\nu}\,,\,q_{-1/3}+\ell^+}{\quad} \nu \qquad (33)$$

Leptoquark vector bosons of this type are expected in the class of Abbott-Farhi initiated composite models [33], characterized by the global 'hyper'-flavor symmetry $[\, \alpha_c(\Lambda_F), \alpha \sim 0\,]$

$$SU(4)_{Pati-Salam} \supset SU(3)_{color} \times U(1)_Y. \qquad (34)$$

It remains to be investigated in detail whether the constraints from rare decays involving $\Delta S \neq 0$ neutral currents (e.g., $K_L \rightarrow \bar{\mu}e, ...$) allow composite J = 1 leptoquark bosons to be sufficiently light [34,35].

As a further possible source of monojets let me finally discuss spinless leptoquark bosons χ of the (pseudo-) Goldstone boson type [36]. They arise in schemes of compositeness (technicolor) involving a spontaneous breaking of the Pati-Salam type 'hyper'-flavor symmetry [34]

$$SU(4)_{Pati-Salam} \xrightarrow{SSB} SU(3)_{color} \times U(1)_Y . \qquad (35)$$

As their vector partners (32) they are color triplets, have charge $Q = 2/3$ and decay as in (32). In nearby compositeness they are expected to be light [36],

$$m_\chi \sim \sqrt{\frac{\alpha_c(\Lambda_F)}{\pi}} \, \Lambda_F \sim \mathcal{O}(40\,GeV) \ll \Lambda_F , \qquad (36)$$

(give or take a factor two).

For spinless composite leptoquarks of the Goldstone boson type there is virtually no problem with rare decays since they typically couple very weakly to quarks and leptons of one generation[36]

$$g_{\chi q \bar{\ell}} \sim \langle m_{q,\ell} \rangle / \Lambda_F \ll 1 . \qquad (37)$$

They are, however, copiously pair-produced via color and electromagnetic gauge interactions, just like squarks ($\tilde{q}$)

$$\text{gluon gluon} \longrightarrow \chi\bar{\chi} \quad \text{in } \bar{p}p \text{ collisions} \qquad (38)$$

$$\text{and} \quad e^+e^- \longrightarrow \gamma \longrightarrow \chi\bar{\chi} \quad \text{in } e^+e^- \text{ collisions.} \qquad (39)$$

As to monojets, it is amusing to note, that the signals due to such leptoquarks are indistinguishable from those due to squarks [36] (eq. (24)). The production cross sections and expected mass ranges are virtually identical. Due to (37), the dominant decay of a leptoquark, associated e.g. with the second q-l generation is

$$\chi^{(2)} \longrightarrow c + \bar{\nu} \quad (\gg \chi^{(2)} \longrightarrow s + \mu^+) , \qquad (40)$$

which is indistinguishable from squark decay in SUSY schemes

$$\tilde{q} \longrightarrow q + \tilde{\gamma} . \qquad (41)$$

Hence, to the extent that monojets are evidence for SUSY [24] they are also evidence for leptoquarks. However, in contrast to squarks, leptoquarks also decay subdominantly into charged leptons, eq. (40), giving rise to anomalous 1^+jet $- 1^-$jet events [36,38] (Fig. 3c). This brings me to

iii) Anomalous μ^+jet $- \mu^-$jet events (CELLO/UA1)
One strikingly planar μ^+jet $- \mu^-$jet event was reported by CELLO [15] at PETRA and very recently two quite similar ones [37] by UA1 (Figs.3a,b).

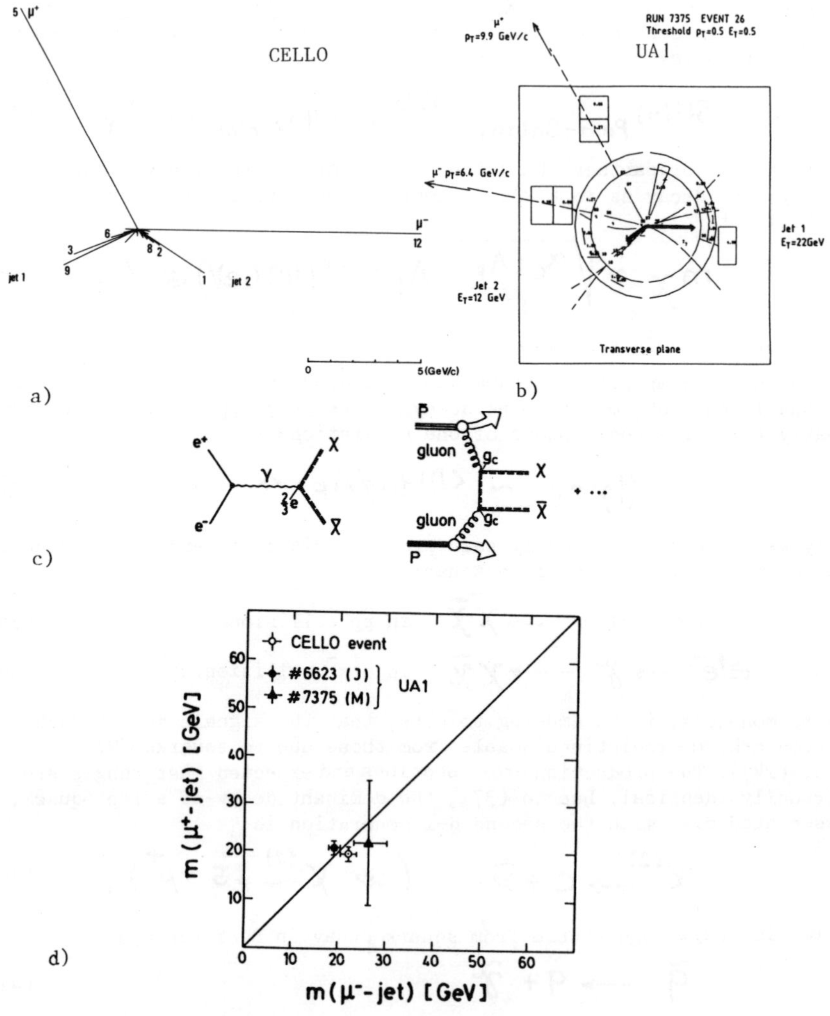

Fig. 3: Evidence for spinless leptoquark bosons X from μ^+jet — μ^-jet events
(a): CELLO event (ref.15)
(b): one of two UA1 events (ref.37)
(c): diagrams for X pair production in $\bar{e}e$ and $\bar{p}p$ collisions
(d): consistency of $\mu^{\pm}$jet invariant mass with $m_X \approx 20 \div 22$ GeV according to fig. 3c).

As displayed in Fig. 3d, all three events are compatible [36] with

$$m(\mu^+ jet_1) \sim m(\mu^- jet_2) \simeq 20 \div 22 \text{ GeV.} \qquad (42)$$

Hence it is tempting to associate them with pair production [38] of the above spinless leptoquarks $\chi^{(2)}$. This means [15]

$$m_{\chi^{(2)}} \simeq 20.5 \pm 1 \text{ GeV.} \qquad (43)$$

Such a mass is still compatible with the rough estimate (36) and also marginally within the mass range (25) needed to explain monojets along the lines of Ref. 24). The kinematics of the CELLO event ($\sqrt{s}$ = 43.45 GeV) would correspond to a $\chi\bar{\chi}$ pair being produced almost at rest. Then each χ decays into a s-μ^+ pair approximately back to back. Since two intersecting lines form a plane this mechanism naturally accounts for the almost planarity of the CELLO event [36]. Overall compatibility of the CELLO and UA1 μ^+jet - μ^-jet event rates is found for [36]

$$Br(\chi^{(2)} \rightarrow s\mu^+) \sim 0.2 , \qquad (44)$$

leaving

$$Br(\chi^{(2)} \rightarrow c\bar{\nu}) \sim 0.8 \qquad (45)$$

for monojet type signals and acoplanar two-jet events [in e^+e^- collisions].

iv) <u>A narrow state $\zeta(8.3)$ in $\Upsilon(1S) \rightarrow \gamma X$ (CRYSTAL BALL)</u>
The $\zeta(8.3)$ was announced at the Leipzig conference [16] by the CRYSTAL BALL collaboration. It had caused great excitement, e.g. as a possible candidate for a (non-standard) Higgs. Unfortunately, further runs with comparable statistics could not confirm this state [39]. I have correspondingly skipped the discussion on its implications for two-photon physics from the written version of my talk.

In conclusion, I feel it is certainly a good time to think about new particles. But, beware of inflation! Presently, more new particles have been proposed than there are 'anomalous' events. The results from the fall-run of the pp collider are certainly awaited with excitement.

3. New Particles in Photon-Photon Collisions

The problematic aspects of two-photon collisions in the context of new particle searches are quite obvious: relatively small cross sections and a strongly decreasing flux at large values of the $\gamma\gamma$ cm energy W.

As outlined in the preceeding sections, theoretical arguments lead us to expect new particles with masses somehow related to the Fermi scale Λ_F. A pessimistic guess would be

$$m_{new} > m_{W,Z}. \qquad (46)$$

In such a case there seems to be very little chance for two-photon physics in the foreseeable future. Even at the level of a disfavoured subprocess, e.g.

$$\text{gluon + gluon} \longrightarrow \text{new particles} \tag{47}$$

the $\bar{p}p$ collider, in contrast, reaches gg energies

$$W \sim O(100 \text{ GeV}). \tag{48}$$

It may be instructive to compare the 'potential for new physics' in the gg subprocess (47) at the $\bar{p}p$ collider with that in two-photon collisions

$$\gamma + \gamma \longrightarrow \text{new particles} \tag{49}$$

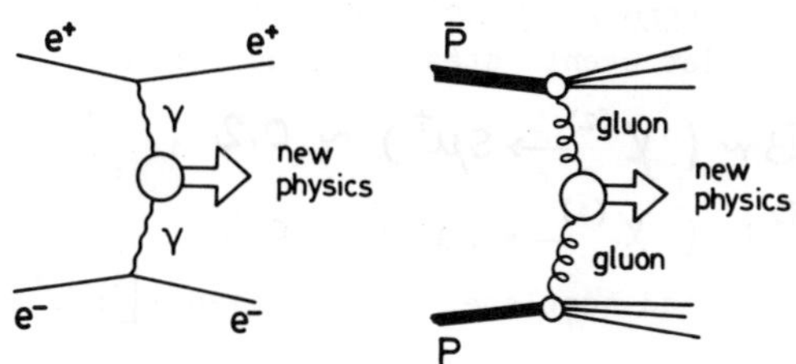

Fig. 4: $\gamma\gamma$ versus gluon gluon collisions

at PETRA/PEP and LEP in a more quantitative way (Fig. 4). For this purpose, consider the $\gamma\gamma$ and gg luminosities $d\mathcal{L}/dz$

$$\frac{d\sigma}{dz}\left(\frac{e^+e^-}{\bar{P}P}\right) = \frac{d\mathcal{L}_{gg}^{\gamma\gamma}}{dz} \cdot \hat{\sigma}_{gg}^{\gamma\gamma}(W), \tag{50}$$

$$\text{with} \quad z = W/\sqrt{s}, \tag{51}$$

computed via the equivalent photon approximation [40] and the gluon structure functions of Ref. 41, respectively. Since one expects

$$\hat{\sigma}_{\gamma\gamma}(W) \propto \frac{\alpha^2}{W^2} \quad \text{and} \quad \hat{\sigma}_{gg}(W) \propto \frac{\alpha_c^2}{W^2} \tag{52}$$

it seems fair to compare the quantities

$$\frac{\alpha^2}{W^2}\frac{d\mathcal{L}^{\gamma\gamma}}{dz} \quad \text{with} \quad \frac{\alpha_c^2}{W^2}\frac{d\mathcal{L}^{gg}}{dz}, \tag{53}$$

with the dimensions of a cross section as function of W. The result is displayed in **Fig.** 5 and looks somewhat depressing for two-photon

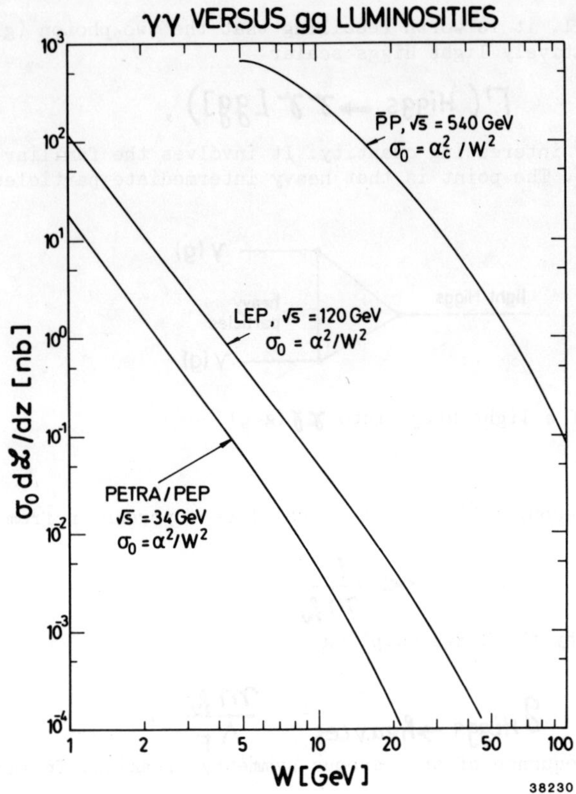

Fig. 5: Comparison of $\gamma\gamma$ and gluon gluon luminosities

physics, even at LEP.

On the other hand, both in SUSY and (nearby) compositeness it is quite possible that some of the new particles are exceptionally light

$$m \ll \Lambda_F , \qquad (54)$$

and – most importantly – that their two-photon couplings are substantially _enhanced_. For obvious reasons (Fig. 5) I shall concentrate on such possibilities.

186

3.1 Enhanced Non-standard Higgs Scalars (SUSY)

First of all, it is worth recalling that the two photon (gluon) width of a relatively light Higgs scalar

$$\Gamma(\,\text{Higgs} \to \gamma\gamma\,[gg]\,), \tag{55}$$

is an extremely interesting quantity. It involves the familiar triangle graph in Fig. 6. The point is that heavy intermediate particles excepti-

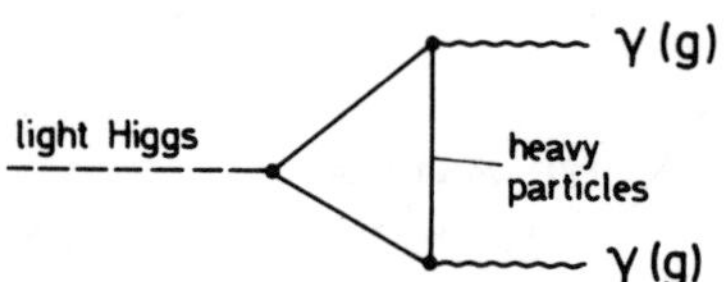

Fig. 6: Decay of a light Higgs into $\gamma\gamma$(g g)

onally do <u>not</u> decouple [42,43]. The usual loop suppression from heavies of mass m_h

$$\propto \frac{1}{m_h} \,, \tag{56}$$

is compensated by the Higgs coupling

$$g_{\text{Higgs} \to \text{heavies}} = \frac{m_h}{\Lambda_F} \,, \tag{57}$$

which is a consequence of spontaneous symmetry breaking. In fact [42,43]

$$\Gamma\left(\text{Higgs} \to \begin{matrix}\gamma\gamma\\gg\end{matrix}\right) \quad \text{'counts' the number}^2 \text{ of } \left\{\begin{matrix}\text{charged}\\\text{colored}\end{matrix}\right\} \tag{58}$$

heavy particles ($m_h \gg m_{\text{Higgs}}$) in the theory which is remarkable.

The problem is that for a standard GSW Higgs $\Gamma(\text{Higgs} \to \gamma\gamma)$ turns out to be hopelessly small [42,43]. The result for 3 generations is displayed in Fig. 7, to which I shall return repeatedly . Also shown are, for comparison, realistic lower limits achievable at PETRA/PEP and LEP for the quantity

$$\frac{\Gamma(R \to \gamma\gamma)\ \text{Br}(R \to X)}{M_R^3} \left[\frac{\text{keV}}{\text{GeV}^3}\right] \,, \tag{59}$$

as a function of the mass M_R of a given spinless particle. The region above the limits corresponds to more than 20 events at an assumed integrated luminosity

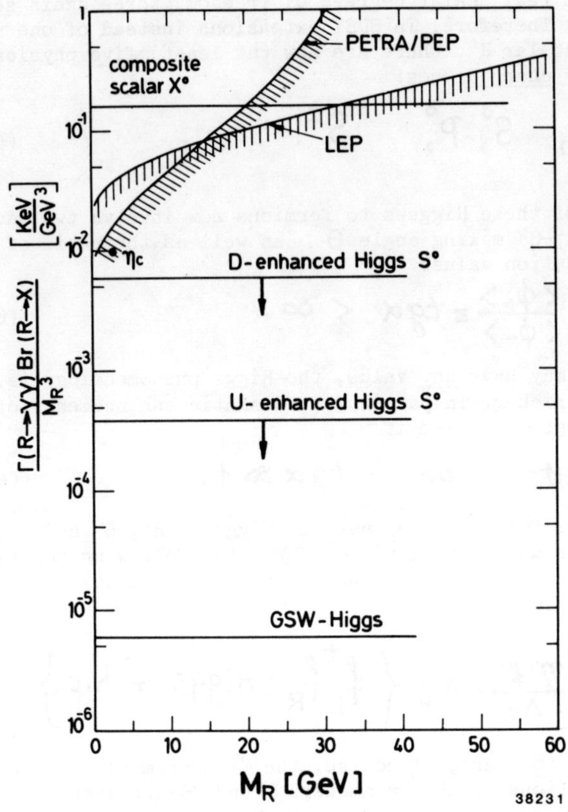

Fig. 7: Predictions for $\gamma\gamma$ widths of GSW-Higgs, non-standard U- and D-enhanced Higgs and composite scalar X^o compared to lower limits at PETRA/PEP and LEP (see text).

$$\int \mathcal{L}\, dt = \begin{cases} 100\ pb^{-1} & at\ PETRA/PEP; \\ 20\ pb^{-1} & at\ LEP.;\ E_b = 60\ GeV \end{cases} \tag{60}$$

and a typical no-tag efficiency of $\varepsilon = 5\%$ for a certain final state X. Again the equivalent photon approximation was used.

The purpose of this section is to examine, how the situation may change for Higgs particles in SUSY extensions of the Standard Model.

In the minimal, non-supersymmetric case with one complex Higgs doublet ϕ , ϕ gives mass to up quarks, whereas ϕ^* gives mass to down quarks and charged leptons via spontaneous symmetry breaking $\langle\phi\rangle \neq 0$. In SUSY, however, ϕ and ϕ^* cannot be both members of SUSY multiplets and hence at least <u>two</u> (complex) scalar doublets ϕ_+ and ϕ_- must exist [6,44]

$$\phi \longrightarrow \phi_+ \quad ; \quad \phi^* \longrightarrow \phi_- \tag{61}$$

188

Of these 2 x 2 x 2 = 8 real scalar degrees of freedom, three again serve
to give mass to $W^{\pm}, Z^{o}$. Therefore, in SUSY extensions instead of one neutral, physical Higgs scalar H^{o}, there are now (at least) _five_ physical
Higgses, including two _charged_ ones!

$$H^{o}, \quad S^{o}, P^{o}, \quad S^{\pm}. \tag{62}$$

The Yukawa couplings of these Higgses to fermions now involve two additional parameters, a H^{o}-S^{o} mixing angle Θ as well as the ratio of
the two vacuum expectation values

$$0 < \frac{\langle \phi_{+} \rangle}{\langle \phi_{-} \rangle} \equiv tg\,\alpha < \infty. \tag{63}$$

Since $tg\,\alpha$, a priori may have _any_ value, the Higgs phenomenology is,
correspondingly, much richer. In particular, dramatic enhancements of
certain Yukawa couplings may arise if

$$tg\,\alpha \ll 1 \qquad \text{or} \qquad tg\,\alpha \gg 1. \tag{64}$$

Specifically, one finds for the three neutral Higgses, H^{o}, S^{o} and P^{o},
contributing via the triangle (Fig. 6) to $\gamma\gamma$, the following general
couplings [44] to up quarks U = (u,c,t), down quarks D = (d,s,b) and
charged leptons L^{-} = (e,μ,τ)

$$\mathcal{L}_{Yuk.} = \sum_{f=U,D,L^{-}} \frac{m_{f}}{\Lambda_{F}} \cdot X_{f} \left\{ f_{L}^{\dagger} f_{R} \cdot Higgs + h.c. \right\}. \tag{65}$$

Of central interest in (65) are, of course, the 'enhancement' factors
$X_{U,D,L^{-}}$ which, as functions of the parameters α and Θ are displayed
in table 2. Note, in particular, that $X_{L^{-}} = X_{D} \gg 1$ implies $X_{U} < 1$ and
vice versa.

Table 2   Enhancement factors for Yukawa couplings of neutral Higgses in
minimal SUSY extensions, compared to the single Higgs case
(H^{o}) in the Standard Model (GSW)

SUSY-GSW	$X_{L^{-}} = X_{D}$	X_{U}
H^{o}	$\cos\Theta / \cos\alpha$	$\sin\Theta / \sin\alpha$
S^{o}	$\sin\Theta / \cos\alpha$	$-\cos\Theta / \sin\alpha$
P^{o}	$-i\,\sin\alpha / \cos\alpha$	$-i\,\cos\alpha / \sin\alpha$
GSW		
H^{o}	1	1

In addition, the masses of the Higgs particles also depend on
$\text{tg}\,\alpha = \langle\phi_+\rangle/\langle\phi_-\rangle$ and they are related among each other [44]. For large
classes of (broken) SUSY schemes one finds in case of enhancement of
either the U-sector ($X_U \gg 1$) or the D-sector ($X_D = X_{L^-} \gg 1$)

$$m_{S^o} \sim m_{P^o} \quad \text{arbitrary, but probably } \underline{\text{light}} \tag{66}$$

$$\text{and} \quad m_{H^o} > m_{Z^o} \,, \; m_{S^\pm} > m_W. \tag{67}$$

Note that the mass of (S^o, P^o) is not even bounded from below by the
Weinberg–Linde bound [45]

$$m_{\text{Higgs}} \gtrsim 10 \text{ GeV}, \tag{68}$$

since it only applies to the single Higgs case!

Hence, it seems, there is a characteristic signature of SUSY extensions of the Standard Model, worth being looked for in $\gamma\gamma$ collisions.
The (S^o, P^o) Higgses could both be possibly light and strongly enhanced in
$\gamma\gamma$ via the fermions in the triangle (Fig. 6). For instance

$$\frac{\Gamma(S^o \to \gamma\gamma)}{m_{S^o}{}^3} \approx (GSW)_{\substack{fermionic \\ contribution}} \cdot X_f^2. \tag{69}$$

Furthermore, in SUSY, all the charged (heavy) SUSY partners $(\tilde{q}, \tilde{l}, \text{etc.})$
contribute via the triangle. According to (58), a measurement of $\Gamma(S_i^o P^o \to \gamma\gamma)$
could thus in addition provide information on the existence of SUSY
degrees of freedom, too heavy to be accessible directly!

An enhancement of the type

$$X_{L^-} = X_D \gg 1 \tag{70}$$

would be most favorable for two-photon collisions since it implies a
large branching ratio of (S^o, P^o) into charged leptons and thus a good
detection efficiently, $\varepsilon \gg 5\%$: On the other hand, (70) requires

$$m_{S_i^o P^o} > m_\gamma \simeq 9.46 \text{ GeV}, \tag{71}$$

else (S^o, P^o) would probably have been seen in

$$\Upsilon \longrightarrow \gamma X \tag{72}$$

(see Fig. 8). The $\Upsilon(8.3)$, which was unfortunately not confirmed [16,39]
(see sect. 2), could have been, of course, an exciting application for
the type of exercise presented in this section.

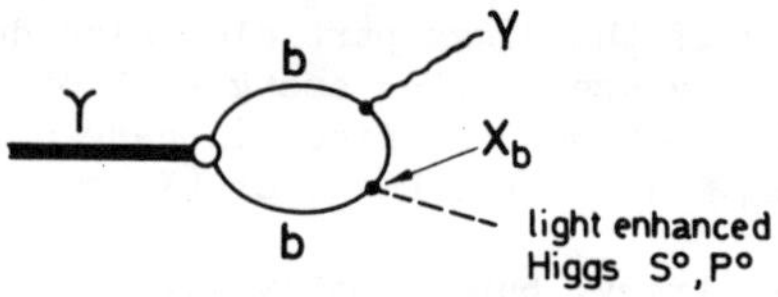

190

Υ
b
Υ
X_b
b
light enhanced
Higgs S°, P°

Fig. 8: A light D-enhanced Higgs in Υ decay

Finally, as a warning against too much optimism, let me point out that the crucial enhancement factors X_f cannot be too large if one insists on perturbation theory to remain valid. Then, in analogy to the perturbative unitarity bound [46] for the GSW Higgs mass

$$m_H \lesssim \sqrt{8\pi} \cdot \Lambda_F \simeq 1.2 \text{ TeV} \tag{73}$$

from

$$W_L \overline{W}_L \longrightarrow (\text{Higgs}) \rightarrow W_L \overline{W}_L \ , \tag{74}$$

there is a corresponding limit [47]

$$m_f \cdot X_f \lesssim \sqrt{2\pi} \cdot \Lambda_F \simeq 640 \text{ GeV} \tag{75}$$

from

$$f\bar{f} \longrightarrow (\text{Higgs}) \rightarrow f\bar{f}, W\overline{W} . \tag{76}$$

Using $m_b \simeq 5$ GeV and $m_t \simeq 42$ GeV, one finds

$$X_{L^-} = X_D \lesssim 130 \quad and \quad X_U \lesssim 15 . \tag{77}$$

The corresponding upper limits for $\frac{\Gamma_{S^\circ \to \gamma\gamma}}{m_{S^\circ}}$ from (77) (ignoring contributions from SUSY partners) are displayed again in Fig. 7. Unfortunately they are still small. It should be emphasized, however, that the perturbative bounds (73), (75) are not sacred. In fact, related to the anomalous $\bar{p}p$ collider events the possibility of a strongly interacting Higgs sector (violating (73) and/or (75)) has received much attention recently [48].

3.2 Enhanced Scalars X° in Nearby Compositeness

In this section, I want to consider 'nearby' compositeness, i.e. the weak W and Z° bosons are viewed as prominent composite spin 1 bosons along with composite quarks and leptons [12]. The issue is then to look for possible composite $J = 0$ partners X° of the W,Z vector bosons. By playing the role of $J = 0$ 'ground states', they might well be lighter than their $J = 1$ counter parts W and Z°. As a basis for the discussion, let us consider a tentative mass range

$$30 \div 50 \text{ GeV} \lesssim m_{X^\circ} < m_W \tag{78}$$

If such X° bosons were to exist in form of (pseudo) Goldstone bosons (in analogy to π°) they could of course be even lighter.

The essential point I want to make is that the $\gamma\gamma$ and $Z^\circ\gamma$ channels offer some unique possibilities to detect such bosons – if they exist – even though they are a priori not expected to be very light.

(1) If X° is a member of a weak isospin <u>triplet</u> like $W^\pm, Z^\circ$, it does <u>not</u> couple to two (or more) gluons since those are iso-singlets.

(2) If X° is an isospin <u>singlet</u> its coupling to two gluons is still very weak if its constituents carry no color, like it is the case for $W^\pm, Z^\circ$ in a variety of popular models.

(3) The coupling of X° to light quarks and leptons is probably strongly suppressed

$$g_{X^\circ f\bar{f}} \propto \frac{m_f}{\Lambda_F} \ll 1, \tag{79}$$

for reasons of chiral symmetry [49,50).

The two photon channel then remains as a dominant [50) decay mode of X°!

Next I want to argue that the $X^\circ\gamma\gamma$ and $Z^\circ X^\circ\gamma$ couplings are strongly enhanced for a composite X° particle as compared to an elementary scalar [51). The crucial difference becomes qualitatively obvious from Fig. 9.   Whereas an elementary Higgs-type scalar couples to $\gamma\gamma$ and $Z^\circ\gamma$

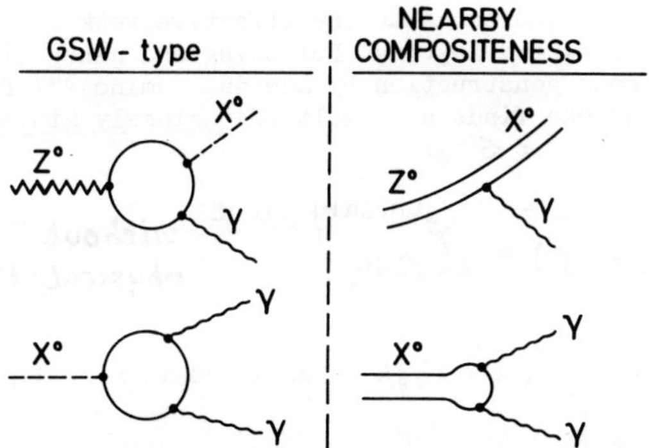

Fig. 9: $X^\circ\gamma\gamma$ and $Z^\circ X^\circ\gamma$ couplings for a composite scalar X° and an elementary, Higgs-type scalar X°, respectively

only via strongly suppressed loops (ignoring possible enhancements à la sect. 3.1), there is a <u>direct coupling of photons to the common charged</u> <u>constituents of X° and Z°</u> in case of compositeness! In order to obtain estimates of these couplings one may employ either methods familiar from onium-bound states [51) or the concept of W-dominance [52) (Fig. 10) which I prefer. The results happen to be quite similar, though.

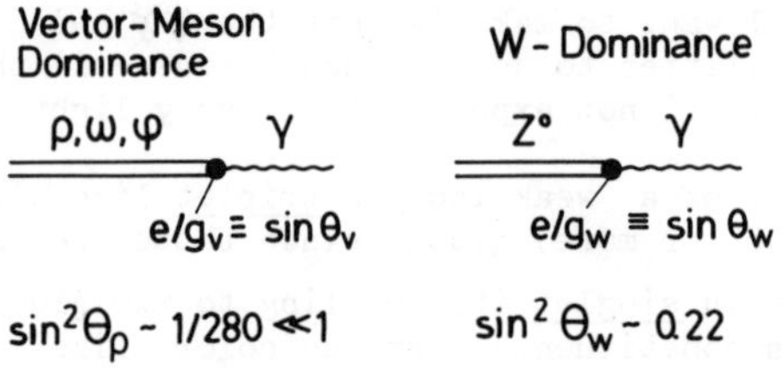

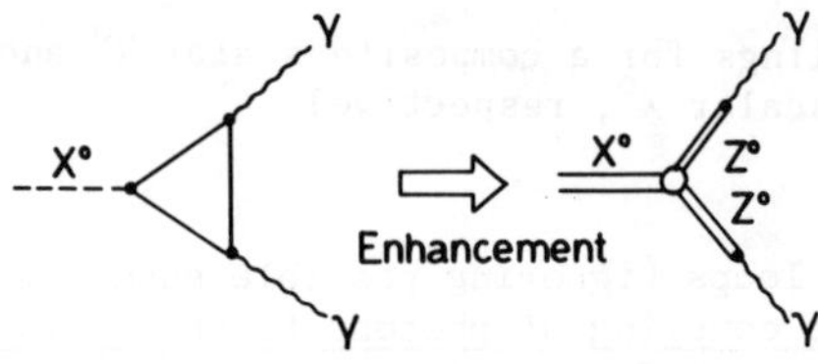

Fig. 10: Analogy of W- dominance to vector - meson dominance

W-dominance is pictured in direct analogy to vector-meson dominance [53] of the electromagnetic current in strong interactions (Fig. 10). The corresponding direct coupling of the photon to the composite Z° is

$$e/g_w = \sin\theta_w \simeq \sqrt{0.22} \tag{80}$$

It is comforting that the requirement of W-dominance in its stronger, operator form [28] ('current-field identity')

$$J_\mu^{e.m} = \frac{m_W^2}{g_W} \cdot W_\mu^3 + J_\mu^Y \tag{81}$$

gives very sensible results as to the effective weak interactions of composite $\vec{W}$'s, quarks and leptons. Following and generalizing the classical, analogous construction by Lee and Zumino [54] for composite ρ 's and nucleons one finds a result [28] <u>closely mimicking</u> the standard GSW model for $E \lesssim m_W$!

$$\mathcal{L}_{eff}^{weak}(\vec{W}, q, \ell, \gamma) = \mathcal{L}_{GSW}^{unitary\ gauge}\ {}^{without}_{physical\ Higgs} {}^{+\cdots} \tag{82}$$

Now, let us estimate $\Gamma(X^\circ \to \gamma\gamma)$. As depicted in Fig. 11, W-dominance

Fig. 11: Enhancement of the $X^\circ \gamma\gamma$ coupling in 'nearby' compositeness

relates $g_{X^o\gamma\gamma}$ to an effective $X^o ZZ$ coupling among composites for which one may naturally assume

$$g_{X^o ZZ} \sim g_{Zf\bar{f}} = g_{ZWW} = g_W \simeq 0.64 . \tag{83}$$

One obtains a very large width

$$\frac{\Gamma(X^o \to \gamma\gamma)}{m_{X^o}{}^3} \simeq \frac{4}{3} \frac{\Gamma(Z \to e^+e^-)}{m_Z{}^3} \simeq 0.16 \ \frac{keV}{GeV^3} . \tag{84}$$

As becomes apparent from Fig. 7, there seems indeed to be a good chance for two-photon physics, in particular at LEP, up to masses

$$m_{X^o} \sim 50 \text{ GeV.} \tag{85}$$

Even, if nothing is found, the resulting limits could be of great importance. Here is an instructive illustration, involving experiments at PETRA in particular also the two photon channel

$$e^+e^- \longrightarrow \gamma\gamma . \tag{86}$$

We return to the proposal mentioned in sect. 2, that the anomalous radiative Z^o decays

$$Z^o \longrightarrow e^+e^- \gamma_{hard} , \tag{87}$$

seen by UA1/UA2 are mediated by a spinless composite boson [18] X^o (Fig. 12) of mass $m_X \simeq m_{e^+e^-} \simeq$ 45–50 GeV.

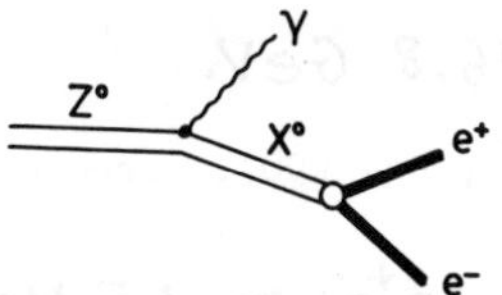

Fig. 12: A composite scalar X^o mediating radiative Z^o decays

The UA1/UA2 rate (16) then implies

$$\Gamma(Z^o \to e^+e^-\gamma) = \varepsilon \ \frac{\Gamma(Z^o \to X^o\gamma) \ \Gamma(X^o \to e^+e^-)}{\Gamma_{X^o}} \sim \mathcal{O}(15 \text{ MeV}) \tag{88}$$

In eq. (88) $\varepsilon = 1$ applies for a single spinless boson and $\varepsilon = 2$ for a (degenerate) parity doublet[19].
W-dominance gives[19] (Fig. 13)

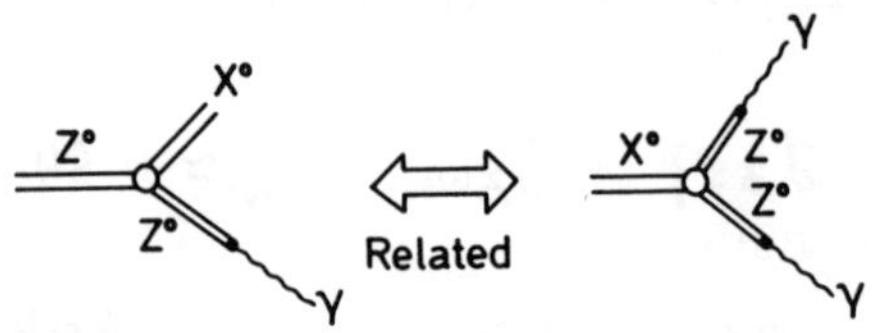

Fig. 13: Relation of $Z^0 X^0 \gamma$ and $X^0 \gamma \gamma$ couplings from W- dominance

$$\frac{\Gamma(Z^0 \to X^0 \gamma)}{\Gamma(X^0 \to \gamma\gamma)} = \frac{2}{3}\left(\frac{m_Z^2 - m_X^2}{m_Z m_X}\right)^3 \frac{1}{\sin^2\theta_W} \equiv r \sim \mathcal{O}(10). \tag{89}$$

Thus, the radiative Z^0 decay rate directly constrains searches for the X^0 boson in the $e^+ e^- \to \gamma\gamma$ channel via the relation following from eqs. (88), (89)

$$\frac{\varepsilon\, \Gamma(X^0 \to \gamma\gamma)\, \Gamma(X^0 \to e^+ e^-)}{\Gamma_{X^0}} \sim \frac{15\ MeV}{r} \simeq 1.5\ MeV. \tag{90}$$

A search was performed by all PETRA groups, but no comparable signal was seen (Fig. 14a), implying [55-58]

$$m_{X^0} > 46.8\ GeV. \tag{91}$$

Since $\Gamma(X^0 \to \gamma\gamma)/\Gamma_{X^0} \lesssim 1$, eq. (90), moreover provides the lower bound[19]

$$\varepsilon\, \Gamma(X^0 \to e^+ e^-) \gtrsim 1.5\ MeV, \tag{92}$$

involving only W-dominance and the UA1/UA2 results. It turned out to be in strong contradiction to the upper bounds from all PETRA groups[55-58] for $M_{X^0} \lesssim 46.8$ GeV (Fig. 14b). In view of the bound (92), Bhabha scattering,

$$e^+ e^- \longrightarrow e^+ e^- \tag{93}$$

is also very restrictive (Fig. 14c).

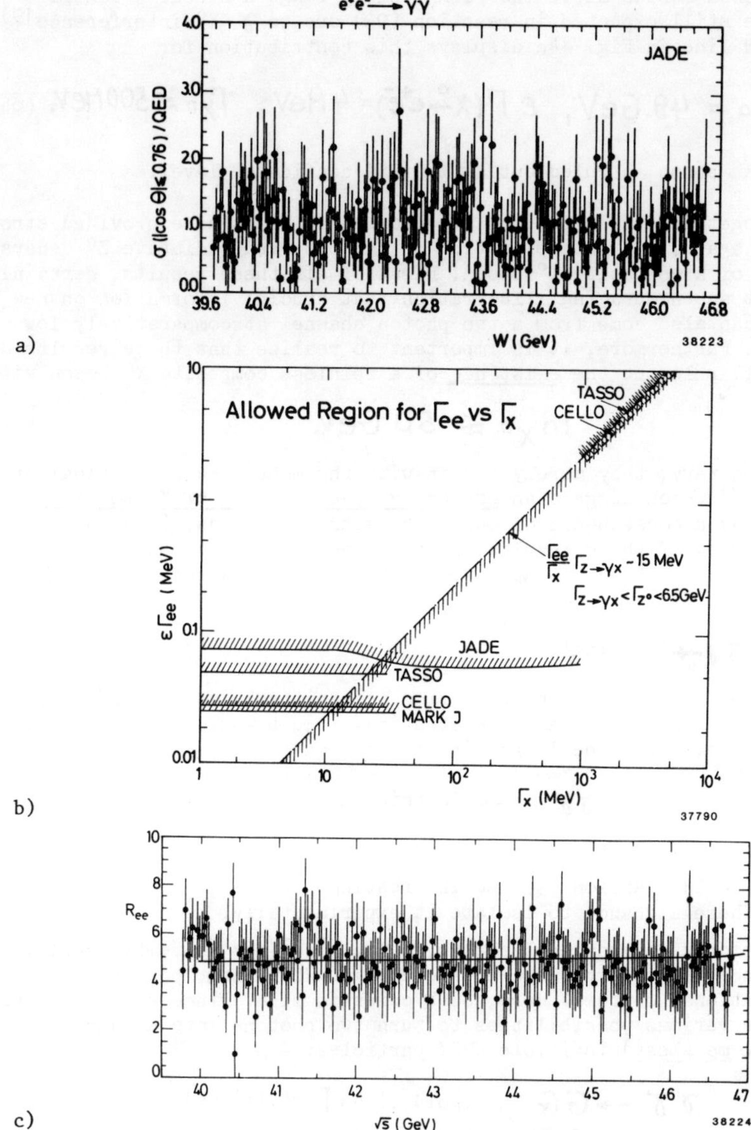

Fig. 14: Typical results of PETRA searches for a composite scalar X^O.
(a): e+e- → $\gamma\gamma$ data from ref.55,
(b): summary of PETRA bounds for $\Gamma(X^O \to$ e+e-) vs. Γ_Xo (ref.55),
(c): e+e- → e+e- data from ref.56.

For X^0 boson masses above the PETRA energy range a sizeable contribution is still expected in reaction (93) due to γ-X^0 interference[19]. The solid line in Fig. 14c displays this contribution for

$$m_{X^0} = 49 \text{ GeV}, \quad \varepsilon\,\Gamma(X^0 \to e^+ e^-) = 4 \text{ MeV}, \quad \Gamma_{X^0} \lesssim 500 \text{ MeV}. \tag{94}$$

Such an X^0 boson is ruled out at the 95% confidence level.

Alltogether, the experiments performed at PETRA have provided strong evidence against an interpretation of the observed radiative Z^0 decays in terms of a composite X^0 boson. First of all these results, certainly, represent an encouraging illustration that crucial information on new physics can also come from a two-photon channel at comparatively low energies. Furthermore, it is important to realize that these results do not at all rule out the <u>existence</u> of a spinless composite X^0 boson with

$$m_{X^0} \lesssim 50 \text{ GeV}. \tag{95}$$

On the contrary, they merely tie in with the mentioned theoretical arguments[49,50] which suggest an <u>extremely weak coupling of X^0 to light fermions</u> as a consequence of chiral symmetry (c.f. eqs. (73) and (92)). Thus, in view of these results and the expected large coupling of X^0 to two photons, eq. (84), it appears worthwhile to look for it in $\gamma\gamma$ collisions (LEP).

3.3 $\gamma\gamma \to$ 'nothing'?

The content of this short section is admittedly very speculative, but nevertheless potentially interesting. I just would like to stimulate some thinking about the 'impossible' reaction

$$\gamma\gamma \longrightarrow \text{'nothing'}. \tag{96}$$

Here are some reflections on two obvious questions

i)   Why could reaction (96) be interesting?
ii) Is there a chance to isolate it experimentally?

(i):   The only conventional source, $\gamma\gamma \to \nu\bar{\nu}$, is extremely small, since ν's carry no electric charge. What about possible unconventional sources? Here, SUSY comes to mind, since a priori there are various possibilities to turn two photons into a pair of ($\sim$ <u>massless</u>) invisible SUSY particles:

$$\gamma\gamma \to GG \qquad \text{(goldstinos [gravitinos])} \tag{97}$$
$$\gamma\gamma \to \tilde{\gamma}\tilde{\gamma} \qquad \text{(photinos)}. \tag{98}$$

As discussed in sect. 2, goldstino pair production within a nonlinear realization[26] of (spontaneously broken) SUSY has been suggested in Ref. 25 as a possible explanation of the monojets and monophotons observed by UA1 at the $\bar{p}p$ collider (c.f. eqs. (26)). In this scheme one may even compute the rate (97) without free parameters, which seems worthwhile. On general grounds (low-energy theorem) one expects a very strong dependence of $\sigma(\gamma\gamma\to GG)$ on the $\gamma\gamma$ energy W, analogously to eq. (27).

(ii): If possible at all, an experimental isolation of reaction (96), is certainly going to be difficult. In e^+e^- collisions, it corresponds to events of the type

$$e^+e^- \longrightarrow e^+e^- + \text{'nothing'}. \tag{99}$$

Double tagging at large angles should at least strongly reduce the probability that photons radiated from the scattered electrons are lost in the beam pipe and fake reaction (99). However, selectron ($\tilde{e}$) pair production with subsequent decay

$$\tilde{e} \longrightarrow e + \tilde{\gamma}, \tag{100}$$

represents another possible SUSY 'background' to relation (96) with a similar signature (99). More promising seems to be an attempt to extract reaction (96) from direct $\gamma e^\pm$ collisions involving __real__ photons. As has been discussed in the literature[59], $\gamma_{real} e^\pm$ collisions might be realized at the Stanford Linear Collider (SLC) with γe energies $\lesssim$ 90 GeV and good luminosity by means of a laser beam.

4 New Particles in $e^\pm \gamma$ Collisions

4.1 Photon Structure Functions

Deep inelastic scattering of electrons on a (quasi-real) photon 'target' represents the standard process from which the photon structure functions $F_2^{\gamma}(x,Q^2)$ and $F_L^{\gamma}(x,Q^2)$ are extracted.

Modifications of the familiar QCD predictions for F_2^{γ} and F_L^{γ} will arise, if new particles exist.

For the case of SUSY, the effects due to squark ($\tilde{q}$) and gluino ($\tilde{g}$) production have been calculated explicitely for the structure functions of the photon[60-62] (LEP), [as well as of the nucleon[63] (HERA)] . In general, of course, the results strongly depend on the masses of the new particles under consideration.

Let me remind you here of a characteristic effect related to their __spins__[61,62]. The point is that the __longitudinal__ structure function of the photon $F_L^{\gamma}(x,Q^2)$ turns out to be quite sensitive to the presence of __spinless__ particles such as

$$\begin{array}{lll} & \text{squarks } \tilde{q} & \text{in SUSY (Fig. 15)} \\ \text{or} & & \\ & \text{leptoquarks } \chi & \text{in 'nearby' compositeness .} \\ & \text{X}^\circ \text{ bosons} & \text{(c.f. sects. 2 and 3.2)} \end{array} \tag{101}$$

It is well known that in QCD, with only $J = 1/2$ quarks and $J = 1$ gluons, F_L^{γ} exhibits <u>no</u> scaling violations both at the parton <u>and</u> $O(\alpha_s)$ levels for large $Q^2 \gg m_q^2$

$$F_L^{\gamma}(x,Q^2) = \frac{\alpha}{2\pi} f(x). \tag{102}$$

Transient Q^2 dependences will, of course, arise if new, heavy quark thresholds are crossed, e.g. around

$$Q^2 \approx 4\, m_{top}^2. \tag{103}$$

In contrast, new $J = 0$ particles give rise to an <u>asymptotically surviving</u>, qualitatively different behaviour[61,62)]

$$F_L^{\gamma}/_{J=0} \propto \log Q^2, \tag{104}$$

as a consequence of simple helicity conservation arguments (Fig. 15).

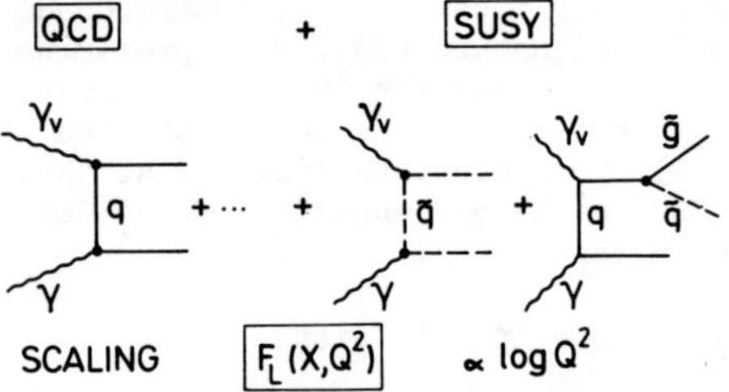

Fig. 15: Spinless squarks $(\tilde{q})$ causing a $\log Q^2$ scaling violation in the longitudinal photon structure function $F_L(x,Q^2)$

However, in order to be able to explore this discriminating effect much effort will have to go into an experimental isolation of F_L^{γ} at LEP energies.

4.2 Hunting for Selectrons $(\tilde{e})$

A detailed computation of cross sections with general couplings [and polarization states] for <u>exclusive</u> $e^{\pm}\gamma$ collisions

$$e^{\pm}\gamma \rightarrow \text{boson} + \text{fermion} \tag{105}$$

was performed in ref. 64.

An exclusive SUSY process of special interest is (Fig. 16)

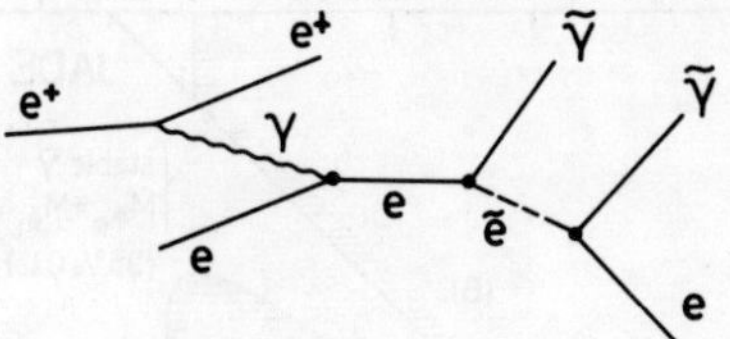

Fig. 16: Single selectron ($\tilde{e}$) production in $e\gamma$ scattering

$$e\gamma \longrightarrow \tilde{e}\,\tilde{\gamma} \atop \hookrightarrow e\tilde{\gamma} \qquad\qquad (106)$$

In contrast to <u>pair</u> production of selectrons ($\tilde{e}$) in e^+e^- collisions (Fig. 17)

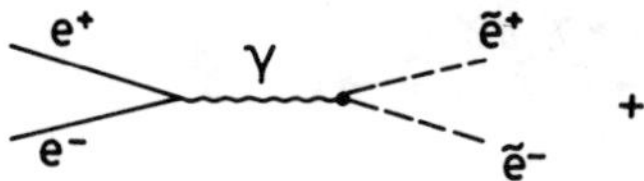

Fig. 17: Selectron ($\tilde{e}$) pair production

$$e^+e^- \longrightarrow \tilde{e}^+\tilde{e}^- \atop \hookrightarrow e^+e^- + \;'nothing'\;, \qquad (107)$$

it offers an opportunity to <u>search for selectrons with mass higher than the e^+e^- beam energy</u>. The $e^\pm$ which radiates the quasi-real photon in (106) goes mostly along the beam direction, and is not observed. Thus, the final state to be observed in (106) consists of a single electron coming from the $\tilde{e} \to e + \tilde{\gamma}$ decay , with no other visible particles. The cross section for this process (106) was first calculated in ref. 65 for a massless photino. A generalization to massive photinos may be found in ref. 66. Fig. 18, taken from JADE[67] illustrates the type of bounds on selectron and photino masses one may achieve at PETRA. Note, the selectron mass bound

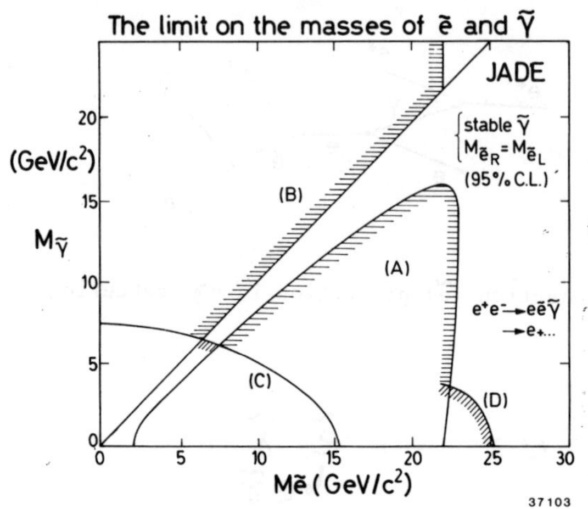

Fig. 18: Lower limits on selectron ($\tilde{e}$) and photino ($\tilde{\gamma}$) masses (ref.67)
   (A): from e+e- $\to \tilde{e}+\tilde{e}-$ with m($\tilde{e}$) > m($\tilde{\gamma}$),
   (B): from e+e- $\to \tilde{e}+\tilde{e}-$ with stable $\tilde{e}$, m($\tilde{e}$) < m($\tilde{\gamma}$),
   (C): from e+e- $\to \gamma \tilde{\gamma}\tilde{\gamma}$
   (D): from e$\gamma \to \tilde{e}\,\tilde{\gamma}$, $\tilde{e} \to$ e + $\tilde{\gamma}$

$$m_{\tilde{e}} \gtrsim 25.2 \text{ GeV for } m_{\tilde{\gamma}} = 0, \qquad (108)$$

essentially coming from the eγ process (106). For comparison, the
MARK-II and MAC collaborations obtained lower $\tilde{e}$ mass limits of 22.2
and 22.4 GeV, respectively[68].

Acknowledgements

I wish to thank Hermann Kolanoski for helpful discussions on experimental aspects of two-photon physics, and Richard Lander and colleagues for organizing an enjoyable and fruitful meeting.

References

1) S.L. Glashow, Nucl. Phys. 22 (1961) 579;
    S. Weinberg, Phys. Rev. Lett. 19 (1967) 1264;
    A. Salam, Proc. of the 8th Nobel Symposium, ed. N. Svartholm
    (Amquist and Wiksells, Stockholm, 1969), p. 367
2) UA1 collaboration: G. Arnison et al., Phys. Lett. 122B (1983) 103;
                                                       126B (1983) 398
    UA2 collaboration: M. Banner et al., Phys. Lett. 122B (1983) 476;
                       P. Bagnaia et al., Phys. Lett. 129B (1983) 130

3) UA1 collaboration: G. Arnison et al., CERN-EP/84-134 (1984)
4) K. Wilson as quoted in L. Susskind, Phys. Rev. $\underline{D20}$ (1979) 2619
5) G. 't Hooft in Recent Developments in Gauge Theories, Proc. NATO
    Advanced Study Institute, Cargèse 1979, ed. G. 't Hooft et al.,
    (Plenum, New York, 1980)
6) J. Wess and B. Zumino, Nucl. Phys. $\underline{B70}$ (1974) 39;
                            Phys. Lett. $\underline{49B}$ (1974) 52;
    A. Salam and B. Strathdee, Phys. Rev. $\underline{D11}$ (1975) 1521;
    P. Fayet and S. Ferrara, Phys. Rep. $\underline{32C}$ (1977) 249
7) S. Dimopoulos and H. Georgi, Nucl. Phys. $\underline{B193}$ (1981) 150;
    N. Sakai, Z. Phys. $\underline{C11}$ (1982) 153
8) see e.g.
    D.V. Nanopoulos, Rapporteur Talk at the XXII Int. Conf. on High
    Energy Physics, Leipzig, July 1984, CERN-TH-3995 (1984)
9) L. Susskind, Phys. Rev. $\underline{D20}$ (1979) 2619;
    S. Weinberg, Phys. Rev. $\underline{D13}$ (1976) 974; $\underline{D19}$ (1979) 1277
10) S. Dimopoulos, Nucl. Phys. $\underline{B168}$ (1980) 69
11) see e.g.
    M. Peskin, Proc. 1981 Int. Symp. on Lepton and Photon Interactions,
    Bonn, ed. W. Pfeil, p. 880;
    H. Harari, 'Composite Models for Quarks and Leptons', Weizmann
    Institute report WIS-82/60 Dec-Ph (1982)
12) H. Harari and N. Seiberg, Phys. Lett. $\underline{98B}$ (1981) 269;
    O.W. Greenberg and J. Sucher, Phys. Lett. $\underline{99B}$ (1981) 339;
    L.F. Abbott and E. Farhi, Phys. Lett. $\underline{101B}$ (1981) 69;
                            Nucl. Phys. $\underline{B189}$ (1981) 547;
    H. Fritzsch and G. Mandelbaum, Phys. Lett. $\underline{102B}$ (1981) 319;
    B. Schrempp and F. Schrempp, Nucl. Phys. $\underline{B231}$ (1984) 109;
                                            $\underline{B242}$ (1984) 203;
    W. Buchmüller, R. Peccei and T. Yanagida, Nucl. Phys. $\underline{B231}$ (1984)
    53; $\underline{B244}$ (1984) 186
13) UA1 collaboration: G. Arnison et al., Phys. Lett. $\underline{139B}$ (1984) 115
14) UA2 collaboration: P. Bagnaia et al., Phys. Lett. $\underline{139B}$ (1984) 105
15) CELLO collaboration: H.J. Behrend et al., Phys. Lett. $\underline{141B}$ (1984)
    145
16) CRYSTAL BALL collaboration: H.J. Trost, Talk at the XXII Int. Conf.
    on High Energy Physics, Leipzig, July 1984, DESY 84-064/
    SLAC-PUB-3380 (1984)
17) see e.g.
    J. Fleischer and F. Jegerlehner, Bielefeld University preprint
    BI-TP 1984/6 (1984)
18) U. Baur, H. Fritzsch and H. Faissner, Phys. Lett. $\underline{135B}$ (1984) 313;
    R.D. Peccei, Phys. Lett. $\underline{136B}$ (1984) 121;
    F.M. Renard, Phys. Lett. $\underline{139B}$ (1984) 449
19) W. Hollik, B. Schrempp and F. Schrempp, Phys. Lett. $\underline{140B}$ (1984)
    424;
    F.W. Bopp, S. Brandt, H.D. Dahmen, D.H. Schiller and D. Wähner,
    Univ. Siegen preprint SI-84-3 (1984)
20) N. Cabibbo, L. Maiani and Y. Srivastava, Phys. Lett. $\underline{139B}$ (1984)
    459;
    K. Enquist and J. Maalampi, Phys. Lett. $\underline{135B}$ (1984) 329;
    F.M. Renard in Ref. 18

21) W. Marciano, Phys. Rev. Lett. $\underline{53}$ (1984) 975;
M. Matsuda and T. Matsuoka, Phys. Lett. $\underline{144B}$ (1984) 443;
B. Holdom, Phys. Lett. $\underline{143B}$ (1984) 241

22) F.M. Renard in Ref. 18;
V. Barger, H. Baer and K. Hagiwara, Phys. Rev. $\underline{D30}$ (1984) 1513

23) J. Ellis and H. Kowalski, Phys. Lett. $\underline{142B}$ (1984) 441;
E. Reya and D.P. Roy, Phys. Rev. Lett. $\underline{53}$ (1984) 881

24) J. Ellis and H. Kowalski, Nucl. Phys. $\underline{B246}$ (1984) 189;
V. Barger, K. Hagiwara and W.Y. Keung, Phys. Lett. $\underline{145B}$ (1984) 147;
A.R. Allan, E.W.N. Glover and A.D. Martin, Phys. Lett. $\underline{146B}$ (1984) 247;
A.R. Allan, E.W.N. Glover and S.L. Grayson, Durham preprint DTP/84/28 (1984)

25) O. Nachtmann, A. Reiter and M. Wirbel, Univ. of Heidelberg preprint HD-THEP-84-11 (1984)

26) J. Wess in "Quantum Theory of Particles and Fields", ed.
B. Jancewicz, J. Lukierski (World Scientific Publishers, Singapore, 1983);
S. Samuel and J. Wess, Nucl. Phys. $\underline{B221}$ (1983) 153

27) A. De Rujula, L. Maiani and R. Petronzio, Phys. Lett. $\underline{140B}$ (1984) 253;
G. Pancheri and Y. Srivastava, Frascati preprint LNF-84(10)P (1984);
J. Kühn and P. Zerwas, Phys. Lett. $\underline{147B}$ (1984) 189

28) B. Schrempp and F. Schrempp, DESY-84-055 (1984)

29) H. Fritzsch and G. Mandelbaum in Ref. 12;
H. Fritzsch, Proc. of the Workshop on Feasibility of Hadron Colliders in the LEP Tunnel, Lausanne-CERN, March 1984;
U. Baur and K.H. Streng, MPI-Munich preprint, MPI/PAE/PTh 50/84 (1984);
G. Gounaris and A. Nicolaidis, Phys. Lett. $\underline{148B}$ (1984) 239

31) UA1 collaboration: J. Rohlf, Plenary Talk at the XXII Int. Conf. on High Energy Physics, Leipzig, July 1984, CERN-EP/84-126 (1984)

32) W. Buchmüller, CERN-TH-3873 (1984); Phys. Lett. $\underline{145B}$ (1984) 151

33) L.F. Abbott and E. Farhi in Ref. 12

34) W. Buchmüller, in preparation;
B. Schrempp and F. Schrempp, in preparation

35) O.W. Greenberg, R.N. Mohapatra and S. Nussinov, Univ. of Maryland preprint No 85-26 (1984)

36) B. Schrempp and F. Schrempp, DESY-84-117 (1984)

37) UA1 collaboration: G. Arnison et al., 'Intermediate Mass Dimuon Events at the CERN $\bar{p}$p Collider at $\sqrt{s}$ = 540 GeV', CERN-EP-report (1984) in print;
K. Eggert, Talk at the Int. Conf. on Cosmic Ray and Particle Physics, March 1984, Tokyo, Japan, preprint (Nov. 1984)

38) R.N. Mohapatra, G. Segré and L. Wolfenstein, Phys. Lett. $\underline{145B}$ (1984) 433

39) CRYSTAL BALL collaboration: I. Brock, Talk at the APS-DPF meeting, Santa Fe, Nov. 1984, to be published in the proceedings

40) see e.g.
V.M. Budnev, I.F. Ginzburg, G.V. Meledin and V.G. Serbo, Phys. Rep. $\underline{15}$ (1975) 181

41) D.W. Duke and J.F. Owens, Florida State University preprint
    FSU-HEP-831115 (1983)

42) F. Wilczek, Phys. Rev. Lett. $\underline{39}$ (1977) 1304;
    M.A. Shifman, A.I. Vainshtein and V.I. Zakharov, Phys. Lett. $\underline{78B}$
    (1978) 443

43) J. Ellis, M.K. Gaillard and D.V. Nanopoulos, Nucl. Phys. $\underline{B106}$ (1976)
    292;
    M.A. Shifman, A.I. Vainshtein, M.B. Voloshin and V.I. Zakharov,
    ITEP-preprint, ITEP-42 (1979);
    L.H. Chan and T. Hagiwara, Phys. Rev. $\underline{D20}$ (1979) 1698

44) R.A. Flores and M. Sher, Ann. Phys. $\underline{148}$ (1983) 95;
    R.M. Barnett, G. Senjanović and D. Wyler, Univ. of California,
    Santa Barbara preprint NSF-ITP-84-45 (1984)

45) A.D. Linde, JETP Lett. $\underline{23}$ (1976); Phys. Lett. $\underline{70B}$ (1977) 306;
    S. Weinberg, Phys. Rev. Lett. $\underline{36}$ (1976) 294

46) B.W. Lee, C. Quigg and H.B. Thacker, Phys. Rev. Lett. $\underline{38}$ (1977) 883;
    Phys. Rev. $\underline{D16}$ (1977) 1519

47) M.S. Chanowitz, M.A. Furman and I. Hinchliffe, Nucl. Phys. $\underline{B153}$
    (1979) 402

48) see e.g.
    R.D. Peccei, in Proc. of the XIth Int. Conf. on Neutrino Physics
    and Astrophysics, Nordkirchen/W.-Germany, June 1984 (MPI-Munich
    preprint MPI-PAE/PTh 65/84 (1984));
    S. Nussinov, Phys. Rev. Lett. $\underline{52}$ (1984) 963

49) M. Leurer, Phys. Lett. $\underline{144B}$ (1984) 273

50) J.H. Kühn and P.M. Zerwas, Phys. Lett. $\underline{142B}$ (1984) 221

51) F.M. Renard, Phys. Lett. $\underline{126B}$ (1983) 59

52) R. Kögerler and D. Schildknecht, CERN-TH-3231 (1982)

53) J.J. Sakurai, "Currents and Mesons", 1969 (Univ. of Chicago Press)

54) T.D. Lee and B. Zumino, Phys. Rev. $\underline{163}$ (1967) 1667

55) JADE collaboration: S. Yamada, Talk at the XXII Int. Conf. on High
    Energy Physics, Leipzig, July 1984

56) MARK J collaboration: B. Adeva et al., Phys. Rev. $\underline{53}$ (1984) 134;
    Min Chen, Talk at Rencontre de Moriond 1984, MIT report # 139 (1984)

57) CELLO collaboration: H.J. Behrend et al., Phys. Lett. $\underline{140B}$ (1984)
    130

58) TASSO collaboration: M. Althoff et al., DESY-85/1 (1985)

59) C. Akerlof, Univ. Michigan preprint UM-HE 81-59 (1981)

60) E. Reya, Phys. Lett. $\underline{124B}$ (1983) 424

61) D.M. Scott, Proc. 5th Int. Workshop on Photon Photon Collisions,
    Aachen 1983, ed. Ch. Berger (Springer-Verlag) p. 358;
    D.M. Scott and W.J. Stirling, Univ. Cambridge report DAMTP 83/13
    (1983)

62) M. Drees, M. Glück and E. Reya, Univ. of Dortmund preprint
    DO-TH 84/02 (1984)

63) S.K. Jones and C.H. Llewellyn Smith, Nucl. Phys. $\underline{B217}$ (1983) 145;
    M. Drees and K. Grassie, Univ. of Dortmund preprint DO-TH-84/13
    (1984)

64) F.M. Renard, Z. Phys. $\underline{C14}$ (1982) 209

65) M.K. Gaillard, L. Hall and I. Hinchliffe, Phys. Lett. $\underline{116B}$ (1982)
    279

66) T. Kobayashi and M. Kuroda, Phys. Lett. 134B (1984) 271
67) JADE collaboration: W. Bartel et al., DESY-84-112 (1984)
68) L. Gladney et al., Phys. Rev. Lett. 51 (1983) 2253;
     E. Fernandez et al., Phys. Rev. Lett. 52 (1984) 22

UNFOLDING TECHNIQUES

Andreas Bäcker

Siegen University
Fachbereich 7
P.O. Box 101240
5900 Siegen
FRG

PLUTO Collaboration [1]

ABSTRACT

A general method to unfold experimental results for measurement effects is described. The method regularizes the data and presents them with minimal correlations.

Introduction

Measurements of inclusive reactions in the field of two photon physics confront us with a special but known problem of data analysis. This problem is due to the fact that usually the center of mass energy of an individual event has to be measured from the observed final state system. Because of limited detector acceptance and finite detector resolutions this quantity however is measured only to a very limited accuracy. This means that for example the measurement of the total hadronic cross section or of the photon structure function yields distributions of observed variables. Instead of the sought distribution one thus measures a different distribution of visible quantities. The problem addressed in this talk is, how can one recover the true distribution from the measured one?

The folding integral

Formally the detector effect can be described by the folding integral, which is also known as the Fredholm integral of the first kind:

$$g(y) = \int B(x,y) \, f(x) \, dx \qquad (1)$$

where $f(x)$ is the true and $g(y)$ the measured distribution. The two-dimensional function $B(x,y)$ describes all measurement effects, it might also incorporate a model for the manifestation of the investigated process in observable quantities as event multiplicities and so on; the properties and parameters of the model have to be found from different measurements. Usually this model is then used to generate Monte Carlo events, which are passed through a detector simulation. The final results of these Monte Carlo calculations is the response function $B(x,y)$, which now is known, but still might be model dependent. The detector effects are the detector acceptance, the detector resolution and a transformation $x{\to}y$ (fig. 1); in general these effects are correlated. The remaining problem thus is the mathematical task of inverting the folding integral to find $f(x)$. This, as we shall see in a moment, is a non trivial problem, mainly because the solution of the inversion is not unique.

First discretisation

Measured distributions are usually given in the form of histograms. This is a first discretisation, which however does not add new errors as soon as the bin width is chosen small compared to the resolution. This same discretisation can be applied to the true function $f(x)$, so that both the functions $f(x)$ and $g(y)$ are represented by vectors:

$$g_i = \int_{y_{i-1}}^{y_i} g(y)\, dy \left/ \int_{y_{i-1}}^{y_i} dy \right. \qquad i = 1 \ldots n \qquad (2)$$

$$f_j = \int_{x_{j-1}}^{x_j} f(x)\, dx \left/ \int_{x_{j-1}}^{x_j} dx \right. \qquad j = 1 \ldots m$$

Now B is not any longer a two dimensional function of x and y, but a nxm matrix. The integration is then converted to a summation and the integral equation becomes a matrix equation

$$g = B \cdot f \qquad (3)$$

Unobservable Oscillations

The direct solution to obtain f by inverting B and computing $f = B^{-1} \cdot g$ suffers from an instability inherent to the solution of the basic integral equation:

The response of the given system described by B to a fast oscillating function tends to zero:

$$\int A(x,y) \cdot \sin(\nu x) \, dx \to 0 \qquad (4)$$
$$\nu \to \infty$$

Consequently, if there are statistical fluctuations ε, oscillations become unobservable, if, from some frequency ν_c on, the condition

$$\Delta g = \left| \int A(x,y) \cdot f(x) \, dx \right. \qquad (5)$$
$$\left. - \int A(x,y) \cdot \{f(x) + \sin(\nu_c x)\} \, dx \right| < \varepsilon$$

is fulfilled. The inversion method generally infers these unobservable oscillations by amplifying the statistical fluctuations. Moreover, due to the in general non diagonal character of the matrix B, the data points f_i are highly correlated obscuring the interpretation of the measurement errors.

As an example figure 2 shows an assumed true distribution (histogram) together with a simulation of measured points. For this example a well behaved detector was assumed measuring the true value i with a probability of 50%, the value $i \pm 1$ with 20% each and $i \pm 2$ with 5% each. Figure 2 shows the result, which is obtained by applying B^{-1} to the

208

measurements. The unfolded result obviously is close to the true
function, but superposed by a non negligible component oscillating more
rapidly than measurable due to the detector resolution as given in B.
This oscillation can also be interpreted as a negative point to point
correlation, illustrating, that a direct presentation of the measurement errors is impossible. Smoothing algorithms, which are often applied
to cure these problems, can make things worse, if there is no unbiased
ansatz for this procedure.

Outline of the general method

In the remainder of this talk I shall describe a method to invert
the folding integral, which avoids the difficulties mentioned before.
This method has been developed by V. Blobel [2] and has first been used
by the PLUTO collaboration for the determination of the photon structure
function and the total hadronic cross section in $\gamma\gamma$ interactions [3]. The
procedure consists of the following steps, which are described in
detail:

- An ansatz for the true function is made, which decomposes the
  unknown function into B-splines with a priori unknown coefficients.
- By two consecutive transformations of the coefficient space we
  obtain an orthonormal convariance matrix of the new coefficients and a decomposition of the unknown function into
  orthogonal functions.
- By making use of the orthonormal covariance matrix a well
  defined noise analysis will be performed leading to an
  negligible bias smoothing algorithm.
- Analysing the degrees of freedom the number of bins for the
  presentation of the result is defined and remaining point to
  point correlations are suppressed by a special choice of
  binning.

Decomposition into B-Splines

A general ansatz for an unknown function is an expansion into a
set of basic functions. We therefore use a decomposition

$$f(x) = \sum_j P_j(x) \cdot a_j, \tag{6}$$

where the a_j are coefficients and the $P_j(x)$ is a set of basic functions,
for which we take cubic B-splines.

A cubic B-spline is a set of four cubic polynomials joined together in such a way that it is continuous up to the second derivative
at the junctions, which are also called knots (fig. 4). For a decomposition of a function $f(x)$ into splines the x-axis is separated by
knots into pieces small compared to the estimated resolution in x. At
each knot the definition interval of exactly one B-spline starts, so
that $f(x)$ has nonzero contributions of exactly four splines at all x
(see fig. 4). The spacing of the knots is arbitrary, but has to be
monotonous; in this context however we shall only deal with constant
knot spacing. We note the following properties

$$P_j(x) \geq 0 \quad \text{at all } x$$
$$\tag{7}$$
$$\sum_j P_j(x) = 1 \quad \text{at all } x,$$

where the second statement is definition dependent. In calculations the
use of splines is very convenient because of the closed support and the
simple computation of the integral and derivatives. Moreover splines
have optimal interpolation properties.

Second discretisation

With expansion (6) the folding integral reads

210

$$g(y) = \int B(x,y) \, f(x) \, dx$$

$$= \int B(x,y) \sum_j P_j(x) \, a_j \, dx$$

$$g(y) = \sum_j a_j \cdot \int B(x,y) \, P_j(x) \, dx \tag{8}$$

This is a second discretisation; no error however is inferred by it, if the number of knots is chosen to be large enough.

The integral on the right side of (8) is a function of y and the parameter j only, so that we can write:

$$g(y) = \sum_j a_j \cdot A_j(y) \, , \tag{9}$$

where we have defined

$$A_j(y) = \int B(x,y) \, P_j(x) \, dx. \tag{10}$$

Discretisation in y-space according to definition (2) $(g(y) \to g_i$ and $A_j(y) \to A_{ij})$ yields

$$g_i = \sum_j A_{ij} \, a_j \tag{11}$$

or in matrix notation

$$g = A \, a \, . \tag{12}$$

Here g is the n-dimensional vector representing the n bins of the histogram of the expected distribution and a is a m-dimensional vector containing the m coefficients for the B-splines P_j. The matrix A is of the dimension n x m; each column of A represents the response of the detector to the according spline. An example of response functions is shown in fig. 5; each line of the three dimensional plot shows the expected measured distribution, if the true distribution is one of the splines. The total expected distribution is the sum of these response functions weighted with the coefficients a_j.

Apart from the known B-splines the only element necessary to calculate the matrix A is a precise knowledge of the measurement properties. Latter, as mentioned before, is usually obtained by analysing Monte Carlo events which are passed through a detector simulator. Here special attention has to be paid to the requirement, that each location in A is filled with sufficient Monte Carlo statistics. Therefore sometimes large numbers of Monte Carlo events have to be generated, which, due to limited computer capacity, may limit the granularity of the matrix; typical computing times required for such Monte Carlo generations are of the order of several CPU hours on fast computers.

Determination of the coefficients

The vector of the expected histogram thus is

$$g = A\,a$$

while the measured histogram is disturbed by statistical fluctuations $\varepsilon = (\varepsilon_i)$

$$\hat{g} = g + \varepsilon \tag{13}$$

where $\hat{g}$ is understood to be background subtracted (the $^\wedge$ denotes measured quantities). The coefficient vector a can be determined by the least square method minimising the expression $S(a)$

$$S(a) = (\hat{g}-g)^T W (\hat{g}-g)$$

$$= \hat{g}^T W\hat{g} - 2\,a^T A^T W\hat{g} + a^T A^T WAa \rightarrow min, \tag{14}$$

here W is the inverted covariance matrix $W = V^{-1}(\hat{g})$. Minimisation is done be setting the derivative to zero:

$$\frac{1}{2}\frac{\partial S}{\partial a} = -A^T W\hat{g} + A^T WA\hat{a} = 0 \tag{15}$$

resulting in

$$\hat{a} = H^{-1}\,h \tag{16}$$

212

with the calculable quantities

$$H = A^T W A \quad \text{and} \quad h = A^T W \hat{g}. \tag{17}$$

The complete covariance matrices are obtained by error propagation

$$V(h) = A^T W \, V(\hat{g}) \, (A^T W)^T = H \tag{18}$$

$$V(\hat{a}) = H^{-1} \, V(h) \, (H^{-1})^T = H^{-1}. \tag{19}$$

Regularisation

In the introduction it was shown that the inversion of the unfolding integral does not lead to an unique solution in that sense, that one can always add a fast oscillating function to the true distribution $f(x)$ without changing the expected distribution $g(y)$. These fast oscillating functions thus are not observable and one has to require that the result does not contain such oscillations. Knowing the reason for the ambiguity in the result we now have at hand an unbiased description for a smoothing procedure: Contributions to $f(x)$, which are so fast oscillating, that they due to the limited detector resolution cannot be observed, have to be suppressed. These are exactly those contributions, which fulfill the condition stated in equation 5.

The requirement of a smooth result has to be handled as additional condition in the determination of the coefficient vector $\hat{a}$. This is incorporated by extending the expression, which has to be minimised, by an expression of smoothness $r(a)$. Minimisation is then done with respect to a new expression

$$R(a) = S(a) + \tau \cdot r(a). \tag{20}$$

This will yield a result, which is a compromise between a best fit to the measured distribution and a smooth function. The parameter τ controls the degree of smoothness and is still unknown. A good measure for smoothness obviously is the expression

$$r(a) = \int (f''(x))^2 \, dx \tag{21}$$

For the case of cubic B-splines one can explicitely express $r(a)$ by

$$r(a) = a^T C a \tag{22}$$

with the matrix

$$C = \begin{pmatrix}
2 & -3 & 0 & 1 & 0 & 0 & 0 & 0 & 0 & 0 & . \\
-3 & 8 & -6 & 0 & 1 & 0 & 0 & 0 & 0 & 0 & . \\
0 & -6 & 14 & -9 & 0 & 1 & 0 & 0 & 0 & 0 & . \\
1 & 0 & -9 & 16 & -9 & 0 & 1 & 0 & 0 & 0 & . \\
0 & 1 & 0 & -9 & 16 & -9 & 0 & 1 & 0 & 0 & . \\
0 & 0 & 1 & 0 & -9 & 16 & -9 & 0 & 1 & 0 & . \\
0 & 0 & 0 & 1 & 0 & -9 & 16 & -9 & 0 & 1 & . \\
. & . & . & . & . & . & . & . & . & . & .
\end{pmatrix}$$

The modified condition for $\hat{a}$ then is

$$R(a) = a^T H a - 2a^T h + \tau \cdot a^T C a \rightarrow \min \tag{23}$$

This again can be solved by linear algebra, if τ is known.
In order to determine τ, it is necessary to perform a transformation
of the coefficient vector $\hat{a}$ into a new system. The aim of this is to
obtain an expression, which has no contributions of the form $\hat{a}_{2,i} \cdot \hat{a}_{2,j}$,
with $i \neq j$, where the $\hat{a}_{2,1}$ are the components of $\hat{a}$ in the new
space. To this end two transformations are applied:

The first transformation is

$$a = U_1 \cdot a_1 \tag{24}$$

with the definition

$$U_1^T \cdot H \cdot U_1 = 1 \tag{25}$$

This yields

$$R(a_1) = a_1^T a_1 - 2 a_1^T h_1 + \tau\, a_1^T C_1 a_1 \tag{26}$$

with the new matrix

$$C_1 = U_1^T C U_1 \tag{27}$$

The second transformation again is a diagonalisation

$$C_1 = U_2 D U_2^T , \tag{28}$$

where D is diagonal and the elements are ordered wrt. increasing diagonal elements representing squares of the curvatures. With

$$a_2 = U_2^T a_1 \tag{29}$$

we have

$$R(a_2) = a_2^T a_2 - 2 a_2^T h_2 + \tau\, a_2^T D a_2 \tag{30}$$

The minimisation yields

$$\frac{\partial R}{\partial a_2} = 2\,\hat{a}_2 - 2\,h_2 + 2\,\tau \cdot D\,\hat{a}_2 = 0 \tag{31}$$

$$\hat{a}_2 + \tau \cdot D\,\hat{a}_2 = h_2 \tag{32}$$

The advantages of the result in this form are evident. Due to the diagonal character of D the matrix equation (32) represents a set of m decoupled linear equations for the m components of $\hat{a}_2$. Consequently the elements of $\hat{a}_2$ are independent from each other and the covariance matrix $V(\hat{a}_2)$ is diagonal.

Moreover, using the relations (18) and (19) one finds for the unregularised case ($\tau = 0$), that the covariance matrix is unity

$$V(\hat{a}_2^0) = 1 \tag{33}$$

In this case the elements of $\hat{a}_2^0$ are independent and have standard deviation 1. (The upper index o denotes the unregularised case.)

In the new parameter space the decomposition of $f(x)$ is

$$f(x) = \sum_j P_j(x) \cdot a_j$$

$$= \sum_j P'_j(x) a_{2,j}$$

The $P'_j(x)$ are a new set of spline functions, which are linear combinations of our B-splines and, most important, the $P'_j(x)$ are a set of orthogonal functions.

As an example figure 6 shows an assumed true function $f(x)$ (solid line), which after detector effects is observed like shown by the data points. One finds that the detector significantly mismeasures the distribution at the high end of the scale. Using simulated events the response matrix A was calculated and the described formalism was used to calculate the coefficient vector $\hat{a}_2$ and the functions $P'_j(x)$, which are shown in figure 7, the oscillating character of the new basic functions P'_j is evident.

Determination of the Regularisation Parameter

The remaining problem is the determination of the regularisation parameter τ. To do so we have to go back to equation 32:

$$\hat{a}_2 + \tau \cdot D \, \hat{a}_2 = h_2.$$

For the unregularised case this results in

$$\hat{a}_2^0 = h_2. \tag{35}$$

Using this result and making use of the diagonal form of D, we obtain from equation 32 a set of m linear equations for the coefficients $\hat{a}_{2,j}$:

$$\hat{a}_{2,j} = \hat{a}_{2,j}^0 \frac{1}{(1 + \tau \cdot D_{jj})} \qquad j = 1 \ldots m \tag{36}$$

Without regularisation the number of coefficients m is equivalent to the number of degrees of freedom. By the regularisation however we effectively decrease this number and thus remain with an effective number of degrees of freedom given by

$$j_0 = \sum_{j=1}^{m} \frac{1}{(1 + \tau \cdot D_{jj})} \tag{37}$$

This is the best estimate for the number of components in the decomposition of $f(x)$. That means, because the D_{jj} are ordered with respect to increasing values and thus the p'_j are ordered with respect to increasing frequencies, $P'_{j_0}(x)$ is the fastest oscillating observable component of $f(x)$.

Instead of identifying this component we will look for a definition of the unobservable oscillations. To answer this question we note that these oscillations are created by statistical noise and therefore their contribution has to be compatible with zero. It is here, that we make use of the result (37), that the covariance matrix of the unregularised coefficients $\hat{a}^0_{2,j}$ is unity, saying that all coefficients with a value smaller than one are not significant and the contributions of the according P'_j are compatible with zero. We consequently have to find that value of $j = j_0$, from where on there is no significant deviation of $\hat{a}^0_{2,j}$ from zero with unit standard deviation. With that value of j_0 the parameter τ can be calculated using equation 37 and by 36 the final regularised values for $\hat{a}_{2,j}$ are found.

Figure 9 shows for our example the result for the unregularised coefficients $\hat{a}^0_{2,j}$ (bars) and illustrates a good method to find a value for j. The expected behaviour is very obvious; for low frequencies equivalent to low values of j the coefficients are significantly different from zero, while for values of $j = 10$ and higher all values are, with standard deviation one, consistent with zero. A value of $j_0 = 10$ therefore would be reasonable, to be not more restrictive however than necessary we set j_0 to 12. With that a damping factor as function of j as shown in figure 8 is obtained. The regularised coefficients are plotted in figure 9 (circles). Obviously, with the chosen value of j_0, the regularisation does not affect the significant amplitudes, but only the noise; this is the most important criterion, one has to regard when selecting j_0.

Presentation of the Result, Binning

In conclusion we so far have found a decomposition of the true function into a set of orthogonal functions with the additional condition that the result is free from unobservable oscillations. Also the error matrix of the coefficients in known. We finally have to define a method to present this result in such a way that each data point has a maximum information content, because only then a proper interpretation of the result is possible to the reader. That is, the number of bins and their location have to be determined.

First of all the number of data points has to be chosen so that all significant oscillations are represented, the maximum given by j_0. For this we use the fact that $P'_{j_0}(x)$ has j_0-1 zeros, bracketing j_0-2 extrema, each of which has to be presented by one data point. Two more datapoints are necessary to show the behaviour of the result to the left and right of the zeros leading to a total number of j_0 necessary bins.

The locations of the bins we select with the requirement that correlations between the datapoints are minimised. To do so we have to minimise the contribution of the first not significant component j_0+1, effects of higher order components can be neglected. The contribution of P'_{j_0+1} is almost completely cancelled, if the bin limits of the presentation are put to the extrema of this function, because in that case on average the integral over P'_{j_0+1} between adjacent bin limits is minimal. This, together with the endpoints of the scale, leaves us with j_0+1 bin limits, dividing the scale into j_0 bins as required above.

Figure 10 shows, again for our example, the first not significant component $P'_{13}(x)$. The bin limits according to the given scheme are indicated by the vertical lines. For one of the bins the contribution

218

of P'_{13} has been indicated by the hatched area. Obviously this contribution is almost cancelled, because the area to the left of the zero is close to the one on the right.

The final result of the complete method applied to the example is shown in figure 11 (data points). For comparison the true distribution is added (solid line). One finds immediately that the serious distortion of the true function towards the upper end of the scale has been properly corrected for, the measured and unfolded data points are a good representation of the true function $f(x)$ and no artificial oscillations are observed.

Conclusions

For the reader, when finding a result given in the form as described, there are three important statements to keep in mind, which conclude my talk:

- The data have been corrected for all known detector effects
- There is a minimal correlation between the data points
- There is no hint in the data for any additional structure of the true function, which is not explicitely visible from the given data points.

The first remark is almost trivial, this can under special conditions also be achieved with other simpler methods. The second statement is not trivial in that sense that it is often assumed by a reader of a publication, but sometimes not true. The third remark finally is an important note to everybody, who wants to read something out of data, where there is no evidence for.

Acknowledgement I would like to thank Mr. V. Blobel for introducing me into the field and for many helpful discussions. The organizers of the workshop I thank for their kind hospitality and the manifold help during the Tahoe conference.

List of References

1. The members of the PLUTO-group are: Ch. Berger, A. Deuter, H. Genzel, W. Lackas, J. Pielorz, F. Raupach, W. Wagner, I. Phys. Institut der RWTH Aachen, Federal Republic of Germany; A. Klovning, E. Lillestöl, University of Bergen, Norway; J. Bürger, L. Criegee, F. Ferrarotto, G. Franke, M. Gaspero, Ch. Gerke, G. Knies, B. Lewendel, J. Meyer, U. Michelsen, K.H. Pape, B. Stella, U. Timm, G.G. Winter, M. Zachara, W. Zimmermann, DESY, Hamburg, Federal Republic of Germany; P.J. Bussey, S.L. Cartwright, J.B. Dainton, B.T. King, C. Raine, J.M. Scarr, I.O. Skillicorn, K.M. Smith, J.C. Thomson, University of Glasgow, U.K.; O. Achterberg, V. Blobel, D. Burkart, K. Diehlmann, M. Feindt, H. Kapitza, B. Koppitz, M. Krüger, M. Poppe, H. Spitzer, R. van Staa, II. Institut für Experimentalphysik der Universität Hamburg, Federal Republic of Germany; C.Y. Chang, R.G. Glasser, R.G. Kellogg, S.J. Maxfield, R.O. Polvado, B. Sechi-Zorn, J.A. Skard, A. Skuja, A.J. Tylka, G.E. Welch, G.T. Zorn, University of Maryland, U.S.A.; F. Almeida, A. Bäcker, F. Barreiro, S. Brandt, K. Derikum, C. Grupen, H.J. Meyer, H. Müller, B. Neumann, M. Rost, K. Stupperich, G. Zech, Universität-Gesamthochschule Siegen, Federal Republic of Germany; G. Alexander, G. Bella, Y. Gnat, J. Grunhaus, Tel-Aviv University, Israel; H. Junge, K. Kraski, C. Maxeiner, H. Maxeiner, H. Meyer, D. Schmidt, Universität-Gesamthochschule Wuppertal, Federal Republic of Germany.

2. Unfolding Methods in High Energy Physics Experiments, V. Blobel, 1984 CERN School of Computing, Aiguablava (Spain)

3. Measurement of Deep Inelastic Electron Scattering off Virtual Photons, PLUTO Collaboration, Physics Letters 142B (1984) 119-124; Measurement of the Photon Structure Function $F_2(x,Q^2)$, PLUTO collaboration, Physics Letters 142B (1984) 111-118; Measurement of the Total Photon-Photon Cross Section for the Production of Hadrons at Small Q^2, PLUTO collaboration, DESY 84/080, to be

published; A Measurement of the Q^2 and W Dependence of the $\gamma\gamma$ Total Cross Section for Hadron Production, PLUTO collaboration, DESY 84/081, to be published

Figure Captions

1. Schematic presentation of detector effects, $f(x)$ (top graph) is the true distribution and $g(y)$ (bottom graph) the observed distribution

2. Assumed true distribution (histogram) with a simulated measurement (points)

3. Corrected data obtained from the measured points shown in figure 2 and the inverse response matrix B^{-1}

4. Decomposition of a given function into cubic B-splines, for one of the splines the knots are indicated

5. Graphical presentation of the detector response matrix A (see text)

6. Example of a true distribution (line) with measured data

7. Graphic presentation of the orthogonal functions $P'(x)$

8. The damping coefficient as function of the ordering parameter j

9. Undamped (bars) and damped (circles) coefficients for the orthogonal basic functions $P'(x)$

10. First non observable component of the orthogonal basic functions. The bin limits are marked (bars). For one of the bins the cancellation of the contribution from this component is also indicated, both the hatched areas are of almost equal size, so that

the integral over the bin is small.

11.   The unfolded measurement obtained from the data points given in
       fig. 6 compared with the true function f(x).

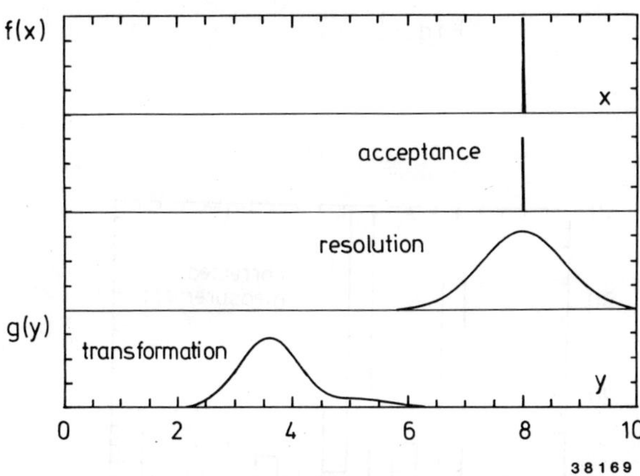

Fig. 1

40
30
20
10
0
true f(x)
measurement
0   5   10   15   20
38171

Fig. 2

40
30
20
10
0
-10
corrected
measurement
0   5   10   15   20
38172

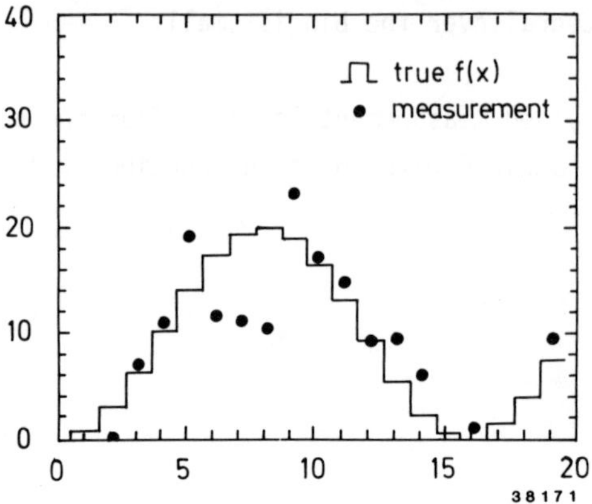

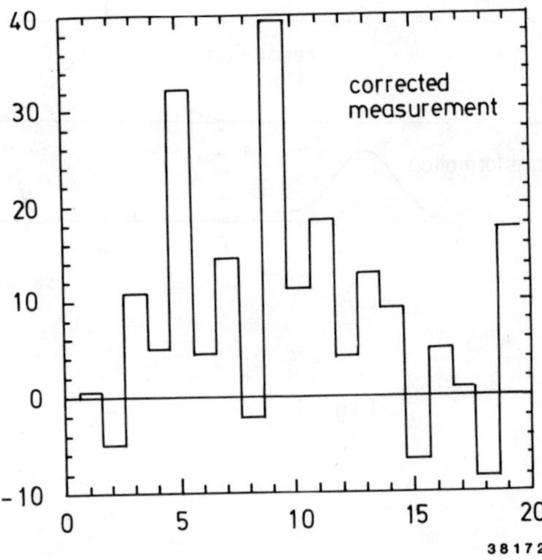

Fig. 3

2.0
1.5
1.0
0.5
0
$f(x) = 1 + (1-x)^2 = \sum a_j p_j(x)$
$a_j p_j(x)$
knot
0   0.2   0.4   0.6   0.8   1.0
x
38170

Fig. 4

38167

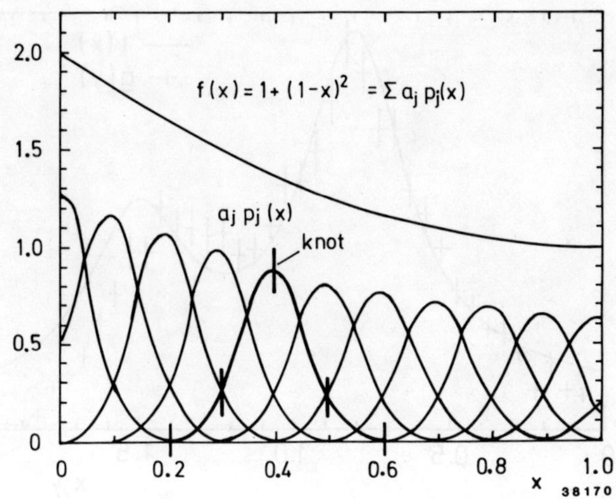

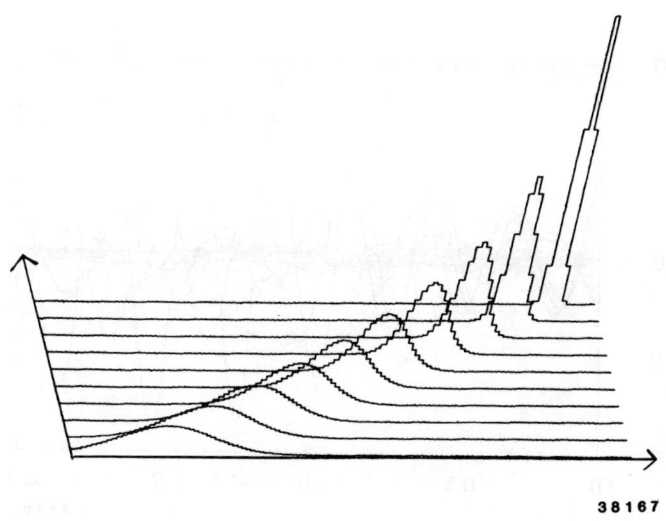

Fig. 5

6000
4000
2000
0
f(x)
g(y)
0   0.5   1.0   1.5   2.0
x,y
38168

Fig. 6

4.0
2.0
0.0
-2.0
-4.0
0.0   0.5   1.0   1.5   2.0
38166

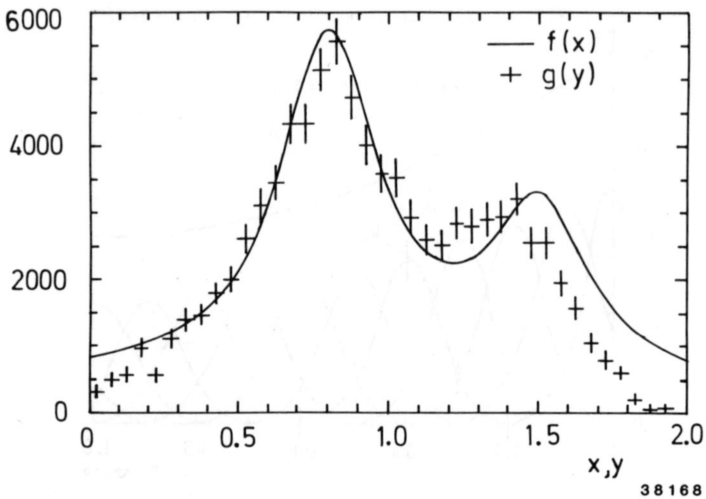

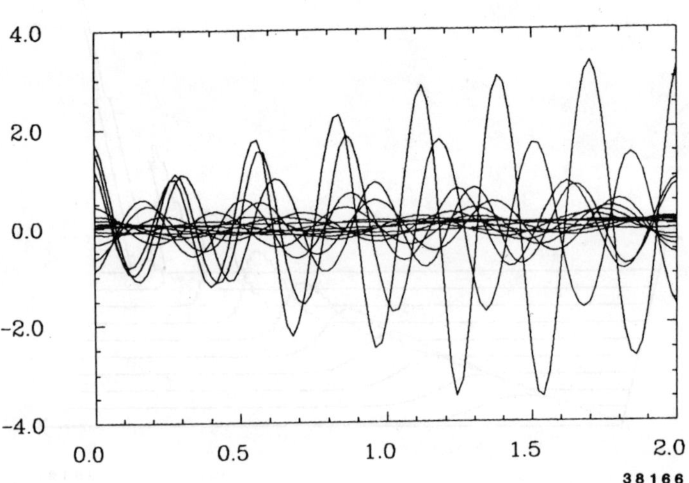

Fig. 7

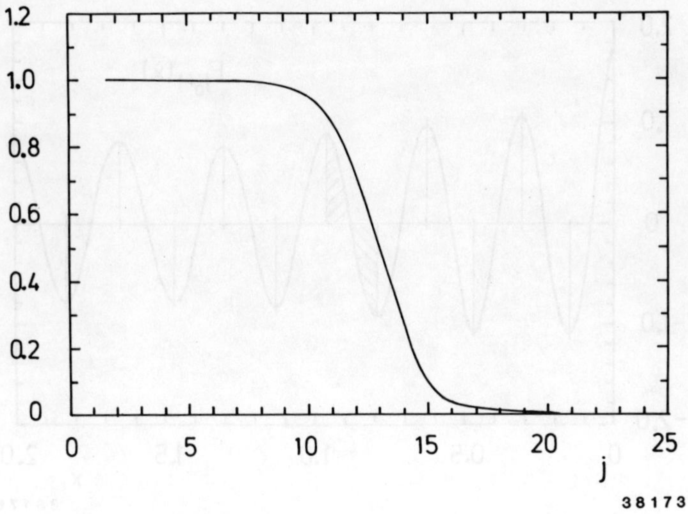

Fig. 8

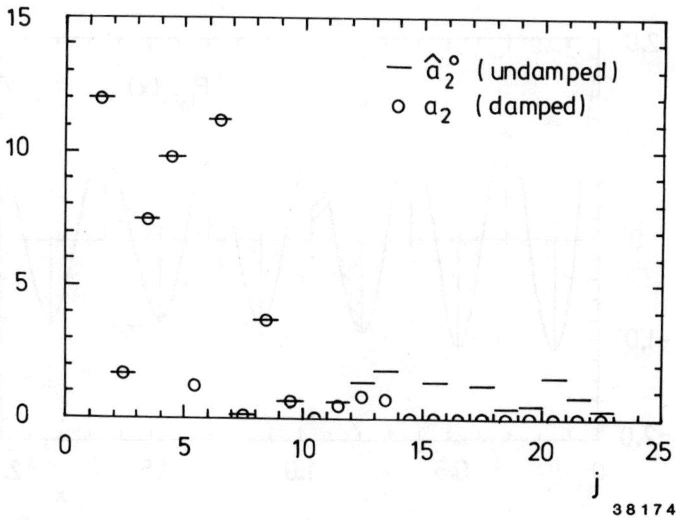

Fig. 9

$P_{j_0+1}(x)$
2.0
1.0
0
-1.0
-2.0
0
0.5
1.0
1.5
2.0
x
38176

Fig. 10

$P_{j_0+1}(x)$
2.0
1.0
0
-1.0
-2.0
0
0.5
1.0
1.5
2.0
x
38176

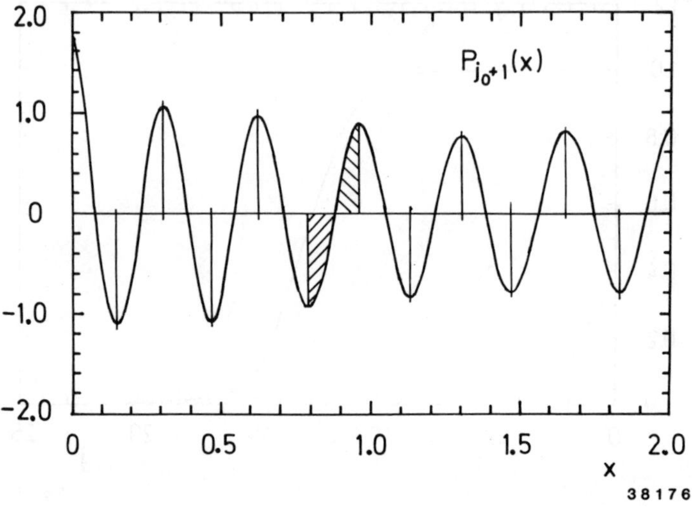

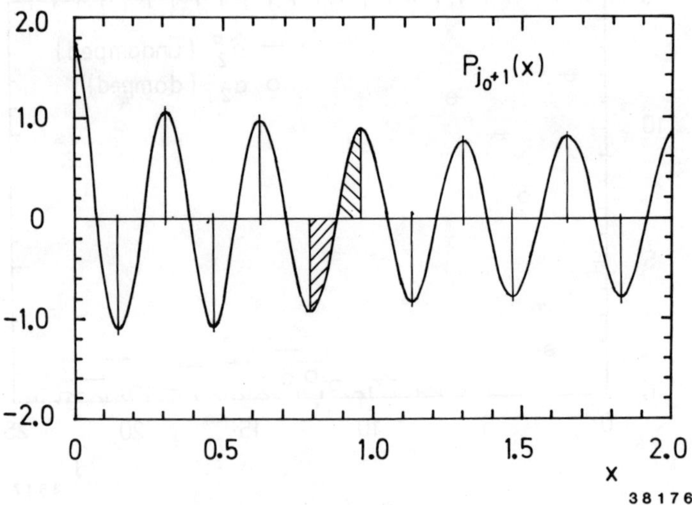

Fig. 11

The Photon Structure Function - Theory

William A. Bardeen

Fermilab
P.O. Box 500
Batavia, IL 60510

ABSTRACT

The theoretical status of the photon structure function
is reviewed.   Particular attention is paid to the hadronic
mixing problem and the ability of perturbative QCD to make
definitive predictions for the photon structure function.

1.   Introduction.

Deep inelastic scattering provides a unique probe of the pointlike
structure of matter. The structure of the photon has special interest due
to its two component nature where it can interact directly through its
pointlike couplings or indirectly through its hadronic component.   Initial
interest[1] in the photon structure function was based on the parton model.
The parton model predicts that the virtual photon can interact directly with
the target photon through the exchange of charged pointlike partons. The
parton model prediction for the photon structure function becomes

$$F_2^{\gamma}(x,Q^2) = \langle e^4 \rangle \times \left\{ P(x) \times \log(Q^2/m^2) + B(x) \right\} \tag{1}$$

where we see the sensitivity to the fourth moment of the parton charge
and the nonscaling Q^2 dependence.   The parton x distribution, P(x), reflects
the direct coupling to the photon.   The parton mass sets the scale of the
logarithm and reflects the infrared sensitivity of the parton structure
function.   In the following we will study the photon structure function
within the context of the theory of perturbative quantum chromodynamics.
We are particularly concerned with the separation of the direct pointlike

couplings of the photon from the effects of the quark and gluon hadronic constituents.

2. Leading order QCD.

The application of perturbative QCD to the photon structure function is similar to its application to hadronic processes. The reaction can be factorized into a hard scattering cross-section of the constituents times their target probability,

$$F_2^\gamma(x,Q^2) = \sum_c F_{2c}^\gamma(Q^2) * A_c \qquad (2)$$

The infrared sensitivity is absorbed in constituent probabilities, A_c. The constituent cross-sections, F_{2c}, are directly computed in perturbative QCD. Witten[2] was the first to observe that the proper treatment of the photon structure function requires that the photon be considered as its own constituent. Witten used operator product expansion and renormalization group methods to compute the hard scattering cross-sections. In leading order, the diagrams for quark and gluon production as shown in Figure 1 are summed to all orders. Of course only the hard scattering parts of these diagrams are correctly predicted by perturbative QCD. I emphasize the association of all Q^2 dependence with the hard scattering cross-sections rather than the Q^2 evolution of the parton distributions. The leading order results were also obtained using a wide variety of methods[3].

Quantum chromodynamics makes a unique prediction for the asymptotic behavior of the photon structure function. Witten presented the results for the moments of the structure function which are summarized by

$$\begin{aligned}
M_n(Q^2) &= \int dx\, x^{n-2} \times F_2^\gamma(x,Q^2) \\
&= a_n/\alpha_s(Q^2) + b_n \qquad \text{(photon)} \\
&\quad + \sum_i [\alpha_s(Q^2)]^{d_{n,i}} \times (1 + ...) \times A_{n,i} \qquad \text{(hadrons)} \qquad (3)
\end{aligned}$$

where $d_{n,i} = \gamma^0_{n,i}/\beta_0 \geq 0$ are the hadronic anomalous dimensions. The

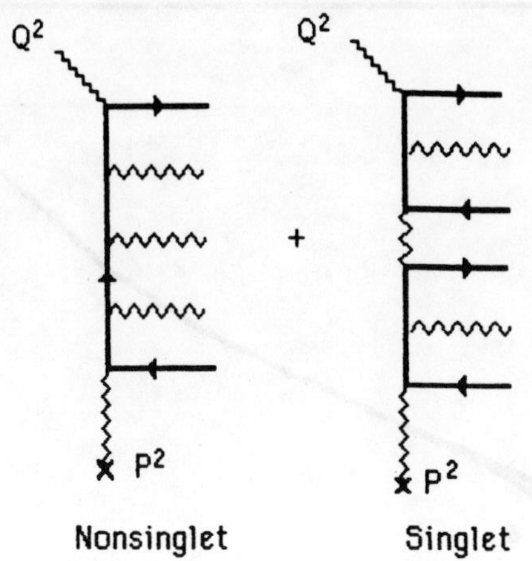

Figure 1.   Leading order perturbative diagrams.

hadronic part has exactly the same structure as hadronic deep inelastic scattering.   In leading order, asymptotic freedom predicts that the effective strong coupling should vanish for large Q^2 as $\alpha_S(Q^2) \rightarrow 4\pi/\beta_0 \times \log(Q^2/\Lambda^2) \rightarrow 0$.   Hence the photon component of the structure function in Eq. 3 dominates asymptotically over the hadronic components and the moments have the behavior,

$$M_n(Q^2) \rightarrow a_n \times (\beta_0/4\pi) \times \log(Q^2/\Lambda^2). \tag{4}$$

The coefficients a_n are computed in perturbative QCD and yield the nonparton but stiff x distribution shown in Figure 2.   The result of Eq. 4 predicts the ultimate asymptotic behavior of the moments as the anomalous dimensions given in Eq. 3 imply that all other corrections to the moments are logarithmically suppressed.

3. Higher order QCD.

The higher order corrections to the moments can also be computed in perturbative QCD.   The next corrections to the photon component are O(1) in an α_S expansion and will continue to aymptotically dominate over the

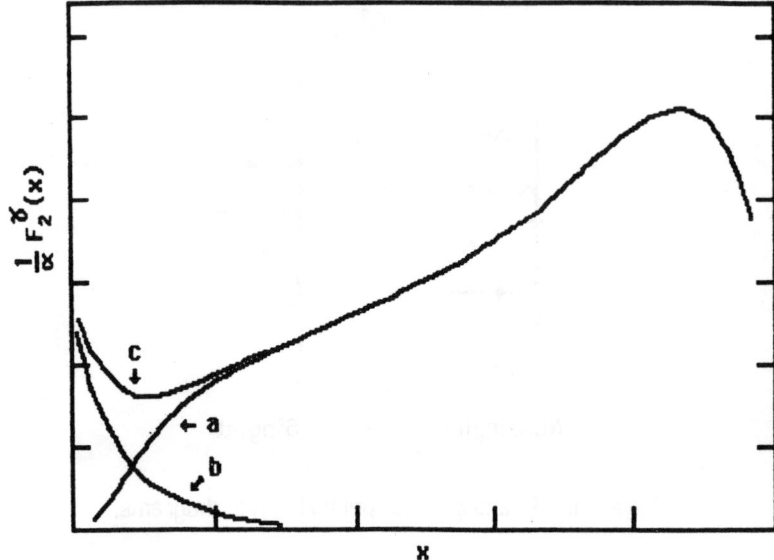

Figure 2.   Leading order structure function.   **a)** Valence
component. **b)** Sea component. **c)** Total.

hadronic components due to the positivity of the hadronic anomalous
dimensions with the exception of the second moment where the anomalous
dimension can vanish.   The coefficients, b_n, were computed[4] for $n > 2$ and
combined with the higher order corrections to α_S to determine the photon
component of the structure function through next leading order.   These
corrections are required for a significant determination of the QCD scale,
Λ_{QCD}, from this process.

   The moments can be directly compared to data or inverted to give
directly the structure function.   For moderate x, the higher order
corrections do not dramatically alter the shape of the x distribution but do
provide the overall scale of the structure function.   The results are shown
in Figure 3.   For small x, the higher order prediction of the photon
component breaks down as it appears to predict a negative cross-section.
As emphasized by Duke and Owens[5], this effect is due to mixing with the
hadronic component which can not be suppressed at small x.   The effect is

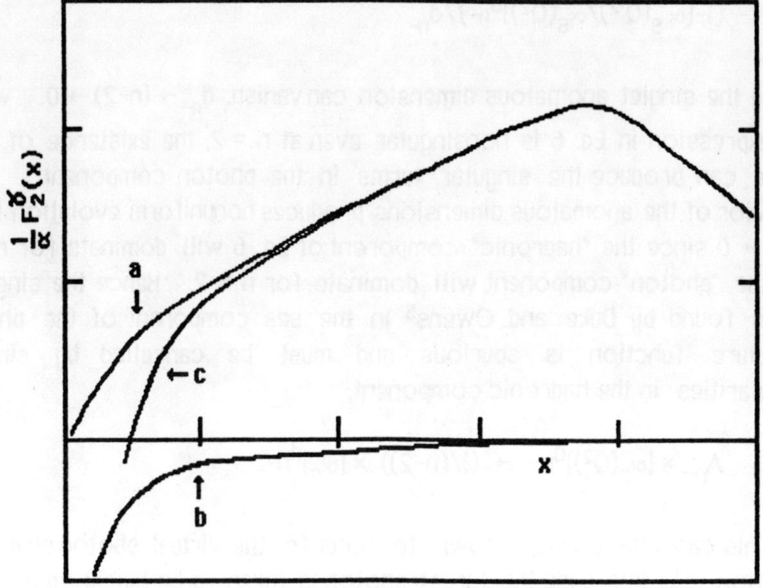

Figure 3.   Higher order structure function.   a) Valence
component.   b) Sea component.   c) Total.

best seen by the separation of the theoretical prediction into the valence
component, $\sim[\langle e^4\rangle - \langle e^2\rangle^2]$, and the sea component, $\sim\langle e^2\rangle^2$.   The negative
terms appear only in the sea component.

4.   Mixing singularities.

The negative contribution to the sea component of the structure
function arises from a pole in the b_n coefficient at $n = 2$, $b_n \to b/(n-2)$.
When inverted this pole generates a singularity at $x = 0$,

$$F_2^{sea}(x,Q^2) \to - (1/x). \tag{5}$$

This singularity arises from the mixing of the photon and hadronic
components and can be seen from the evolution equations for the structure
functions[4,6].   The evolution equations generate terms in the form

$$\left\{1-[\alpha_S(Q'^2)/\alpha_S(Q^2)]^{d_{n-}}\right\}/d_{n-} \tag{6}$$

where the singlet anomalous dimension can vanish, $d_{n-} \to (n-2) \to 0$.   While the expression in Eq. 6 is nonsingular even at $n = 2$, the existence of such terms can produce the singular terms in the photon component.   This behavior of the anomalous dimensions produces nonuniform evolution in Q^2 as $x \to 0$ since the "hadronic" component of Eq. 6 will dominate for $n < 2$ and the "photon" component will dominate for $n > 2$.   Hence the singular terms found by Duke and Owens[5] in the sea component of the photon structure function is spurious and must be cancelled by similar singularities in the hadronic component,

$$A_{n-} \times [\alpha_S(Q^2)]^{d_{n-}} \to (1/(n-2)) \times [\alpha_S]^{d_{n-}}. \tag{7}$$

This cancellation was shown[7] to occur for the virtual photon structure function.   In this case the target photon is taken to be highly virtual and the entire amplitude is calculable in perturbative QCD.   The coefficient A_{n-} can be computed exactly and does contain the pole expected form Eq. 7.   We conclude that the "hadronic" component may not necessarily be ignored even for real photons as the coefficients may be enhanced due to poles even though the terms are suppressed by powers of α_S for $n > 2$.

5.   Regularization.

The singularities discussed in the previous section require that the simple separation of the photon and hadron components be modified.   Much of the predictive power of perturbative QCD may be retained through the proper regularization of these singularities[8].   The basic point is that the singular terms produce a large effect in the sea distribution at small x. However, except for the singularity, the sea component is expected to be small.   Therefore, any reasonable regularization will cancel the singularity and leave a remaining small sea component.   Antoniadis and Grunberg[9] have made an explicit construction of the regularized structure function.   The method first involves the explicit separation of the singular terms.   Then they introduce a new parameter, λ, to tune the strength of the induced terms. The resulting structure function is nonsingular,

$$b_n = b_n^{reg} + b/(n-2), \qquad [\, t = \lambda \times \alpha_S(Q^2) \,],$$

$$M^{\gamma}_n{}^{reg} = a_n/\alpha_S(Q^2) + b_n^{reg} + [b/(n-2)] \times \left\{ 1-[t]^{\,d_n-} \right\}. \qquad (8)$$

For reasonable values of λ, the singular terms are reduced to a true higher order correction with sensitivity to λ only at small x. Their results are shown in Figure 4. The procedure used here is by no means unique but other methods will yield similar results.

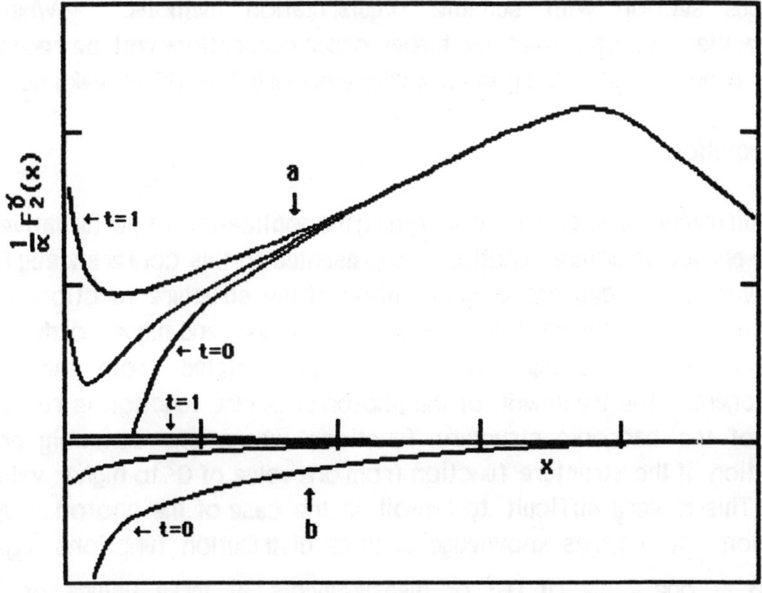

Figure 4. Regularized structure functions. **a)** Total
contribution for various **t** values. **b)** Sea component.

234

6. Higher order singularities.

We have discussed the singularities and the regularization of the next leading contributions to the photon structure function. G. Rossi[10] has made a systematic study of singularities induced by higher order corrections. He finds that the mixings generated in higher orders produce poles which move to larger values of n. Poles at larger values of n correspond to more singular x distributions,

$$M_n \to 1/(n-n_0) \;\Rightarrow\; F_2(x) \to 1/x^{n_0-1}. \tag{9}$$

These poles are a further reflection of the nonanalytic behavior at small x. The singularities must be cancelled by similar singularities in the "hadronic" terms. The poles can be discussed from the perspective of the previous section with suitable regularization methods. When the singularities are cancelled, the higher order corrections will be reduced to higher order except at very small x where perturbative QCD breaks down.

7. Evolution.

A different perspective[11] concerning the application of perturbative QCD to the photon structure function was presented to this conference by Drees. This work concludes that only evolution of the structure functions can be computed due to the mixing singularities in next and higher order. The dominant photon component can not be isolated from the hadron component. The treatment of the photon structure function is reduced to that of the hadronic structure functions where one can only predict evolution of the structure function from one value of Q^2 to higher values of Q^2. This is very difficult to exploit in the case of the photon structure function. It requires knowledge of three distribution functions q_{NS}, q_S, and G at one value of Q_0^2 or measurements at three values of Q^2 to determine the full Q^2 dependence (note there are no sum rules in this case). In this procedure we lose all sensitivity to Λ_{QCD}.

In practice one must make an ansatz, at one Q^2, to relate the singlet quark and the singlet gluon distributions to the nonsinglet quark distribution.   Drees et al choose the following relations,

$$\Sigma^{\gamma}(x,Q_0^2) = \{\langle e^2 \rangle / [\langle e^4 \rangle - \langle e^2 \rangle^2]\} \times q_{NS}(x,Q^2),$$

$$G^{\gamma}(x,Q_0^2) = \{2/\beta_0\} \times P^0_{gq} * \Sigma^{\gamma}. \tag{10}$$

In order to test the evolution predictions, they let $Q_0^2 = 1$ GeV^2 and evolve to fit the data at $Q^2 = 5$ GeV^2 which determines the valence quark distribution, $q_{NS}(x,Q_0^2)$.   Their predictions for the photon structure function at higher Q^2 is shown in Figure 5.   They compare the leading order evolution and higher order evolution and find little difference.   They find no sensitivity to Λ_{QCD} as expected in this approach.   With their parameterization, they find only a slow approach to the asymptotic structure function.

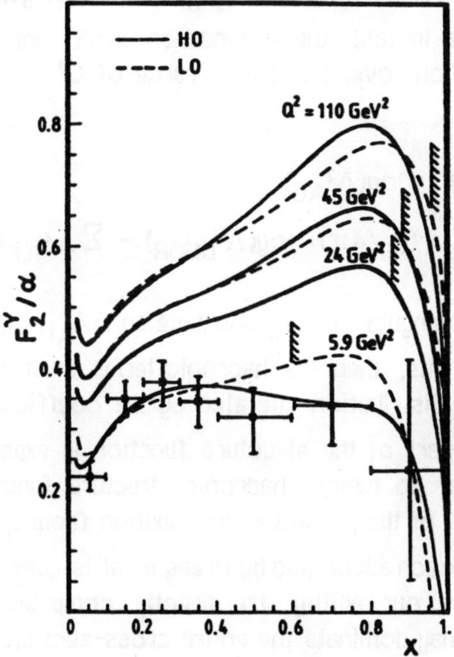

Figure 5.   Comparison of leading and higher order
evolution of the photon structure function.

236

We may try to analyze the results of Drees et al by comparing their results with the results of Antoniadis et al.   We may compare the asymptotic forms for the fitted moments,

$$M_n^\gamma = a_n/\alpha_s(Q^2) + b_n + \sum_{NS,+,-} A^\gamma_{n,i} \times [\alpha_s(Q^2)]^{d_{n,i}} . \qquad (11)$$

The two approaches must agree on the values of a_n and b_n. The coefficient $A_{n,-}$ must have the same $n=2$ singularities.   However the nonsingular parts of the $A_{n,i}$ can be expected to differ as well as the choice of Λ_{QCD}. We see that the two approaches differ only in those terms which are not calculable in perturbative QCD.   Whether the photon component or the hadron component dominates depends on the fitting procedure.

Assuming that both approaches can fit the data, we can address the sensitivity to the determination of Λ_{QCD}. The nonsingular hadronic terms of Drees et al can imitate the dependence on Λ^2 only if they have the correct x distribution over the fitted range of Q^2.   The Λ^2 values are related by

$$a_n \times (\beta_0/4\pi) \times \log(\Lambda^2_{AG})$$
$$= a_n \times (\beta_0/4\pi) \times \log(\Lambda^2_{DGGR}) - \sum \Delta A_{n,i} \times (\alpha_s)^{d_{n,i}} \qquad (12)$$

where only the nonsingular parts contribute to $\Delta A_{n,i}$. If the values of Λ^2 differ in the two fits, then the hadronic terms must have the pointlike structure of the x distribution dictated by the coefficient, a_n. However the true hadronic part of the structure function is expected to have an x distribution similar to typical hadronic structure functions and not the stiff x distribution of the pointlike contribution from a_n. I conclude that the evolution approach advocated by Drees et al is quite conservative. It chooses to ignore our ability to directly compute the large photon component which may dominate the entire cross-section.

8. Conclusions.

In this talk I have only briefly discussed the fundamental QCD analysis of the photon structure function as it has been extensively presented in the literature. Instead I have focussed on the questions related to the hadronic mixing problem. My basic conclusion is that a large pointlike photon component can dominate the photon structure function with a calculable dependence on Λ^2_{QCD}. The mixing singularities discovered in the perturbative analysis must be properly treated. However the analysis of Antoniadis et al does provide a reasonable prescription for regularizing the perturbative singularities but requires the introduction of a new parameter, λ. The structure function is sensitive to the value of λ only at small x. Of course the size of the true hadronic component is not calculable in perturbative QCD. However since this hadronic component is not expected to have the pointlike x distribution, the data can be used to determine whether a large hadronic component is required. If the structure function is pointlike, then we can presume the photon component dominates and use the moderate x range to determine Λ^2_{QCD}.

In fits to the data, a vector meson dominance (VMD) contribution is usually included with the hadronic x distribution. Although only a small effect, this hadronic component should include the expected Q^2 dependence. The λ parameter of Antoniadis et al produces a VMD-like effect with the correct Q^2 dependence.

Strictly perturbative analysis can not be used further to resolve the structure of the hadronic components. Diagram calculations are sensitive to the wrong, perturbative infrared dependence which neglects all confinement effects. In the future we must look to nonperturbative estimates of the hadronic coefficients, $A_{n,i}$. Perhaps the QCD lattice industry can be induced to study the appropriate matrix elements and determine whether anomalously large hadronic components contribute to the photon structure function.

Acknowledgements

This talk was presented at the VIth International Workshop on Photon-Photon Collisions held at Lake Tahoe, September 1984. I wish to thank the organizers of the workshop for providing a stimulating atmosphere for discussions of all aspects of two photon physics.

REFERENCES

1.   S.J. Brodsky, T. Kinoshita, and H. Terazawa, Phys. Rev. Letters 27,280(1971); T.F. Walsh and P. Zerwas, Phys. Letters 44B,195(1973); R.L. Kingsley, Nucl. Phys. B60,45(1973); H. Terazawa, Rev. Mod. Phys. 43,615(1973).

2.   E. Witten, Nucl. Phys. B120,189(1977).

3.   C.H. Llewellyn-Smith, Phys. Letters 79B,83(1978); C.T. Hill and G.G. Ross, Nucl. Phys. B148,373(1979); R. DeWitt, L. Jones, J. Sullivan, D. Willen, and H. Wyld, Phys. Rev. D19,2046(1979); W.R. Frazer and J.F. Gunion, Phys. Rev. D20,147(1979).

4.   W.A. Bardeen and A.J. Buras, Phys. Rev. D22,166(1979); Erratum, D21,2041(1980).

5.   D.W. Duke and J.F. Owens, Phys. Rev. D22,2280(1980).

6.   W.A. Bardeen and A.J. Buras, Phys. Letters 86B,61(1979); A.J. Buras, Rev. Mod. Phys. 52,1(1980).

7.   T. Uematsu and T.F. Walsh, Phys. Letters 101B, 263(1981); Nucl. Phys. B199, 93(1982); C. Peterson, P.M. Zerwas, and T.F. Walsh, Nucl. Phys. B229,301(1983).

8.   W.A. Bardeen, Proceedings of the 1981 International Symposium on Lepton and Photon Interactions, W. Pfiel,ed., Bonn(1981); D.W. Duke, Procedings of Conference of Photon Photon Collisions, C.H. Berger,ed., Aachen(1983).

9.   I. Antoniadis and G. Grunberg, Nucl. Phys. B213,445(1983); I. Antoniadis, L. Baulieu, and R. Lacaze, Phys. Rev. B125,92(1983).

10.   G. Rossi, Phys. Letters 130B,105(1983); Phys. Rev. D29,852(1984).

11.   M. Gluck and E. Reya, Phys. Rev. D28,2749(1983); M. Gluck, Procedings of the Brighton Conference, Brighton(1983); M. Gluck, K. Grassie, and E. Reya, Dortmund Preprint DO-TH 83/23; M. Drees, M. Gluck, K. Grassie, and E. Reya, Dortmund Preprint DO-TH 84/12.

COMMENTS

S. BRODSKY: As you have emphasized, the mass dependence in the QPM result for $F_2^\gamma \sim \log(Q^2/m^2)$ is replaced by the QCD scale Λ^2 in the all-order calculation.   For heavy quarks $m^2 \gg \Lambda^2$, the mass dependence can not be neglected.   Thus shouldn't we include higher dimension operators in the analysis to recover the mass dependence, including the vacuum expectation values which give large consistent quark masses? Neglecting these contributions could affect the determination of $\Lambda_{\underline{MS}}$ from F_2^γ.

J.H. FIELD: For the charm quark contribution to F_2, the quark parton prediction is normally used as the mass scale is set not by Λ_{QCD} but by the charm quark mass.   If however Λ_{QCD} is as small as 100 MeV and the lightest constituent quark mass is $\sim$300 MeV, is to be expected that even for the light quarks, the quark mass will set the energy scale, not Λ_{QCD}. In fact existing data are fitted equally well by the QPM with conventional constituent quark masses or by asymptotic QCD predictions with Λ as an adjustable parameter.

K. GRASSIE: How do you obtain the typical behavior of structure functions as predicted by QCD, namely an increase of F_2 at large x and decrease at small x if you change Λ^2 or Q^2 appropriately? This behavior is predicted by the A_i terms which have been neglected in the semi-asymptotic solution of Antoniadis and Grunberg.

WINSTON KO: If we measure the longitudinal structure function, F_L, which is expected to not have a Q^2 dependence, would it be useful to determine the Q^2 – independent term of F_2?

EXPERIMENTAL RESULTS ON THE PHOTON STRUCTURE FUNCTION

Walter Wagner

Physics Department
University of California, Davis
Davis, CA 95616 U.S.A.

ABSTRACT

The experimental situation of the photon structure function is reviewed. The data are compared to a regularized higher order QCD calculation. Utilizing the total available information the un-calculable hadronic pieces can be kept under control and a value of $\Lambda_{\overline{MS}} = 230 \pm 70$ MeV is obtained.

1.   Introduction

Since the first experimental results on the photon structure function have been presented about four years ago,[1] we have witnessed an almost explosive development of that field. The total amount of data available has crossed the 10000 event limit and the lever arm in Q^2 covers three orders of magnitude. In addition, very solid methods have been developed, which allow to present the data in a detector independent way.[2]

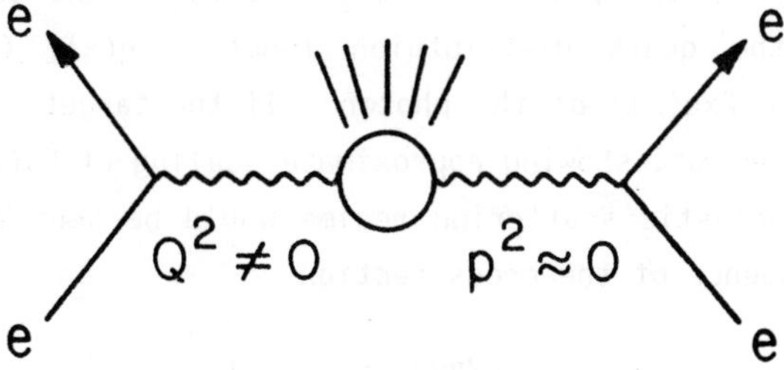

Figure 1.   Two photon diagram.

The process under consideration is hadron production by the two photon mechanism (Fig. 1), with one photon almost real ($p^2 \underset{\sim}{} 0$) and the other

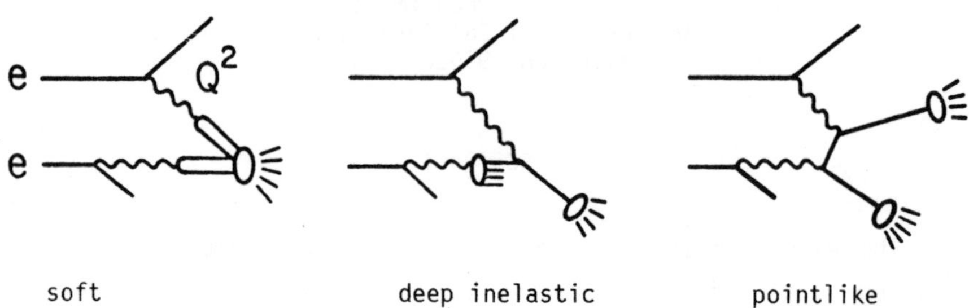

Figure 2.   Different levels of inelasticity for photon photon scattering.

one virtual ($Q^2 \neq 0$).   It can be looked at three levels of inelasticity (Fig. 2a-c).   In the soft regime both photons have enough time to convert into vector mesons, mainly ρ's, and the subsequent process will be 'typical' hadronic reaction.   The ρ propagator gives rise to a $1/Q^4$ behavior of the total cross section

$$\sigma(Q^2) \;=\; \sigma_0 (\frac{1}{1 + \frac{Q^2}{m_\rho^2}})^2 \to \frac{1}{Q^4} \quad . \tag{1}$$

As the probing photon gets off its mass shell it will more likely couple directly to the pointlike constituents of the target photon, thus testing the quark distribtuion function $q(x)$, (or structure function $F_2(x) = 2xq(x)$) of the photon.   If the target photon behaves like a typical hadron, showing approximate scaling of $F_2(x)$, the onset of this deep inelastic scattering regime would be seen as a turnover to a $1/Q^2$ dependence of the cross section

$$\sigma(Q^2) \;=\; \frac{(2\pi\alpha)^2}{Q^2} \frac{F_2}{\alpha} \to \frac{1}{Q^2} \quad . \tag{2}$$

Figure 3 nicely demonstrates[3] that the scale where the turnover occurs is given by the ρ-mass squared. In addition to the hadronic part the photon has a pointlike component, which is expected to show up and even dominate at large values of Q^2. This pointlike piece results in a structure function that rises with x (in contrast to all other known structure functions, which decrease with x) and which shows a strong scale breaking effect $\sim \ell n\ Q^2$. Thus the Q^2 dependence in this regime is given by

$$F_2 \sim \ell n\ Q^2 \ ; \quad \sigma(Q^2) \ \rightarrow \ \frac{1}{Q^2}\ \ell n\ Q^2 \ . \tag{3}$$

An independent way to look for the pointlike component is to investigate the final state structure. As the photon quark coupling is entirely pointlike there is no scale, limiting the transverse momenta p_T of the quarks relative to the photon direction, except for phase space boundaries. That is quite different in a hadron-like

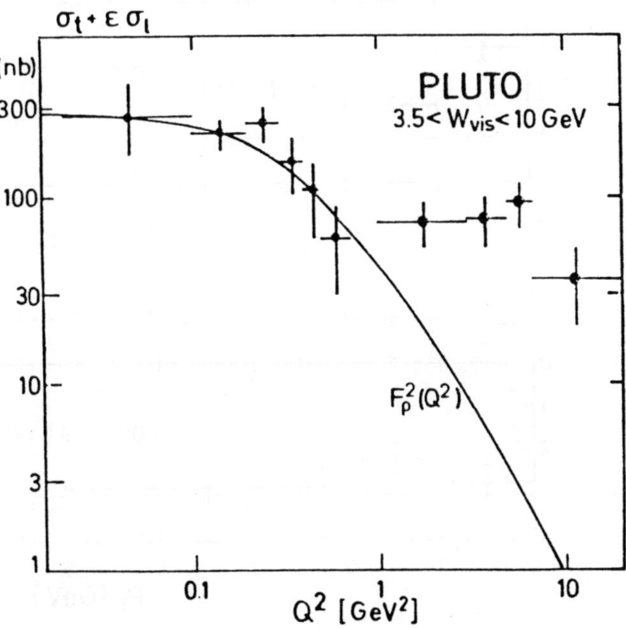

Figure 3. Total cross section as function of Q^2.

photon state, where the transverse momenta are limited to about 300 MeV, corresponding to the size of a typical hadron. The measurement[4] of $\tilde{R}_{\gamma\gamma}(p_T)$ which is the ratio of the measured jet cross section over the predicted pointlike part, nicely demonstrates two things (see Fig. 4).

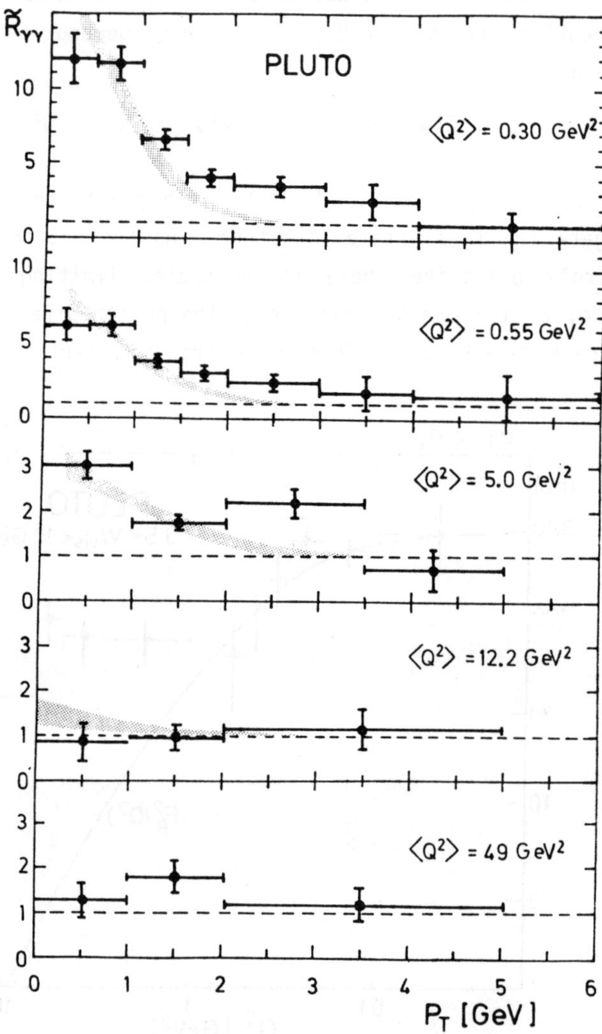

Figure 4. $\tilde{R}_{\gamma\gamma}$ vs. jet p_T in 5 bins of Q^2. The dashed line shows the QPM expectation and the bands give the QPM + VDM prediction. For $Q^2 = 45$ GeV^2 the VDM contribution is negligible.

i) at all Q^2 we see the presence of the pointlike component sticking out at large p_T.

ii) Though the pointlike component plays a minor role for the total cross section at low Q^2, ($< 10\%$), it is quite substantial at moderate Q^2 ($\sim 30\%$ at 5 GeV^2) and completely dominates for $Q^2 > 10$ GeV^2.

It is interesting to note that the cut made in this jet analysis of W>4 GeV^2 favours the events in the low x region, and even there the pointlike piece can be isolated. On the other hand, the very early measurements of the photon structure function[5] already indicate the dominance of the pointlike component at large x (Fig. 5).

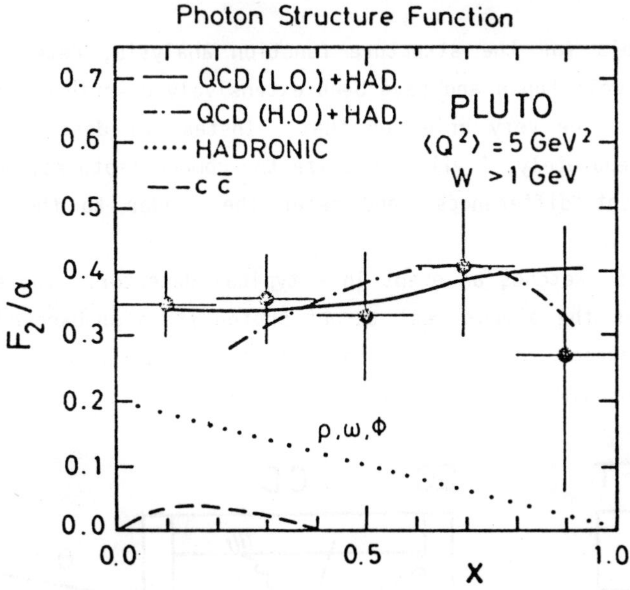

Figure 5.   First result on $F_2^\gamma(x)$.

To conclude this section we can summarize that we enter the deep inelastic regime at Q^2 values above the ρ-mass squared, that the pointlike component is present at all Q^2 and that it even dominates if either Q^2, p_T or x gets large. Thus we may be encouraged to interpret

246

the data in terms of the photon structure function and do some meaningful tests of QCD. The paper will be arranged as follows: in the next chapter I will make some general remarks on how the experiments are done, how the data are presented and to which theoretical models they are compared. In chapter 3 the data are presented and are compared to theory in chapter 4. The guideline for the comparison to theory will be to answer the following questions:

1. Do we see effects of gluons?
2. Do we see higher order effects?
3. Can we do clean QCD tests?
4. Can we determine Λ or α_s?

2. General Remarks

The tools for the structure function analysis, detectors, and methods are well known and have been extensively described elsewhere[6] so that I can be very brief on that. Instead of describing every experiment separately, I will summarize the common features, point out some important differences, and refer the reader to the original publications.

Figure 6 sketches an event in a typical detector. The electron that radiates the almost real target photon stays unobserved in the beam pipe (except for the rare case of the virtual photon structure function measurement in the double tag setup). The other electron is

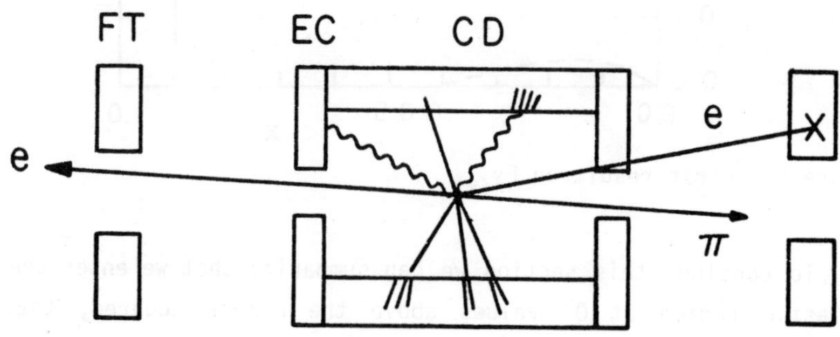

Figure 6. Sketch of a two photon detector.

detected at larger angles in either one of the forward tagging devices (FT), the end cap (EC), or central detector (CD) shower counter devices. The final state hadrons and photons are measured in the central detector and (if available) in forward tracking devices. The kinematic variables of interest are calculated in an obvious way from the scattered electron and the final state particles:

$$Q^2 = 2\ EE'(1-\cos\theta)$$

$$W^2_{vis} = (\sum_{vis}^{i} p_i)^2 < W^2_{true} \tag{4}$$

$$X_{vis} = Q^2/(Q^2 + W^2_{vis}) > X_{true}$$

As the sum goes over the visible particles only, one has to correct for the unseen particles by a Monte Carlo simulation of the detector acceptance. This acceptance simulation depends in principle on the fragmentation model used and it is a major part of the analysis to first find a proper fragmentation model and then estimate the systematic error coming from the remaining uncertainty in the fragmentation. The experiments differ substantially in the angular range for the tagging, corresponding to the Q^2 range available, the detection coverage for the final state particles, and the integrated luminosity. Figure 7 and Table 1 summarize the situation.

The clear signature of the events allows for a quite straight-forward analysis: events with one energetic electron in addition to a multihadronic final state (multiplicity $\geq$ 3) are selected and compared to a Monte Carlo simulation using

$$\frac{d\sigma}{dxdy} = \frac{16\ \pi\alpha^2 EE_\gamma}{Q^4} \{(1-y)\ F_2 + xy^2F_1\}\ N_\gamma(z)dz \tag{5}$$

for the generator. As xy^2 is much smaller than $(1-y)$ within the acceptance of most experiments, the F_1 term is normally neglected.

For the final confrontation of data with theoretical predictions two alternative methods are used

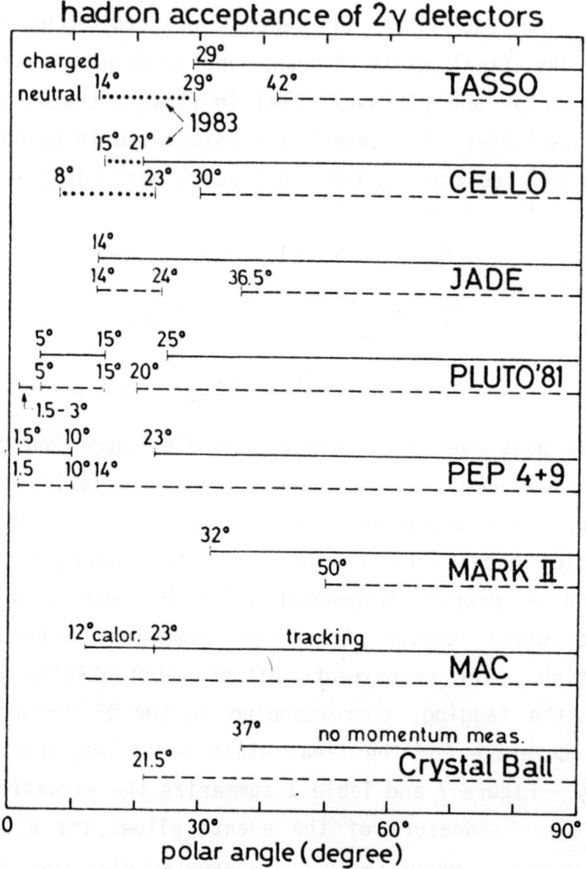

Figure 7. Acceptance for measurement of charged hadrons (full line)
and neutral showering hadrons (dashed line) in existing 2γ
detectors at DESY and SLAC.

i) a whole bunch of different models for $F_2(x,Q^2)$ is used for
the generator (eq. 5) and the resulting X_{vis} distributions
are compared with the data.

Table 1

	ΔQ^2(GeV^2)	$<Q^2>$(GeV^2)	L(pb^{-1})	events
PEP 4/9	.5 - 7	1	50	7000
PEP 4/9	7 - 100	20	70	400
PLUTO	1.5 - 16	5.3	34.2	1400
PLUTO	7 - 100	45	30	100
CELLO	2.5 - 300	9	9.5	200
JADE	10 - 55	24	72.5	400
JADE	30 - 220	100	72.5	50
TASSO	8 - 65	23	52.5	250
MAC	10 - 200	32	224	500

ii)   a generator with $F_2(x,Q^2) = 1$ is used to unfold the data for both, $x \rightarrow x_{vis}$ transformations and x-Q^2 correlations.[2] (The x-Q^2 correlations can be kept small if the data are analyzed in small bins of Q^2 separately). This leads to a detector independent presentation of $F_2(x)$ at a given Q^2 value.

The second method is much more convenient as it allows a direct comparison of different experiments or of experiments with theory (which eases the life of review speakers). The first method, however, is slightly superior if one particular model is to be tested or if a one parameter fit is to be carried out. But the loss of significance in the second method (which is a reflection of the small but still present bin to bin correlations in the unfolded result) is almost negligible as we will see later.

The experiments compare their results to basically four classes of models.

i)   A straightforward QED calculation[7] in the quark parton model (QPM) where the quarks are treated as free particles with an effective mass as the only free parameter:

$$F_2^{QPM}(x,Q^2) = 3 \left(\frac{\alpha}{\pi}\right) \sum e_q^4 \left\{ x(x^2+(1-x)^2) \ln \frac{W^2}{m_q^2} \right.$$

$$\left. + 8 \, x^2(1-x)-x \right\} \tag{6}$$

Some experiments add a VDM part to the QPM which is normally parameterized[8] as

$$F_2^{VDM}(x,Q^2) = 0.2 \cdot \alpha \, (1-x) \tag{7}$$

All experiments use the QPM calculation for the c quark (except for the highest Q^2 point (100 GeV^2) of JADE, where the QCD calculation was extended to the c quark also.

ii) Asymptotic leading order QCD calculations[9] (LO) representing the pointlike part of the photon

$$F_2^{LO}(x,Q^2) = 3 \left(\frac{\alpha}{\pi}\right) \sum e_q^4 \, f(x) \, \ln \frac{Q^2}{\Lambda^2} \tag{8}$$

iii) Regularized asymptotic higher order QCD calculations (HO)

$$F_2^{HO}(x,Q^2) = F_2^R (x,Q^2; \Lambda_{\overline{MS}}) + \Delta(x,t) + F^{VDM}(x) \tag{9}$$

The second term is a regularization term which was calculated by Antoniadis and Grunberg[10] and which removes the $\bar{d}$ pole in the n=2 moment of the pointlike component.[11,12] (For t=0 the first two terms correspond to the non-regularized calculation of Bardeen and Buras[13].) The regularization scheme depends on the non-perturbative parameter t, and there is some interplay and most probably some double counting between the second and the third term in eq. 9 (see discussion below).

iiii) Non-asymptotic QCD prediction in leading and higher order, using parton distribution functions at $Q_o^2 = 1$ GeV^2 as boundary conditions.[14] Either a pure VMD or a superposition of VDM and QPM (using m_u = 200 MeV) were used as inputs for the Q^2 evolution of the structure function.

If one counts the four contributions in the last model separately (LO,HO / VDM,VDM+QPM) this amounts to a total of 7 models to be compared to the data, each of which depends on one or more parameters (m_q, Λ, t, ...). As there are experimental results available at 14 different Q^2 values this amounts to a total of $2 \cdot 7 \cdot 14 = 196$ nice plots, if both methods of presentation are considered. To show them all might slightly exceed both the available space and the patience of the reader.

3.   Presentation of the Results

Figure 8 shows the JADE[15] result at Q^2 = 24 GeV^2 compared to the QPM, using both methods: In Fig. 8a the measured distribution of x_{vis} is compared to the x_{vis} distribution as expected from a QPM generator, and in Fig. 8b the unfolded structure function is compared to the analytic expression (eq. 6) of the same QPM. The conclusion is in both cases the same, namely that at low x the data are above the curve whereas at high x they are slightly below. The significance of the deviation is slightly higher in the direct analysis of Fig. 8a than in

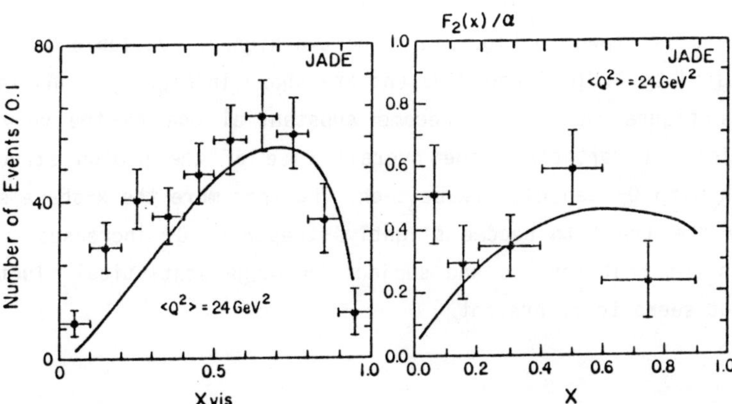

Figure 8. Comparison of the JADE data with the QPM in a) the x_{vis} distribution and b) the unfolded structure function.

the comparison after unfolding. However, as the effect of significance losses is not big, I am going to present the data in terms of unfolded structure functions only. That allows comparing different experiments in one diagram and looking for trends in the data without the bias of theoretical curves to guide the eye.

In Fig. 9 the first set of results[16,17] is shown in 6 Q^2 bins exploiting the Q^2 range from 0.5 - 18 GeV^2. The rise of the structure function with Q^2 at each value of x is observed very clearly and significantly. At $x \sim 0.3$, F_2/α rises from about 0.15 to 0.4. It is very interesting to observe that for low Q^2 ($Q^2 < 2$ GeV^2) $F_2(x)$ bends over to values below $0.2\cdot\alpha$ (1-x) at low x. This might indicate that the parameterization of the VDM part (eq. 7) could need some refinement (see discussion below). The figure also demonstrates nicely the extension of the available x range as Q^2 increases, a simple reflexion of the W cut around 1 GeV.

As the Q^2 ranges of the two experiments overlap, the results can be directly compared at equal (or similar) Q^2. This is shown in Fig. 10a where the high Q^2 PEP4/9 data of Fig. 9 are compared with the low Q^2 PLUTO result. The highest PEP4/9 bin in Q^2 corresponds to the mean Q^2 of the PLUTO data and a comparison is shown in Fig. 10b. Both experiments agree well within the statistical errors.

The TASSO[18] and the low Q^2 JADE[15] result are at substantially higher and similar Q^2. They are in fair agreement with each other and the result is shown in Fig. 11a. The two highest available Q^2 ranges are studied by PLUTO[19] and JADE and are shown in Fig. 11b. As in the previous figure the errors become substantial due to the very low statistics. Nevertheless the overall rise of the photon structure function with Q^2 can clearly be seen. Further more the x-shape seems to have the trend to become slightly steeper as Q^2 increases. This trend is not that obvious and subject to large statistical fluctuations, but seems to be present.

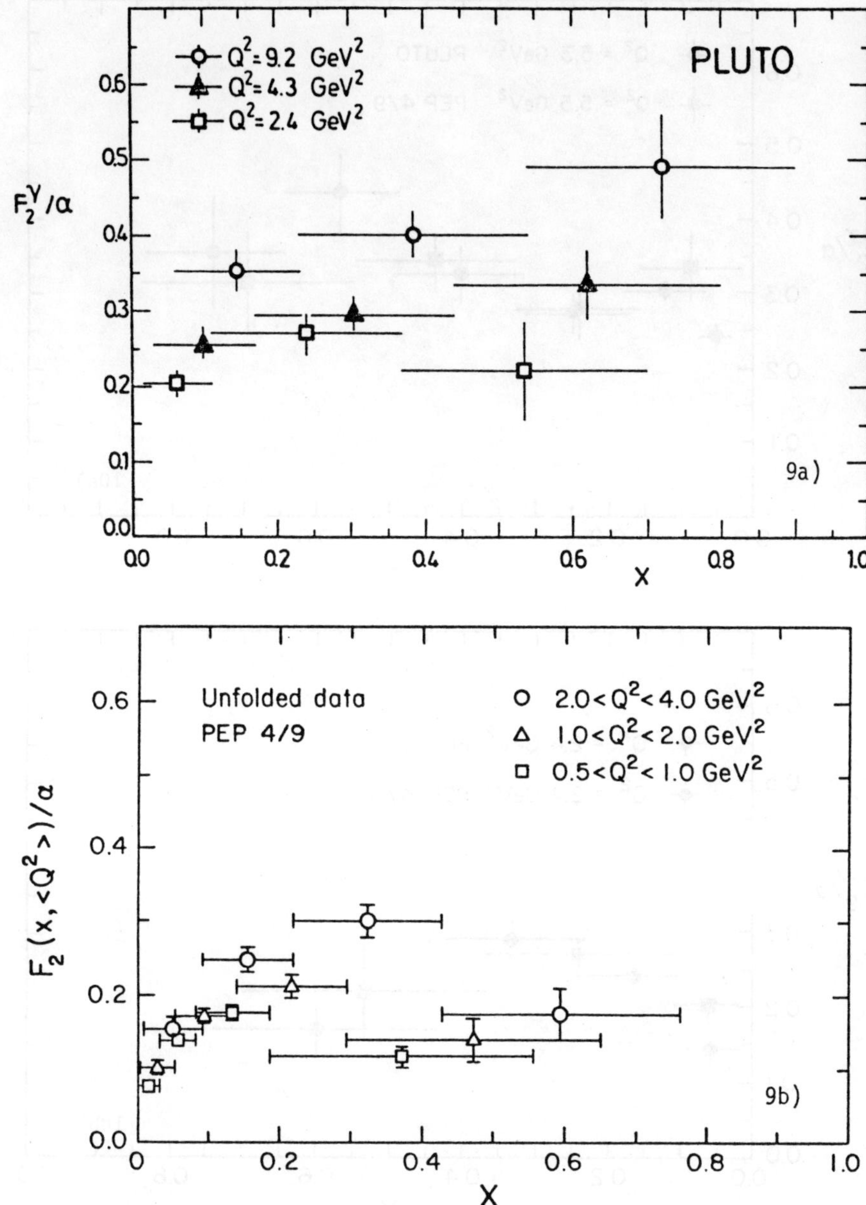

Figure 9. Unfolded structure functions in 6 different bins of Q^2 by a) PLUTO and b) PEP4/9.

254

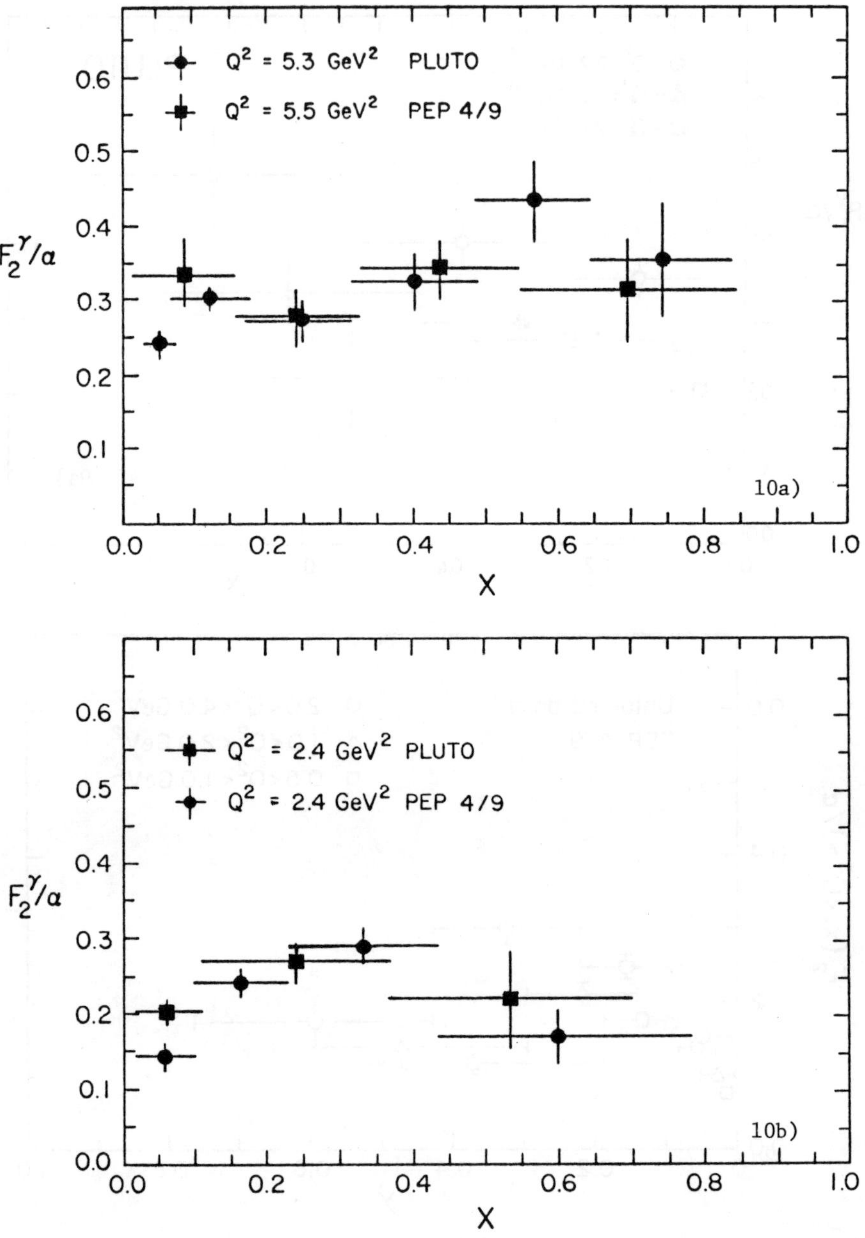

Figure 10.   Comparison of the PEP4/9 and PLUTO results in similar Q^2 ranges.

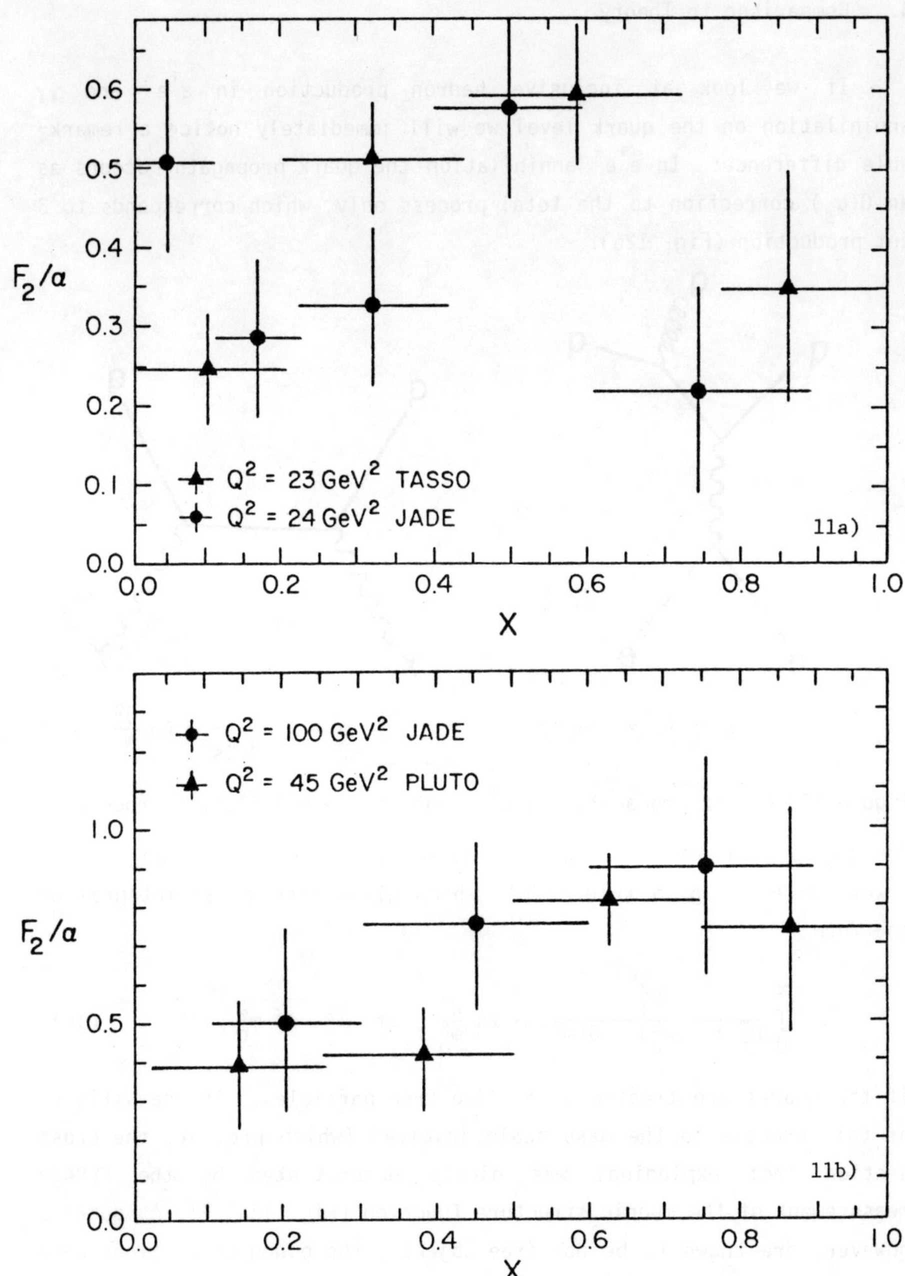

Figure 11.   Results of $F_2(x)$ at high Q^2 by JADE, PLUTO, and TASSO.

4. Comparison to Theory

If we look at inclusive hadron production in e^+e^- and $\gamma\gamma$ annihilation on the quark level we will immediately notice a remarkable difference: In e^+e^- annihilation the quark propagator occurs as an $O(\alpha_s)$ correction to the total process only, which corresponds to 3 jet production (Fig. 12a).

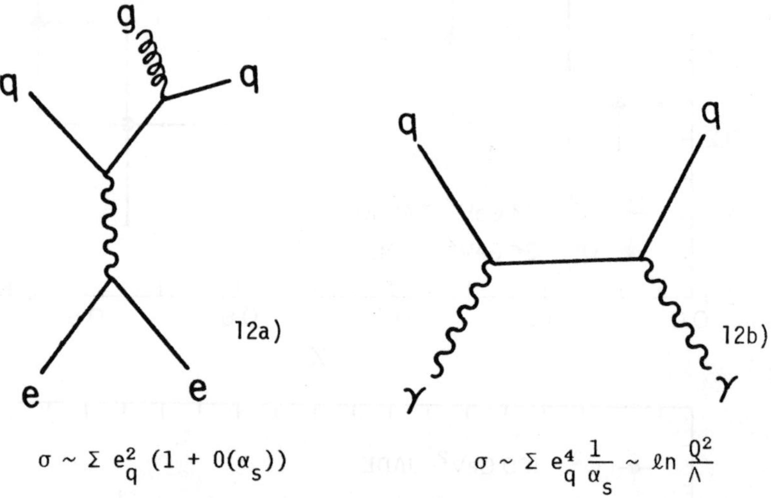

$$\sigma \sim \Sigma\, e_q^2\, (1 + O(\alpha_s)) \qquad\qquad \sigma \sim \Sigma\, e_q^4\, \frac{1}{\alpha_s} \sim \ell n\, \frac{Q^2}{\Lambda}$$

Figure 12. Quark propagator in e^+e^- and in $\gamma\gamma$ annihilation processes.

The two photon process requires the quark propagator already in the lowest order diagram (Fig. 12b), which gives rise to an integral of the form

$$\int_0^{Q^2} \frac{dp_T^2}{p_T^2 + O(m_q^2) + O(m_\gamma^2)} \to \ell n\, \frac{W^2}{m_q^2}\ , \quad \text{if}\quad m_\gamma^2 \ll m_q^2 \tag{10}$$

if the quarks are treated as massive free particles. The sensitivity of this process to the mass scale involved (which prevents the cross section from exploding) was nicely demonstrated by the PEP4/9 measurement of the muonic structure function (Fig. 13a).[21] As quarks, however, are known to be non-free objects, the gluon corrections have to play an extremely important role. Very globally, the effect of the

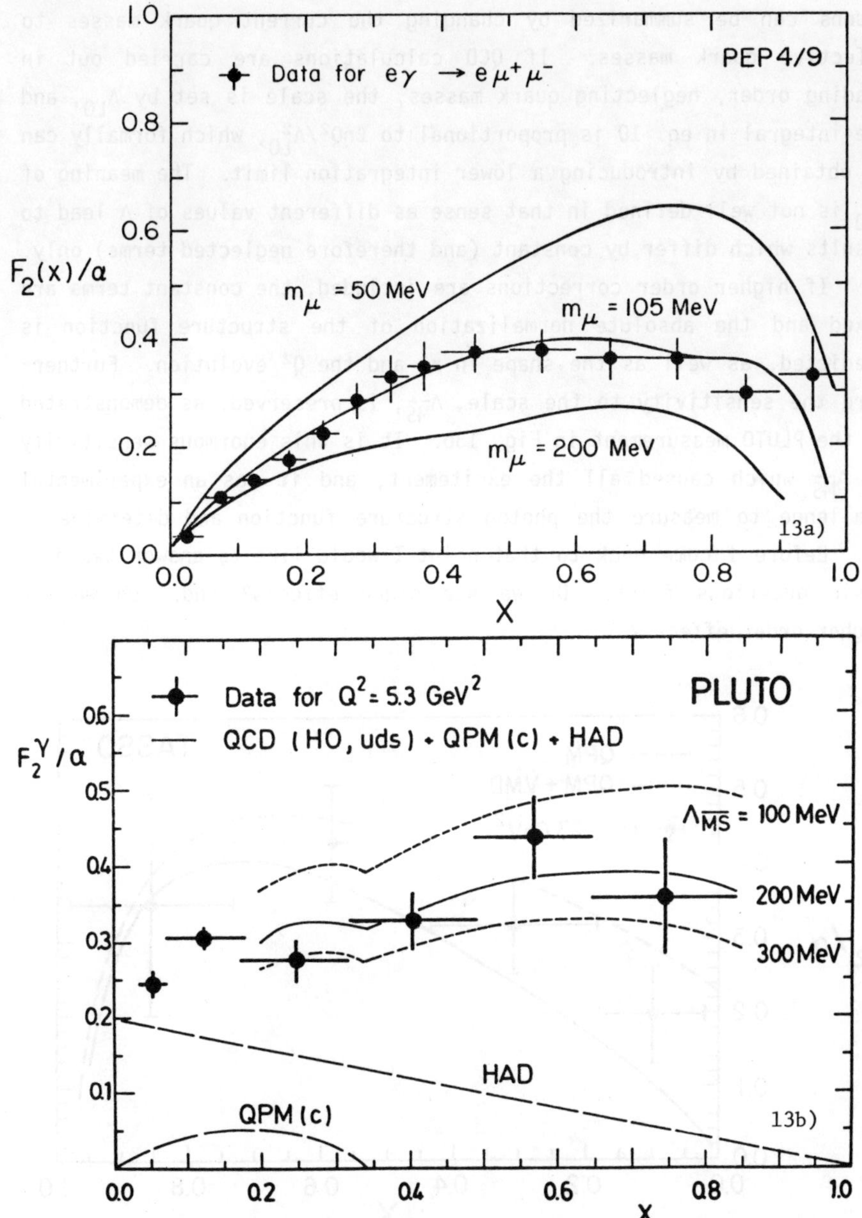

Figure 13.   Sensitivity of a) the leptonic structure function to m_μ and b) the hadronic structure function to $\Lambda_{\overline{MS}}$.

gluons can be summarized by changing the current quark masses to effective quark masses. If QCD calculations are carried out in leading order, neglecting quark masses, the scale is set by Λ_{LO}, and the integral in eq. 10 is proportional to $\ln Q^2/\Lambda_{LO}^2$, which formally can be obtained by introducing a lower integration limit. The meaning of Λ_{LO} is not well defined in that sense as different values of Λ lead to results which differ by constant (and therefore neglected terms) only.

If higher order corrections are included, the constant terms are fixed and the absolute normalization of the structure function is predicted, as well as the shape in x, and the Q^2 evolution. Furthermore the sensitivity to the scale, $\Lambda_{\overline{MS}}$, is preserved, as demonstrated by the PLUTO measurement in Fig. 13b. It is this enormous sensitivity to $\Lambda_{\overline{MS}}$ which caused all the excitement, and it was an experimental challenge to measure the photon structure function and determine Λ.

Before I come back to that point I would like to answer two more basic questions first: Do we see gluon effects? and: Do we see higher order effects?

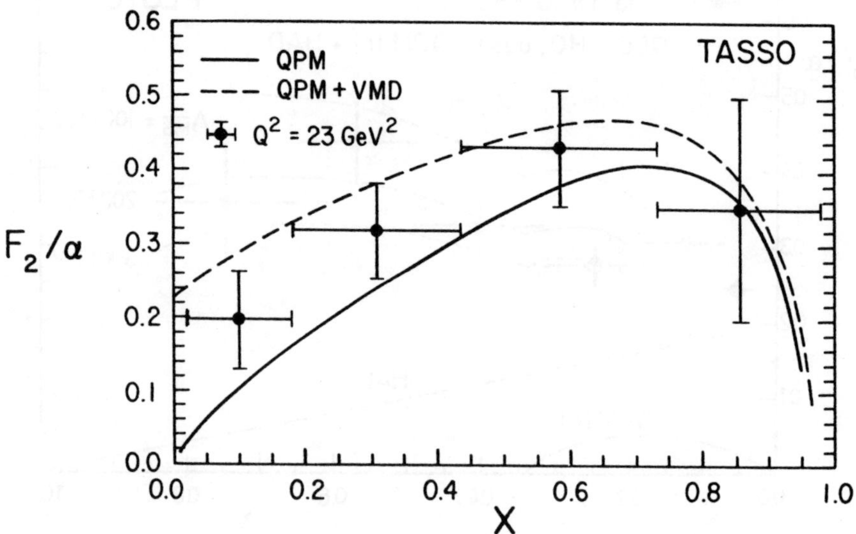

Figure 14. Comparison of the TASSO data to the QPM.

Figure 14 shows as an example the TASSO result at $Q^2 = 24$ GeV^2 compared to the QPM with and without VDM added. Obviously the QPM with effective quark masses of $m_q \sim 300$ MeV describes the data well if some VDM part is added. This statement also holds for all the other

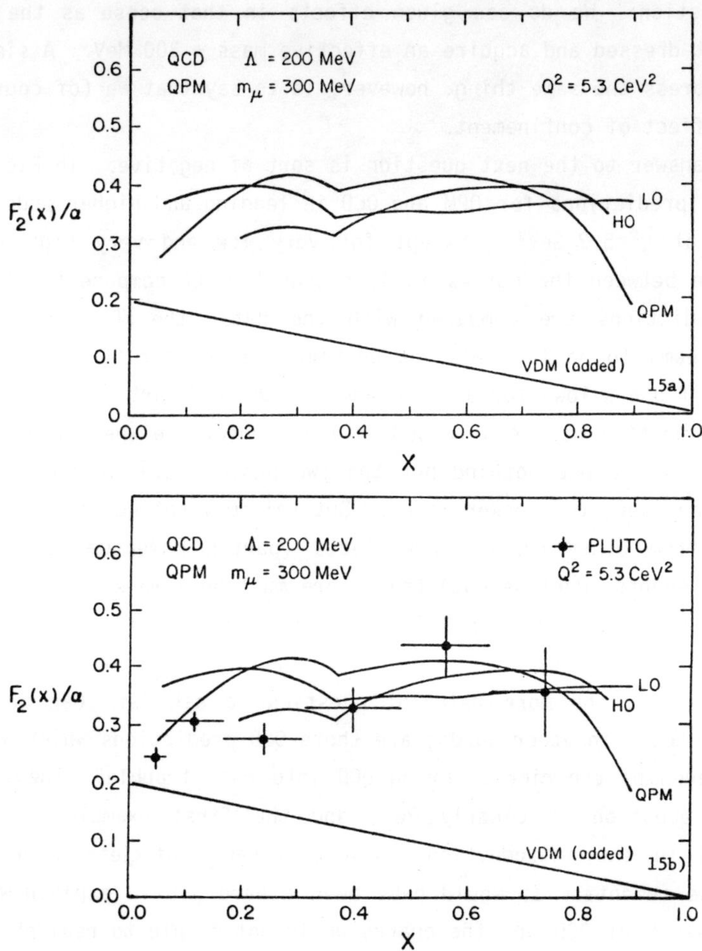

Figure 15.   Comparison of the QPM and QCD in leading and higher order. The PLUTO data favour the higher order (HO) but the differences between the predictions are less than the VDM contribution.

experiments, including the very precise measurements from PLUTO and PEP4/9 (see Fig. 15b). If, however, current quark masses are taken instead of effective masses, the QPM predictions are substantially higher by about a function of 2-3. This can be ruled out un-ambiguously by the experiments (see again Fig. 15b). This answers the first question: We do see gluon effects in that sense as the bare quarks get dressed and acquire an effective mass $\sim$ 300 MeV. A simpler way to express the same thing, however, is to say that we (of course!) see the effect of confinement.

The answer to the next question is sort of negative. In Fig. 15a the three predictions for QPM and QCD in leading and higher order are compared at $Q^2=5.3$ GeV^2. Except for very low and very high x the difference between the curves is less than the VDM component. If the three predictions are compared with the data, the HO calculation, indeed, seems to be favored, but neither the QPM nor QCD, LO can be excluded if one allows for a 30% change in the VDM part.

The question has been raised at this conference whether we would invent QCD if we had nothing but the two photon data to explain. As far as I can see, the answer is no. But this may not be the point, as we know from other experimental facts (parton momentum sum rules, three jet events, confinement) that there must be a more sophisticated theory than just the simple QPM, and QCD is the only theory that passed all tests so far.

Therefore, the more relevant question to ask is: Can we do a clean QCD test, in other words, are there QCD predictions which would, if disproven by experiment, bring QCD into real trouble? The answer to that question is clearly yes, and the first example has been mentioned in the introduction: the occurrence of jets with large transverse momenta. It would have been a hard job to explain within the framework of QCD why the quarks would not couple to real photons. But fortunately they do with about the rate predicted by QCD.

Similarly safe predictions exist for the structure function of the virtual photon if the target mass is much larger than Λ. There is a simple argument why one would anticipate no tremendous problems in this kinematic region: If the integral of eq. 10 is carried out, the large scale $p^2 = -m_\gamma^2$ totally screens the smaller scales Λ^2 and m_q^2 such that the result is $\sim \ell n\ Q^2/p^2$ and very insensitive to Λ and m_q. Thus the effect of the gluon corrections applied to a QPM with zero quark masses is only of the order of a few percent, whereas it is an infinite correction in the case of the real photon structure function. The PLUTO result for Q^2=5 GeV^2 and p^2=0.35 GeV^2 is shown in Fig. 16

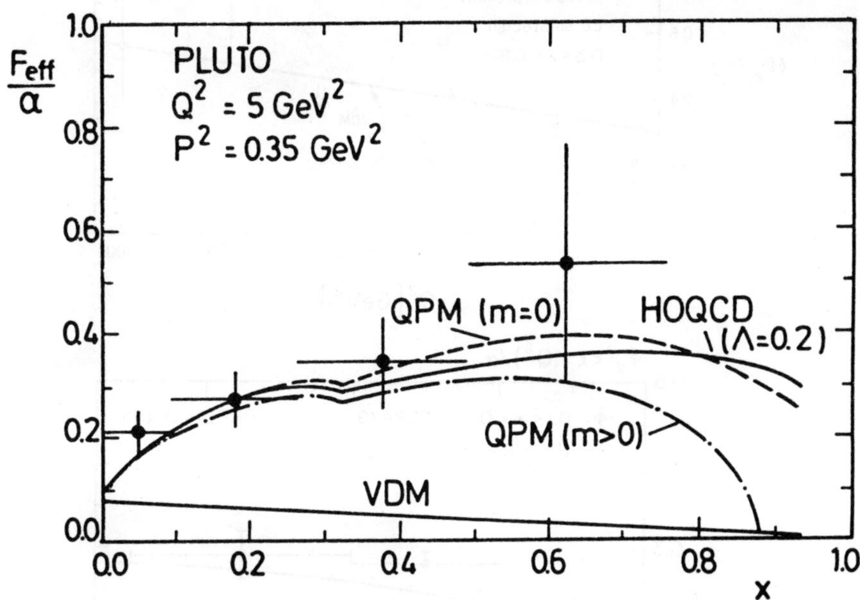

Figure 16.   The measured value of F_{eff}/α as a function of true x.   The data are compared with QCD (solid curve), QPM with constituent quark masses (dash-dotted curve) and QPM with massless quarks (dashed curve).   In each case a QPM c-quark contribution with constituent mass and a VDM contribution (see text) was added.   The VDM contribution is shown separately as the lower diagonal line.

and compared to HO QCD as well as to the QPM, using both effective quark masses and zero quark masses. All three models differ only very little, and are in good agreement with the data.

The third safe QCD prediction is the strong positive scale breaking with ℓnQ^2 at all values of x. This has clearly been observed by all experiments (see Fig. 9-11), and Fig. 17a shows the Q^2 evolution of one common x bin $0.3 < x < 0.8$. The data follow nicely

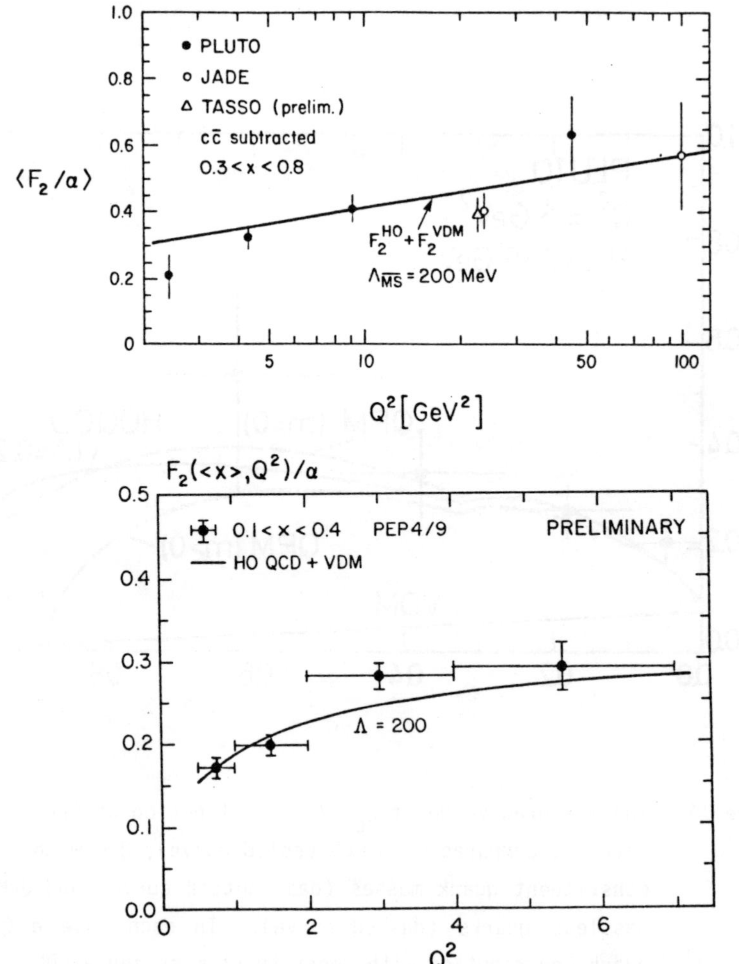

Figure 17. Q^2 evolution of the photon structure function in two bins of x. In both cases the curve represents a superposition of VDM and a higher order QCD calculation.

the predicted a+b ℓn Q^2 rise, with a slope, consistent with QCD. For the data at low Q^2 the high x-range is not accessible (the 2.5 GeV data in Fig. 17a suffer already from that effect) and Fig. 17b shows the low Q^2 PEP4/9 data, averaged over the available x range from 0.1 to 0.4 In the entire Q^2 range from 0.5 to 200 GeV^2 the data do not only follow the predicted slope, but also agree in the absolute normalization with QCD, HO plus VDM, using $\Lambda_{\overline{MS}} \sim 200$ MeV. In all the three cases mentioned here, a substantial deviation from the predictions would have caused a mild disaster for QCD. On the other hand in all three cases the QCD predictions almost coincide with the QPM predicitions, which means that the gluon corrections are small and the calculations safe. Thus these tests are excellent tools to falsify the theory but if they come out o.k. we do not learn more about QCD and their intrinsic parameters.

The obvious next step to do is to try to gain information about QCD from the data, which means a determination of Λ, or α_s respectively. The discussion at this workshop about that point was enthusiastic, partly emotional, but almost always theoretical. The situation is summarized by W. Bardeen[12], and I will repeat only some theoretical points which are needed for the further discussion.

In the framework of the operator product expansion and the renormalization group treatment we get for the moments of the photon structure function

$$F_n^\gamma(Q^2) = \int_0^1 x^{n-2} F_2^\gamma(x,Q^2)dx$$

$$F_n^\gamma(Q^2) = \sum_{\pm,NS} A_n^i(\mu^2) \left[\frac{\alpha_s(Q^2)}{\alpha_s(\mu^2)}\right]^{d_n^i}$$

$$+ \frac{1}{\alpha_s(Q^2)} \sum \frac{a_n^i}{d_n^i+1} \left\{1 - \left[\frac{\alpha_s(Q^2)}{\alpha_s(\mu^2)}\right]^{d_n^i+1}\right\} \tag{11}$$

$$+ \sum \frac{b_n^i}{d_n^i} \left\{1 - \left[\frac{\alpha_s(Q^2)}{\alpha_s(\mu^2)}\right]^{d_n^i}\right\} + C_n$$

This equation predicts the Q^2 evolution from some arbitrary renormalization point μ^2

$$F_n^\gamma(\mu^2) \;=\; \Sigma\; A_n^i(\mu^2) + C_n \tag{12}$$

to infinity, where the dominant term for $n > 2$ is given by

$$F_n^\gamma(Q^2 \to \infty) \;=\; \frac{1}{\alpha_s(Q^2)}\; \Sigma\; \frac{a_n^i}{d_n^i+1} + \Sigma\; \frac{b_n^i}{d_n^i} + C_n \tag{13}$$

Rearranging the terms in eq. 11 according to their large Q^2 behavior

$$F_n^\gamma(Q^2) \;=\; \{\Sigma\; A_n^i(\mu^2) - \frac{1}{\alpha_s(\mu^2)}\; \Sigma\; \frac{a_n^i}{d_n^i+1} - \Sigma\; \frac{b_n^i}{d_n^i}\}\; [\frac{\alpha_s(Q^2)}{\alpha_s(\mu^2)}]^{d_n^i}$$

$$+ \frac{1}{\alpha_s(Q^2)}\; \Sigma\; \frac{a_n^i}{d_n^i+1} + \Sigma\; \frac{b_n^i}{d_n^i} + C_n$$

$$= \; \Sigma\; \hat{A}_n^i(\mu^2) \cdot [\frac{\alpha_s(Q^2)}{\alpha_s(\mu^2)}]^{d_n^i} + \frac{1}{\alpha_s(Q^2)}\; \Sigma\; \frac{a_n^i}{d_n^i+1} + \Sigma\; \frac{b_n^i}{d_n^i} + C_n \tag{14}$$

does not only separate nicely the leading ('asymptotic') and non-leading ('hadronic') parts of the structure function, but, more important, shows a problem that goes along with the separation: Equation 11 is well defined if the anomalous dimensions d_n^i cause zeroes in the denominators at $d_n^i = 0,-1$, as

$$\lim_{d \to o}\; \frac{1}{d}\; (1 - x^d) \;=\; -\,\ell n\; x \;. \tag{15}$$

Separating the two terms $1/d$ and x^d/d, as it is done in eq. 14, artificially introduces poles in both, the asymptotic and the hadronic part, which exactly cancel each other by definition. The situation very much resembles the situation of radiative corrections in QED where the infrared divergency of soft radiation is cancelled by vertex corrections (Fig. 18).

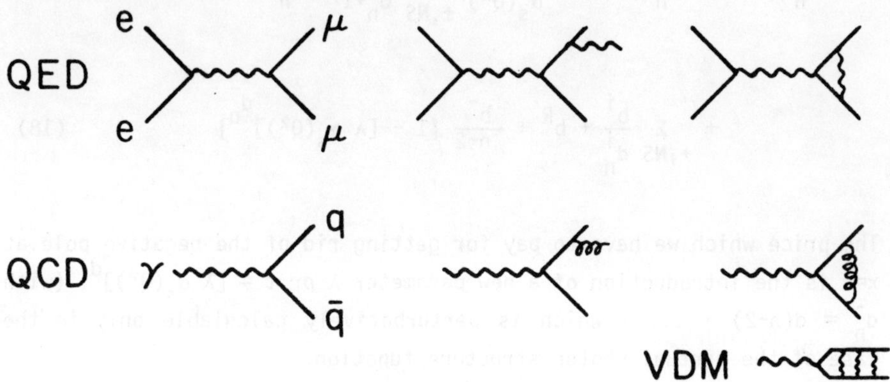

Figure 18.   Radiative corrections in QED and QCD.

A concept for removing the poles from the asymptotic part back to the hadronic part (where they get eaten) has been proposed by Bardeen and worked out by Antoniadis and Grunberg[10] for the most troublesome $\bar{d_2}$ pole, which gives rise to the negative structure function at low x. The basic idea is to split the term $\dfrac{\bar{b_n}}{\bar{d_n}}$, which has the pole structure $\bar{b}/(n-2)$ into a regular and a non-regular piece

$$\frac{\bar{b_n}}{\bar{d_n}} = b_n^R + \frac{\bar{b}}{n-2}$$

$$= b_n^R + \frac{\bar{b}}{n-2} (1 - [\lambda \, \alpha_s(Q^2)]^{\bar{d_n}}) + \frac{\bar{b}}{n-2} [\lambda \, \alpha_s(Q^2)]^{\bar{d_n}} \qquad (16)$$

Reabsorbing the third term into the hadronic part

$$F_n^{HAD}(Q^2) = \{ \sum_{\pm,NS} \hat{A}_n^i(\mu^2) + \frac{\bar{b}}{n-2} [\lambda \, \alpha_s(\mu^2)]^{\bar{d_n}} \} \, [\frac{\alpha_s(Q^2)}{\alpha_s(\mu^2)}]^{d_n^i} \qquad (17)$$

leaves us for each value of λ with the sum of two well-behaving pieces

$$F_n^\gamma(Q^2) = F_n^{HAD}(Q^2) + \frac{1}{\alpha_s(Q^2)} \sum_{\pm,NS} \frac{a_n^i}{d_n^i+1} + C_n$$

$$+ \sum_{+,NS} \frac{b_n^i}{d_n^i} + b^R + \frac{b^-}{n-2} \{1 - [\lambda \, \alpha_s(Q^2)]^{\bar{d}_n}\} \tag{18}$$

The price which we have to pay for getting rid of the negative pole at x=0 is the introduction of a new parameter λ or $t = [\lambda \, \alpha_s(Q^2)]^{\bar{d}}$, (with $\bar{d}_n = d(n-2) + \ldots$) which is perturbatively calculable only in the case of the virtual photon structure function.

At $p^2 = 0$ it has to be regarded as an additional non-perturbative parameter, which has to be determined from the experiment. Figure 19 shows the result for the unregularized case and for two values of t. It is interesting to see that the choice of t mainly influences the low x range and leaves the result for x > 0.3 almost unaffected. This would allow to use the low x data to fit t and the high x data to fit Λ. But we have to be careful with that because the F_n^{HAD} term in eq. 18 contains the (perturbatively) not calculable hadronic matrix elements of the photon. If we are not willing to make any assumptions about the F_2^{HAD} piece we obviously loose the ability to determine Λ. In a way the hadronic part screens Λ, as the main effect in going from one value of Λ to another one is adding a Q^2 independent piece, which can be reabsorbed in the hadronic piece.

One reasonable estimate of the hadronic piece is obtained in the VDM from the pion structure function, which was fitted by the two parameterizations[20]:

$$f_{u/\pi^-}(x) = \begin{cases} 0.52 \, (1-x)^{1.02} \\ 0.9 \, \sqrt{x} \, (1-x)^{1.27} \end{cases} \tag{19}$$

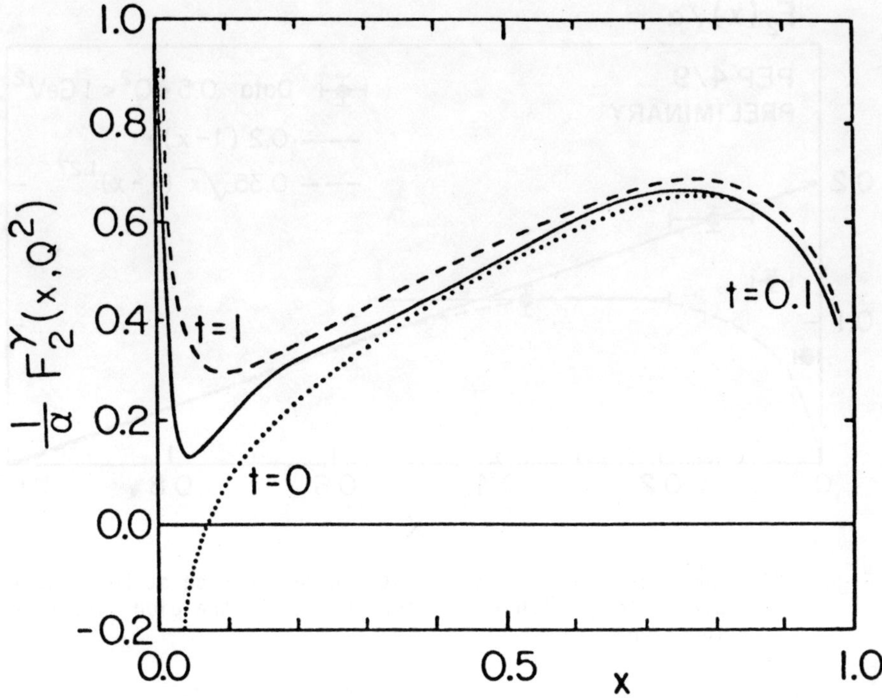

Figure 19.   The result of Antoniadis and Grunberg for the unregularized case (t=0, dotted line) and two values (t=0.1, solid
line and t=1.0, dashed line) of the regularization
parameter.

Applying the proper charge factors this leads to the two photon
structure functions shown in Fig. 20.   The comparison to the data at
the lowest available Q^2 shows that the data are somewhere in between
these two parameterizations.   Thus the first solution of eq. 19 (which
leads to eq. 7) can be regarded to be an upper limit of the hadronic
part.   On the other hand we know from the jet analysis that for Q^2
values below 1 GeV^2 the final state is dominated by a VDM type

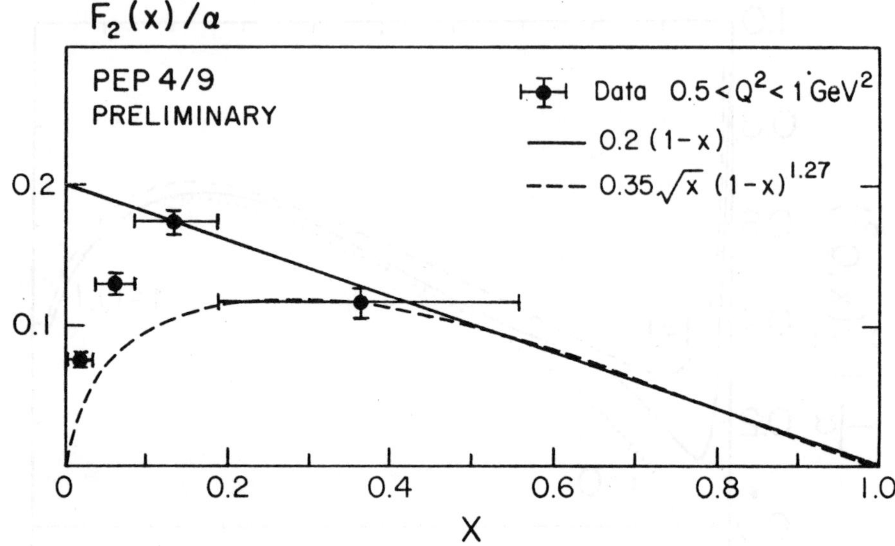

Figure 20.   Comparison of the photon structure function at the lowest available Q^2 with two different fits to the pion structure function.

structure (Fig. 4).   It therefore seems to be not unreasonable to take eq. 7 as an assumption for the hadronic part and fit Λ.   In the x region which is sensitive to Λ, it does not matter which of the two forms of eq. 19 is used.

The results are shown in Fig. 21 for the PLUTO experiment for three values of Q^2, where the parameters t and Λ were simultaneously fitted.   The TASSO data (Fig. 22) are fitted by a one parameter fit only, assuming t=0.   JADE uses t=1 for the Λ determination (Fig. 23). Instead of the fit result JADE shows the predictions for two values of Λ.   All experiments are in good agreement with the predictions and the resulting values for Λ and t are summarized in Table 2.   Taking the average yields

$$\Lambda_{\overline{MS}} = (230 \pm 40) \text{ MeV} \tag{20}$$

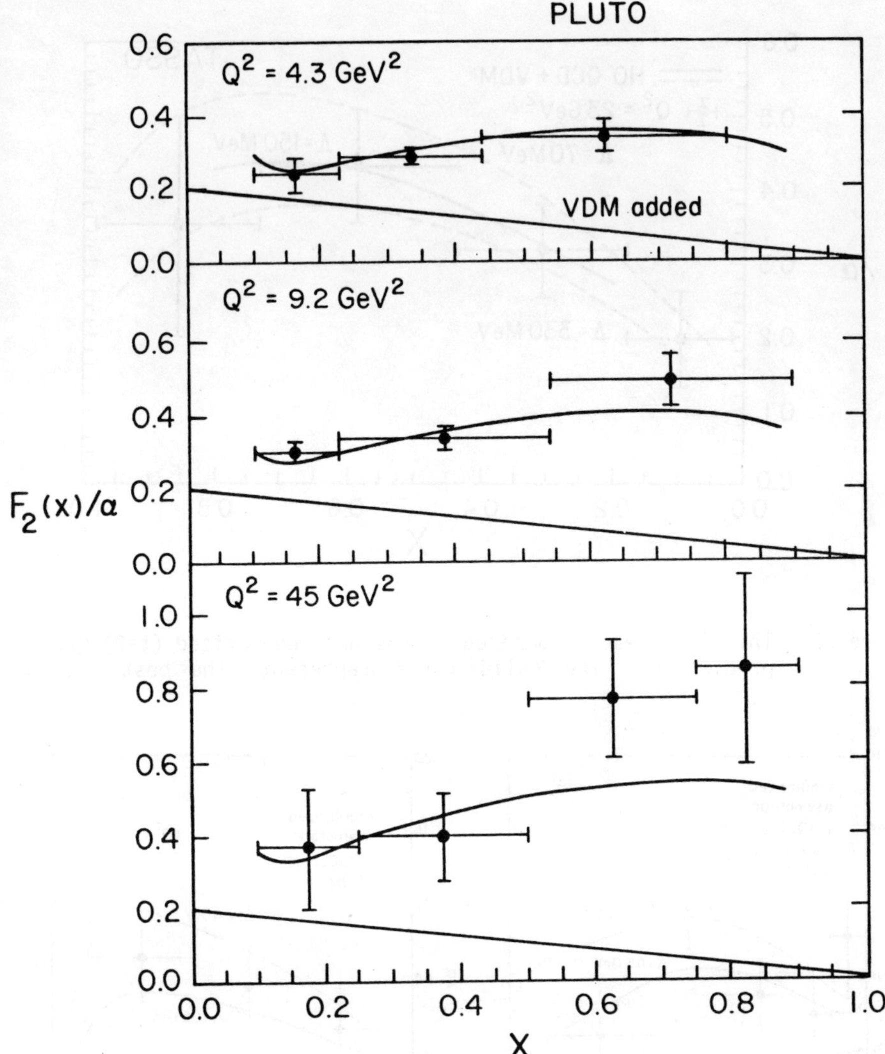

Figure 21. Comparison of the PLUTO results with the regularized higher order QCD prediction, with a VDM contribution added. Λ and t are fitted simultaneously.

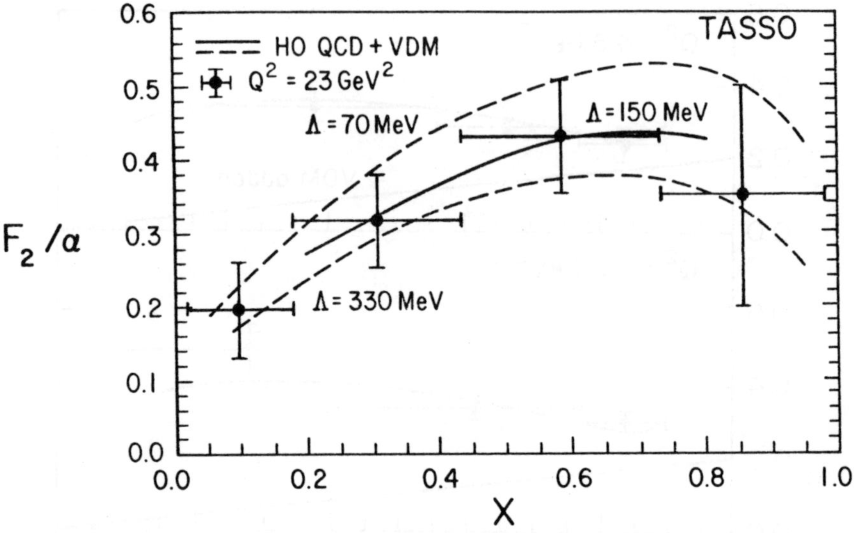

Figure 22.   The TASSO result compared to the not regularized (t=0) QCD prediction.   The solid curve represents the best fit.

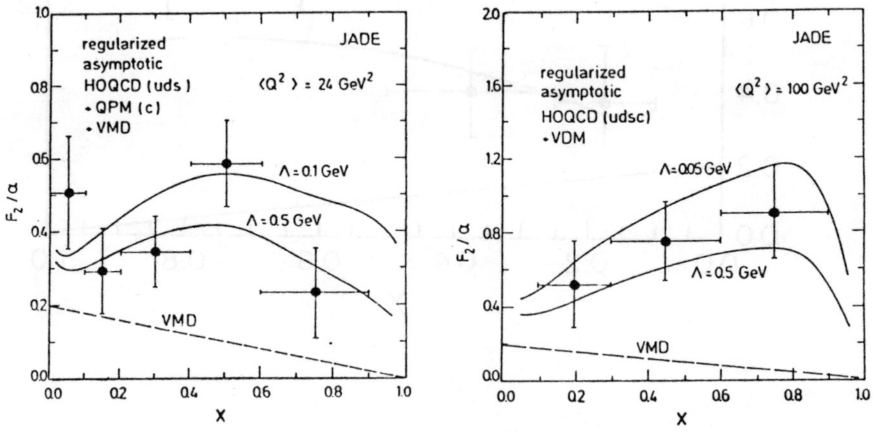

Figure 23.   Comparison of the JADE result with regularized QCD predicttions.   t is not fitted but set to t = 1.

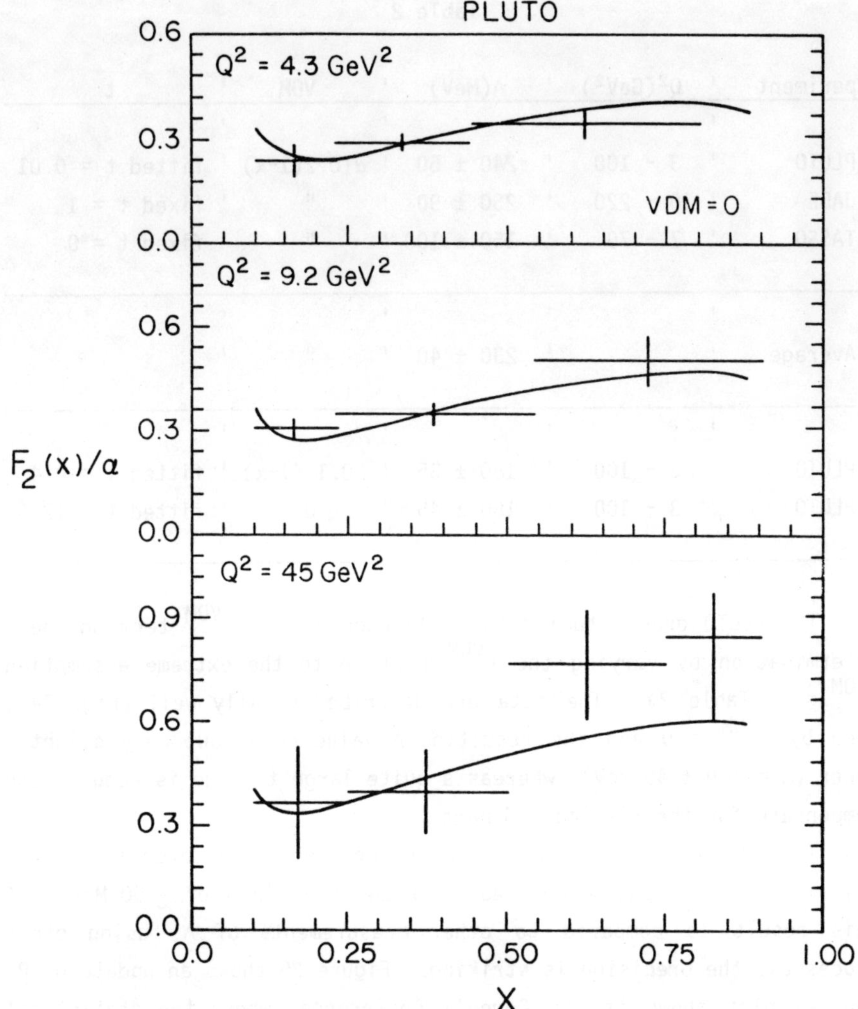

Figure 24. Comparison of the PLUTO results with the regularized higher order QCD prediction. Λ and t are fitted simultaneously.

Table 2

Experiment	Q^2 (GeV^2)	Λ(MeV)	VDM	t
PLUTO	3 - 100	240 ± 50	$\alpha \cdot 0.2(1-x)$	fitted t = 0.01
JADE	10 - 220	250 ± 90	"	fixed t = 1
TASSO	7 - 70	150 ± 100	"	fixed t = 0
Average		230 ± 40	"	
PLUTO	3 - 100	180 ± 35	0.1 (1-x)	fitted t = 0.4
PLUTO	3 - 100	160 ± 45	0	fitted t = 12.6

The PLUTO group studied the influence of the F^{VDM} term on the Λ determination by varying the F^{VDM} part up to the extreme assumption $F^{VDM} = 0$ (Table 2). The data are described equally well (Fig. 24), even by $F^{VDM} = 0$ and the resulting Λ value comes out only slightly lower ($\Lambda = 160 \pm 45$ MeV), whereas a quite large t value is required to compensate for the missing VDM part.

In addition to the statistical error there is an overall systematical error of 10%, which leads to an error in Λ of ~ 50 MeV. If this result is compared to other measurements of Λ, using other processes, the precision is striking. Figure 25 shows an update of P. Lepages plot shown at the Cornell Conference, where the statistical and systematical errors are added in quadrature ($\Lambda_{\overline{MS}} = 230 \pm 70$ MeV).

V. Conclusions

In conclusion, we see deep inelastic electron photon scattering, setting on at $Q^2 \sim m_\rho^2$. We do see the presence of the photons point-like component at <u>all</u> values of Q^2. We even see the dominance of this component at large Q^2 ($Q^2 > 10$ GeV) by looking at the x shape of the

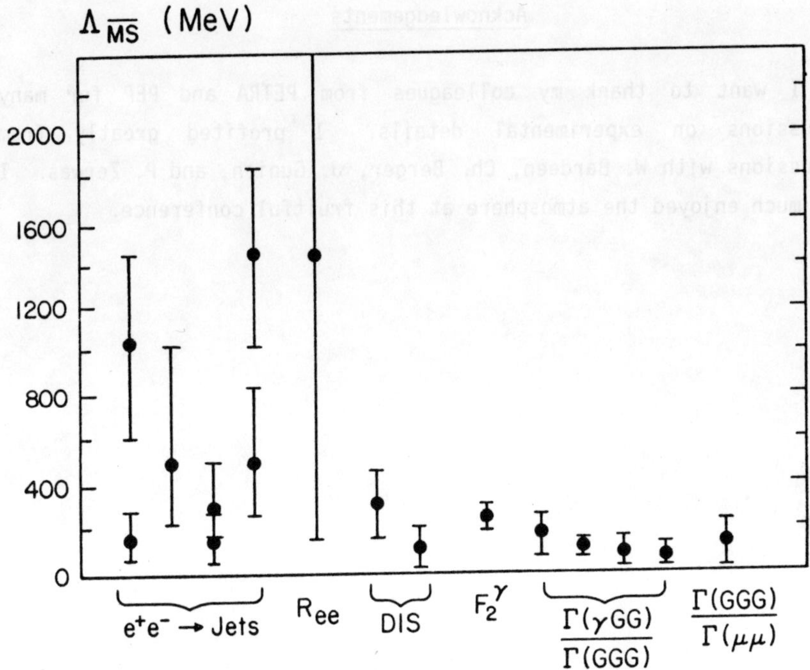

Figure 25.   Comparison of $\Lambda_{\overline{MS}}$ values, determined in different experiments.

structure function as well as the jet structure of the final state. The analysis of the virtual photon structure function and the Q^2 evolution of the real photon structure function offer clear and non-trivial tests of QCD, but with almost no sensitivity to Λ. Using the regularization scheme of Antoniadis and Grunberg we get excellent agreement of the data with next to leading order QCD. Using the data for $Q^2 > 3$ GeV^2, we can extract Λ from the absolute normalization and the x shape of $F_2^{\gamma}(x)$. The result (including systematical errors) of $\Lambda_{\overline{MS}} = (230 \pm 70)$ MeV turned out to be quite insensitive to the assumption of the hadronic component. If, e.g., F^{VDM} is set to zero, Λ changes by only 70 MeV! Thus the problems caused by the non-calculable hadronic pieces seem to be controllable if the whole information, including jet topology and low Q^2 data, are used.

Acknowledgements

I want to thank my colleagues from PETRA and PEP for many discussions on experimental details. I profited greatly from discussions with W. Bardeen, Ch. Berger, J. Gunion, and P. Zerwas. I very much enjoyed the atmosphere at this fruitful conference.

References

1. W. Wagner, Proc. Int. Conf. on High Energy Physics, Madison (1980) Vol. 2, 576.

2. V. Blobel, to be published.
   A. Baecker, invited talk at this conference.

3. PLUTO Coll., Ch. Berger et al., Phys. Lett. 107B (1981) 168.
   G. Knies, invited talk at this conference.

4. F. Foster, invited talk at this conference.

5. PLUTO coll., Ch. Berger et al., Phys. Lett. 107B (1981) 168.

6. See e.g., H. Spitzer, Proc. International Workshop on Photon Photon Collisions, Aachen (1983) 318.

7. T.F. Walsh and P. Zerwas, Phys. Lett. 443 (1973) 195.
   V.M. Budnev et al., Pnys. Rev. C15 (1975) 182.

8. C. Petersen, T.F. Walsh, and P. Zerwas, Nucl. Phys. B229 (1983) 301.
   W. Wagner, Aachen Report PITHA 83/03.

9. E. Witten, Nucl. Phys. B120 (1977) 189.

10. I. Antoniadis and G. Grunberg, Nucl. Phys. B213 (1983) 445.

11. I. Antoniadis, this conference, summarized by P. Zerwas.

12. W.A. Bardeen, invited talk at this conference.

13. W.A. Bardeen and A.J. Buras, Phys. Rev. D20 (1979) 166.

14. M. Glück and E. Reya, Phys. Rev. D28 (1983) 2749.

15. JADE coll., W. Bartel et al., DESY report 84-042 (1984).

16. PLUTO coll., Ch. Berger et al., Phys. Lett. 1423 (1984) 111.

17. J. Smith, this conference, summarized by W. Ko.

18. TASSO coll., F.J. Kirschfink, Aachen internal report (thesis).

19. PLUTO coll., Ch. Berger et al., to be published.

20. C.B. Newman et al., Phys. Rev. Lett. 42 (1979) 951.

21. PEP-9 coll., M. Cain, et al., Phys. Lett. 147B (1984), 232.

Questions and Discussion

H. KOLANOSKI (U. of Bonn): Does the ratio $\tilde{R}_{\gamma\gamma}$ which you showed for different Q^2 ranges correspond to the total cross section or only to the jet cross section? In the former case one needs also information about the event topology, if one wants to conclude consistency with the QPM.

W. WAGNER: The ratio $\tilde{R}_{\gamma\gamma}$ corresponds to the total cross section above 4 GeV invariant mass and charged multiplicity $\geq$ 4. Therefore topological quantities like thrust distributions or jet multiplicities have been studied in addition to get a coherent picture. The implication for the structure function is that, due to the 4 GeV cut in W, we are studying the low x region, where we want to get information about the VDM part.

A. CORDIER (LAL Orsay): Could you evaluate the uncertainty on the Λ measurement due to the unfolding method?

W.W.: The error in Λ of ± 40 MeV contains already the statistical error and the systematics, introduced by the unfolding method. This part of the systematical error is much smaller than the systematics introduced by fragmentation.

J. GUNION: Wouldn't it be possible to bin your $\tilde{R}_{\gamma\gamma}$ plots vs. p_T simultaneously in x and Q^2? Theoretically we expect at small x there must be some excess over pointlike associated with the hadronic regularization of the n=2 singularity in b_n. It would be very interesting to study in this way the x range over which this regularization occurs.

W.W.: That's a very good suggestion. I think it is possible and it will be done.

S. BRODSKY (SLAC): Considering the complexity of hadronic structure functions (multiple Fock states, etc.) it is unlikely that <u>all</u> of the hadronic structure function can be presented by a single parameter, as in the Antoniadis et al., model. In higher order the form of the parameterization may differ and it is not clear at what x these connections can be rigorously neglected. Nevertheless the empirical stability of the parameterization is encouraging.

W.W.: As far as I understand the regularization scheme, there is still a not calculated piece left, coming from the hadronic matrix elements of the photon. I hope that we will get reasonable bounds on this piece in the near future from the data.

J. FIELD (L.P.N.H.E.): Did all the experiments you quoted in your list of Λ determinations take into account radiative corrections?

W.W.: Yes, all experiments correct for target mass effects and radiation. Only for the low Q^2 PLUTO data these two corrections were found to be small and opposite, so that no correction was applied.

P. ZERWAS (RWTH Aachen): How much of the Q^2 evolution is due to $c\bar{c}$ threshold effects?

W.W.: In the plot of the Q^2 evolution in the fixed x range $0.3 < x < 0.8$ the charm contribution has been subtracted. This correction is quite substantial at high Q^2 (30%).

THE TOTAL CROSS SECTION $\gamma\gamma \to$ HADRONS

Gerhard Knies

DESY
2000 Hamburg 52
Germany

ABSTRACT

Measurements of the total cross section for $\gamma\gamma \to$ hadrons are presented. There are new results from double-tagging ($P^2 \neq 0$, $Q^2 \neq 0$, single-tagging ($P^2 \approx 0$, $Q^2 \neq 0$), and no-tagging ($P^2 \approx 0$, $Q^2 \approx 0$) experiments at e^+e^- storage rings. The measurements cover a Q^2 range from 0 to 100 GeV^2, and a W range from 2 to 20 GeV. The significance of these data for the interpretation of the photon as a set of vector mesons (VDM) and as an electromagnetic field quantum coupling to point-like quarks (QPM) is discussed.

OUTLINE:

1.   Introduction

2.   Predictions for $\sigma_{\gamma\gamma}^{tot}$
2.1 The point-like and the VDM nature of the photon
2.2 Duality and double counting
2.3 Is $e\gamma$ like ep?

3.   What can be measured in $e^+e^- \to e^+e^-$ + hadrons?
3.1 Double-tag mode
3.2 Single-tag mode
3.3 No-tag mode

4.   Experiments
4.1 Experimental setups
4.2 Background rejection

5.   Determination of cross section
5.1 Q^2 (P^2) interpolation
5.2 W_{vis} correction
5.3 Acceptance correction
5.4 Hadronic final state model
5.5 The unfolding procedure

6.   Results
6.1 Analysis of ϕ dependence
6.2 W-dependence at different P^2, Q^2
6.3 Point-like and hadron-like contributions

7.   Summary and conclusions

1. Introduction

Much has already been said about the total cross section $\sigma^{tot}($ W, $Q^2 = -\gamma_1^2$, $P^2 = -\gamma_2^2)$ for

$$\gamma_1\gamma_2 \to \text{hadrons}$$

in the two preceeding talks[1,2] in terms of the hadronic structure function F_2^{had} of the photon, via the relation

$$\sigma^{tot} (W, Q^2 \gg 0, P^2 \approx 0) = \sigma_{TT} + \sigma_{LT} = \frac{4\pi^2\alpha}{Q^2} F_2^{had} \quad (x = \frac{Q^2}{Q^2+W^2}, Q^2) \quad (1)$$

Has all been said?

No, since firstly in the kinematic region $(P^2 > 0, Q^2 > 0)$, that is for double-tag data relation (1) is no longer a valid approximation. The total cross section here has the form[3] (see section 3 for definitions and Fig. 1 for notations):

$$\sigma^{tot} = \sigma_{TT} + \sigma_{LT} + \sigma_{TL} + \sigma_{LL} \quad (2)$$

while the structure function F_2 is related[4] to the cross section by

$$F_2(x,Q^2,P^2) = \frac{Q^2}{4\pi^2\alpha} (1-Q^2P^2 \nu^2)^{-1/2} \{\sigma_{TT} + \sigma_{LT} - \frac{1}{2} \sigma_{TL} - \frac{1}{2} \sigma_{LL}\} \quad (3)$$

Secondly, for $Q^2 \ll 1$ GeV^2 and - as always - $P^2 < Q^2$, relation (1) becomes useless, since the predictions for F_2^{had} become meaningless. The reason is that QCD predictions for F_2^{had} are based on perturbative calculations of the photon splitting process[5]:

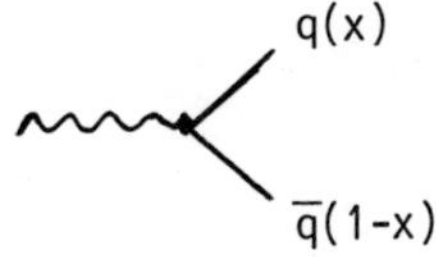

and on a probabilistic interpretation of q(x) as the quark distribution function:

$$F_2(x) = x \cdot q(x) \tag{4}$$

Since this $q^{QCD}(x)$ is proportional to $\ln Q^2/\Lambda^2$, it turns negative for $Q^2 < \Lambda^2$, in obvious conflict with the probabilistic interpretation. Heuristically this means that for $Q^2 \lesssim \Lambda^2$ the probing photon no longer resolves in space the virtual quark states, to an extent that incoherent interaction of the quark and antiquark with the probing photon becomes impossible. However, the total cross section remains a meaningful physics quantity also at $Q^2 \to 0$.

Previous measurements of the total $\gamma\gamma$ cross section[6] at low Q^2 ($Q^2 \lesssim 1$ GeV) are largely in agreement with the VDM expectation[7] for virtual vector meson vector meson collisions and are much larger than predicted by the (perturbative) calculation in the quark parton model (QPM). At high Q^2 ($Q^2 \gtrsim 20$ GeV^2) the cross section predicted through perturbative F_2^{had} calculations[7] accounts approximately for the measured total cross section. The notion of a cross section is therefore more useful than that of a structure function in studying the transition of the photon from a set of vector mesons to an electromagnetic field quantum with point-like couplings to quarks, since there are predictions in the full Q^2 region.

In this talk I will present and discuss results on cross section measurements in double-tag experiments ($P^2 > 0$), and from no-tag and single-tag experiments with $P^2 \approx 0$ and Q^2 from ≈ 0 to ≈ 100 GeV^2. In the latter case, the evolution of σ^{tot} with Q^2 allows us to study the interplay of the vector mesonic and point-like nature of the photons.

2. Predictions for $\sigma_{\gamma\gamma}^{tot}$ (W, Q^2, P^2)

There are two extreme types of predictions, corresponding to treating the photon as an electromagnetic field quantum interacting with the charge of point-like quarks:

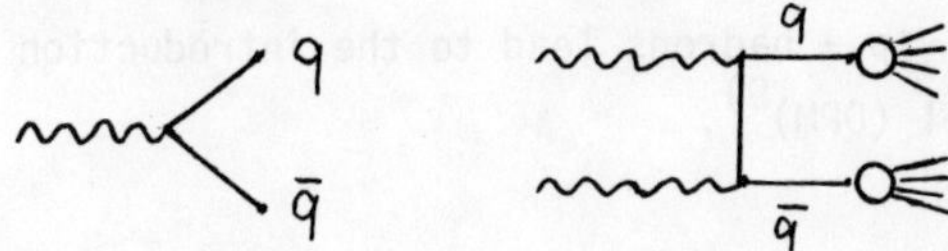

or replacing the photon by a hadronic current, saturated by vector
mesons:

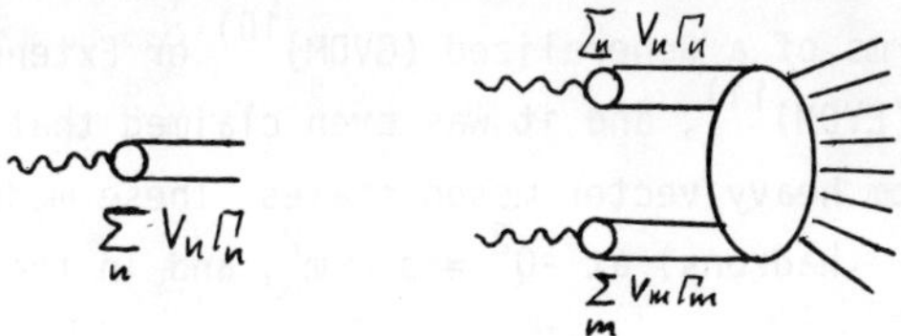

In the first approach $\sigma(W, Q^2, P^2)$ can be evaluated in terms of the
QPM[3] if all gluon radiation corrections are ignored, by introducing
quark masses and charges as the only unknown parameters. In the VDM
approach, in the decomposition of the hadronic current into vector
mesons, the vector meson masses m_V, the γ-V couplings and the cross
section $\sigma^{tot}(VV')$ have to be specified.

As long as full QCD calculation for $\gamma\gamma \to$ hadrons is not available
it is of interest how these two approaches can be combined.

2.1 Brief review of experimental facts in support of the point-like
     and of the VDM hadronic nature of the photon

   a) Q^2 time-like: here the total cross section $\sigma(e^+e^- \to \gamma \to$ hadrons)
      gives support for either picture, as alternatives:
      at $Q^2 \approx m^2_{\rho,\omega,\phi}$: γ like VDM
      at $Q^2 >> m^2_\rho$: γ couples to point-like quarks in the QPM
      approximation
   b) $Q^2 = 0$: Measurements on $\frac{d\sigma}{dt}$ $(\gamma N \to N + \rho,\omega,\phi)$ and σ_{tot}
      $(\gamma p \to$ hadrons) confirmed the Vector Dominance Model (VDM) in
      a quantitative way.

c) Q^2 space like: The deep inelastic inclusive ep scattering experiments i.e. $\gamma^*p \to$ hadrons lead to the introduction of the Quark Parton Model (QPM)[9].

2.2 Duality and double counting

However, it was also possible to describe or parametrize the deep inelastic ep data in terms of a Generalized (GVDM)[10] or Extended Vector Dominance Model (EVDM)[11], and it was even claimed that the Q^2 dependence did arise from heavy vector meson states. These models also described $\sigma^{tot}(e^+e^- \to \gamma \to$ hadrons) at $-Q^2 = s \gg m_\rho^2$, and in the resonance region.

It was obvious that the QPM and the (E or G) VDM parametrizations should not be added, under these circumstances. Sakurai[8] therefore proposed a new duality which is expressed in the following FESR:

$$\int_{4m_\pi^2}^{S_{max}} s \cdot \sigma(e^+e^- \to \gamma^* = \sum_n V_n \cdot \Gamma_n \to \text{hadrons})\, ds$$

$$= \int_{S_o}^{S_{max}} s \cdot (e^+e^- \to \gamma^* = \sum_q q\bar{q} \cdot \alpha_q \to \text{hadrons})\, ds \tag{5}$$

It says that the VDM hadronic and the point-like picture of the photon are equivalent on average. Therefore the combination of both:

$$\sigma^{tot}(e^+e^- \to \text{had}) = \sigma(\text{QPM}) + \sigma((\text{G,E})\text{VDM}) \tag{6}$$

is double counting in this picture.

2.3 Is eγ like ep?

Since ep scattering is well described by the VDM approaches eγ

scattering may be as well. In fact if the transition of a photon into hadrons can be fully described by a coupling of the photon to an infinite set of (extended) vector mesons V_n, with an appropriate choice of couplings Γ_n and masses m_n, then $e\gamma$ scattering can be linked to ep in the following way:

$$\sigma^{tot}(\gamma^*p) = \sigma^{tot}((\sum_n \Gamma_n V_n)p) \tag{7a}$$

$$\sigma^{tot}(\gamma^*\gamma) = \sigma^{tot}((\sum_n \Gamma_n V_n)(\sum_m \Gamma_m V_m)) \tag{7b}$$

Relating the (virtual) vector meson proton cross section $\sigma(V_n P)$ to $\sigma(V_n V_m)$ allows then to predict $\sigma(e\gamma)$ from $\sigma(ep)$.

In ep scattering, the target is an extended hadron, however, and therefore the VDM parametrization of the probing photon may be appropriate to describe the Q^2 dependence of the ep cross section. In $e\gamma$ scattering the target is not necessarily an extended hadron. Therefore, VDM parametrizations of the target and probing photons may not be appropriate. Furthermore, EVDM, GVDM and QPM predict different Q^2 dependences of $\sigma^{tot}_{\gamma\gamma}$ on average. So these approaches <u>cannot</u> be equivalent here in a dual sense. The W dependences of $\sigma(W, Q^2, P^2 \approx 0)$ differ as well, since the QPM predicts no W independent term at small Q^2, but the EVDM does. A measurement of the W and Q^2 dependences of $\sigma^{tot}_{\gamma\gamma}(W, Q^2)$ is therefore suitable to study the roles of the point-like and the VDM nature of the photon.

3. What can be measured in $e^+e^- \to e^+e^- +$ hadrons?

Using the notation of Fig. (1) and following ref. 3 we here briefly summarize the most important cross section expressions for the reaction

$$e^+e^- \to e^+e^- + \text{hadrons} \tag{8}$$

284

3.1 Double-tag mode

$$d\sigma_{ee} = \{\Gamma_{TT} \cdot \sigma^{eff} + \frac{1}{2}\,\varepsilon_1\varepsilon_2\tau_{TT} \cdot \cos 2\phi \tag{9}$$

$$+ \, 2\sqrt{\varepsilon_1(\varepsilon_1+1)\varepsilon_2(\varepsilon_2+1)}\;\tau_{TL}\;\cos\phi\}\;\frac{d^3p_1}{E_1} \cdot \frac{d^3p_2}{E_2}$$

with $\sigma^{eff} = \sigma_{TT} + \varepsilon_2 \cdot \sigma_{TL} + \varepsilon_1\sigma_{LT} + \varepsilon_1\varepsilon_2\sigma_{LL}$ (10)

σ and τ are $\gamma\gamma$ cross sections and interference terms, T,L indicate transverse and longitudinal photons

$$\sigma_{TT} = \frac{1}{2}\,(\sigma_{\parallel} + \sigma_{\perp}) = \frac{1}{2}\,(\sigma_0 + \sigma_2) \tag{11}$$

$$\tau_{TT} = \sigma_{\parallel} - \sigma_{\perp} \tag{12}$$

with $\sigma_{\parallel}$ and $\sigma_{\perp}(\sigma_0$ and $\sigma_2)$ the cross sections for scattering of transverse photons with parallel and perpendicular linear polarization (with helicity 0 and helicity 2). Gauge invariance implies for $q_i^2 \to 0$

$$\sigma_{TL}\sim q_2^2,\; \sigma_{LT} \sim q_1^2,\; \sigma_{LL} \sim q_1^2 q_2^2,\; \tau_{TL}\sim \sqrt{q_1^2 q_2^2} \tag{13}$$

Γ_{TT} is the flux factor for the transverse photons at vertex 1 and 2. Γ_{LT} is the flux factor for longitudinal photons at vertex 1. $\varepsilon_1 = \Gamma_{LT}/\Gamma_{TT}$, $\varepsilon_2 = \Gamma_{TL}/\Gamma_{TT}$ are the polarization parameters.

ϕ is the angle between the electron and positron scattering planes in the $\gamma\gamma$ CMS.

In double tag experiments σ^{eff}, τ_{TT} and τ_{TL} can be measured.

3.2 Single tag mode

Averaging over ψ_2, the azimuthal angle of e_2, and $-q_2^2 = P^2 \approx 0$ lead to $\sigma_{TL} = 0$, $\sigma_{LL} = 0$, $\tau_{TL} = 0$, and $<\cos2\phi> = 0$, which simplifies expressions (9) and (10) for reaction (8):

$$d\sigma_{ee} = \frac{\alpha \cdot E_1 \cdot (1+(1-y)^2)}{2\pi^2 Q^2 y} \{\sigma^{tot} + (\varepsilon_2 - 1)\cdot\sigma_{TL}\}\cdot N_\gamma(z_2,\theta_2^{max}) dz_2 dE_1 d\Omega_1 \quad (14)$$

with $\sigma^{tot} = \sigma_{TT} + \sigma_{TL}$.

$N(z_2,\theta_2^{max})$ describes the flux of quasireal target photons[3] of fractional energy $z_2 = E_2/E_b$ with emission angles smaller than θ_2^{max}:

$$N_\gamma(z,\theta^{max}) = \frac{\alpha}{\pi} \cdot \frac{1}{z} \{1+(1-z)^2\} \ln\frac{E_b(1-z)\theta^{max}}{m_e z} - (1-z)\} \quad (15)$$

y is determined from the tagged electron $y = 1-(E_1/E_b)\cdot\cos^2(\theta_1/2)$. In the actual experiments the tag energy and angle constraints make $\varepsilon_2 \sim 1$ so that the σ_{TL} term in (14) can be neglected.

3.3 No tag mode

The double anti-tagging leads to $P^2 = Q^2 \approx 0$, and eq. (14) reduces to

$$d\sigma_{ee} = \sigma_{TT}(W) \; N(z_2, \theta_2^{max}) \, N(n_1, \theta_1^{max}) \; dz_2 dz_1 \quad (16)$$

4. Experiments

Since the last $\gamma\gamma$ workshop[12] in 1983 there is a substantial improvement of data. Table I gives a summary on the new experimental data.

Table I

Summary on experiments with new results presented to this workshop.

Tagging Mode	Experiment	P^2, Q^2 $(GeV)^2$	E_b GeV	Luminosity nb^{-1}	Events
No tag = double antitag	PLUTO	$P^2=Q^2\approx.01$	17.3	4.8	5223
Single tag $P^2\approx.01$	PLUTO	Q^2 0.1 - 0.9 1.1 - 20 15 - 130	17.3 17.3 17.3	19 37 40	2931 1587 105
Double tag	PEP 4/9	P^2, Q^2 0.1 - 1.6	14.5	50	790

4.1 Experimental setups

Figs. 2a and 2b show the salient features of the PLUTO and the PEP4/9 detectors. The PEP4/9 results are derived from double tag events using the NaI shower counters, and the PLUTO results are derived from no-tag and single-tag data using the small and large angle taggers (SAT and LAT, respectively), and the end cap shower counters (ECT). Table II summarizes the relevant properties of the tagging systems.

Table II

Properties of tagging systems

Experiment	P L U T O			PEP4/9
tagger	SAT	LAT	EC	NaI
polar angle	32-55	90-260	330-680	22-90 mrad
azimuthal angle ψ	2π	180 ± 65 $\quad$ 0 ± 65	2π	2π
tagging energy resolution $\Delta E/\sqrt{E}$	16.5%	25%	28%	$\frac{\Delta E}{E} = 1\%$ at 14.5 GeV
tagging energy trigger threshold	$E > 4$	$E > 4$	$E > 3$	$E_1, E_2 > 0.8$ GeV
other trigger requirements	≥ 1 track in central detector	none	none	$\|\psi_1-\psi_2\| > 0.3$ (>0.55 off line)
Q^2-range (GeV^2)	.14 -.91	1.1-20	15-130	$0.1 < Q^2, P^2 < 1.6$

4.2 Background rejection

The rejection of background is a major task before cross
sections can be inferred from triggered events. The contamination of
$\gamma\gamma$ events by other sources is the more severe the less tags are
required: the no-tag data are heavily contaminated and the double tag
data are almost clean. The separation of the channels $\gamma\gamma \rightarrow$ leptons
and $\gamma\gamma \rightarrow$ hadrons requires further cuts. Table III shows a survey of

the relevant background sources, together with the most efficient cuts
for their remedy.

Table III

Survey on background sources for $\gamma\gamma \rightarrow$ hadrons

Source		Remedy
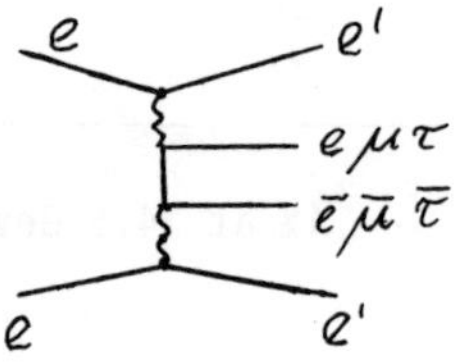 e, e', $e\,\mu\,\tau$, $\bar{e}\,\bar{\mu}\,\bar{\tau}$, e, e'	$\gamma\gamma$ events, but 'QED'	2prg + $\geq$ 1 shower, or $\geq$ 3prg remaining $\tau\bar{\tau}$ subtracted
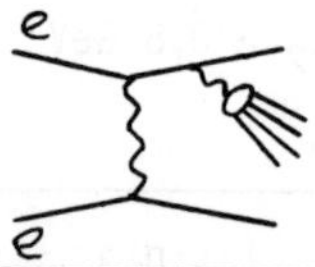 e, e	'inelastic' compton	$W_{vis} \gtrsim 1$ GeV
$eN \rightarrow e' + had$	beam gas	$W_{vis} \gtrsim 1$ GeV tracks with negative charge, side band subtraction
$e^+e^- \rightarrow$ hadrons, 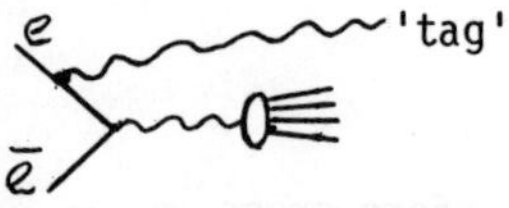 e, 'tag', $\bar{e}$	radiative annihilation	tagging track $W_{vis} \leq \overline{10\ GeV}$ $n_{ch} \leq 10$
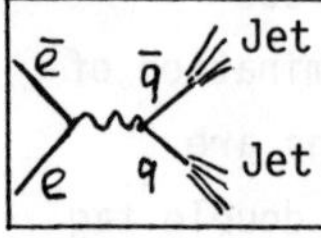 $\bar{e}$, $\bar{q}$, Jet, e, q, Jet at small θ	jet into tagging system	"Isolated" tag p^{miss} cut

In table IV we summarize the cuts applied to the no-tag, single-tag and double-tag samples, respectively, and the remaining background contents.

TABLE IV

Final state cuts which limit the acceptance

Quantity	no-tag	single-tag	double-tag
Anti-tag	$\theta_1,\theta_2 > 30$ mrad $E_1,E_2 > 4$ GeV	$\theta_1 > 30$ mrad $E_1 > 4$ GeV	-
Multiplicity	$3 \leq N_{CH} \leq 10$ at $\theta > 30^o$ not all same charge sign	$N_{CH}=2$, + ≥ 1 sh. $N_{CH} \geq 3$ 1 track$\neq$electr. for $N_{CH} = 2,3$	$N_{CH}=2$, + ≥ 1 sh. $N_{CH} \geq 3$ 1 track$\neq$electr for $N_{CH} = 2,3$
W_{vis}	$1.5<W_{vis}<10$	$1.2<W_{vis}<10$	
net longitudinal momentum		$\|P_2^{miss}\| > 6$ GeV opp. to tagged electron	

Resulting events and background

Events	no tag	SAT	LAT	ECT	double tag
Events	5223	2931	1587	105	~790
Background%					
$\gamma\gamma \to \tau\tau$	1	1.5	5	10	< 5
beam gas	6	4	4	-	-
annihilation	8	-	< 1	9	-

For the no-tag data the background situation is illustrated in Fig. 3. Fig. 3a shows the observed W_{vis} distribution of the no-tag events in the full W_{vis} range. At $W_{vis} < 10$ GeV a clear enhancement due to $\gamma\gamma$ events is seen. At an expanded scale, the data at $W_{vis} < 10$ GeV are compared in Fig. 3b to the sum of the $\gamma\gamma$ MC prediction using the $\sigma_{\gamma\gamma}$ (W) as resulting from the unfolding, the $\gamma\gamma \rightarrow \tau\tau$ and e^+e^- annihilation contamination as estimated by MC simulation, and the beam gas event contribution. The observed W_{vis} distribution is well reproduced by the resulting $\sigma_{\gamma\gamma}$(W) and the estimated background contributions. Fig. 3c shows the share of the contaminations relative to the total. The W_{vis} region above 9 GeV is too much contaminated for a reliable $\gamma\gamma$ cross section determination.

5. Determination of cross section

To evaluate cross sections from the data one has to deal with 3 further problems:

 i) the Q^2 (P^2) interpolation
 ii) the $W_{vis} \rightarrow W$ correction
iii) and acceptance corrections

5.1 The Q^2 (P^2) interpolation

The single- and double-tagged data cover Q^2 and P^2 ranges over which $\sigma(W,Q^2,P^2)$ varies by orders of magnitude. Since event statistics is limited a real multidimensional cross section analysis is not possible, however. Therefore, within limited Q^2 regions (see Table V) factorization is assumed in evaluating $\sigma(W, <Q^2>, <P^2>)$ in these intervals from $\sigma(W,Q^2,P^2)$:

$$\sigma(W, <Q^2>, <P^2>) = \sigma(W, Q^2, P^2) \cdot \frac{F(<Q^2>)}{F(Q^2)} \cdot \frac{F(<P^2>)}{F(P^2)} \qquad (17)$$

The interpolation functions $F(Q^2)$ given in Table V have been checked to describe the data (see Figs. 4a,b for the GVDM form factor).

Table V

Intervals in tagged data, and Q^2 interpolation functions

Tagger	$\cdot$ SAT	LAT	ECT	NaI
Q^2 range	.14 - .91	1.1 - 20	15 - 130	$1 < P^2, Q^2 < 1.6$
$<Q^2>$	.44	5.4	45	0.3
$F(Q^2)$	GVDM	$Q^{-0.9}$	$Q^{-0.9}$	GVDM

5.2 W_{vis} correction

In the no-tag and single-tag data, the effective mass W_{vis} of the observed final state particles is measured instead of the effective mass W of the $\gamma\gamma$ system. In the PLUTO experiment W_{vis} is near 75% of W on average, with a spread of about 25%. Fig. 5 shows the W_{vis}/W ratio for simulated events tagged in the SAT. The original W distribution is reconstructed using the $W - W_{vis}$ correlation from simulated events (see section 5.4), and an unfolding procedure.

In double-tag experiments W is calculated from the colliding photon energies directly:

$$W^2 = 4E_1 E_2^\gamma - 2E_1^\gamma E_2 \sin\theta_1 \sin\theta_2 \cdot \sin \phi/2 \tag{18}$$

At low W (< 3 GeV), however, the resolution is rather poor $\sigma(W) \sim 0.8$ GeV, and initial state radiation pushes W below the true W value. The measured double-tag W distribution (Fig. 6a) has been

corrected for these effects (Fig. 6b). At W < 4 GeV the corrected
distribution is considerably different from the measured one.

5.3 Acceptance corrections

The detection efficiencies for the final state of tagged events
is limited because of the cuts described in section 4. Fig. 7 shows
as examples the acceptance as function of W, for the PLUTO SAT and
for the PEP4/9 double-tagged events. Acceptance corrections are small,
very little model dependent and pretty similar in both experiments at
W > 5 GeV. At lower W values, the PEP4/9 double-tagged data have a
considerably larger acceptance than the PLUTO SAT data. The acceptance
values here depend more on the details of the final state model, in
particular on the charged multiplicities.

5.4 Hadronic final state model

The unfolding procedure requires the use of simulated events for
the reaction $\gamma\gamma \to$ hadrons. To generate such events a model for hadronization is required. The PEP4/9[13] and the PLUTO[14-16] analysis used
the models summarized in Table VI. At the low W end (W < 4 GeV), the
charged multiplicity pattern is particularly important for the event
acceptance calculations. The average multiplicities for charged and
neutral pions differ in the two analyses. The PLUTO analysis uses a
multiplicity parametrization which is close to the e^+e^- annihilation
data[17] in that energy range (Fig. 8), while the PEP4/9 parametrization
is substantially below the e^+e^- data at W < 7 GeV. Both experiments
claim they reproduce their observed multiplicities well. There is,
however, no evidence presented that the PEP4/9 data are inconsistent
with the e^+e^- average together with a KNO distribution.

Table VI

Parametrization of the hadronic final state models

Particles	PLUTO	PEP4/9
Multiplicity		
Average		
$\langle \pi^+ + \pi^- \rangle$	$2 \cdot \sqrt{W}$	$1 + 0.52W$ (GeV)
$\dfrac{\langle \pi^0 \rangle}{\langle \pi^+ + \pi^- \rangle}$	$\dfrac{2}{3}$	$\dfrac{1}{2}$
Distribution	KNO	Poisson
	$\dfrac{\langle N \rangle}{D} = 2.7 \; \pi^{\pm}$ $\phantom{\dfrac{\langle N \rangle}{D} =} 2.4 \; \pi^0$	
Topology: a) invariant phase space (IP) b) limited $P_\perp$ phase space (LP) $\quad LP = IP * \prod\limits_{i=1}^{N} \exp\{-5p_\perp^2\}$ c) $\gamma\gamma \to q\bar{q}$ with QED angular $\quad$ distribution (QM)		$LP = IP * \prod\limits_{i=1}^{N} \exp\{-6.25p_\perp^2\}$

A possible bias of this difference on the acceptance correction should go in the direction that it was estimated too small in the PLUTO and/or too large in the PEP4/9 analysis, relatively. The different multiplicity parametrizations may introduce a systematic difference between the two $\sigma(W)$ evaluations at low W.

There is a further difference in so far as in the PLUTO analysis also the rejected $\pi^+\pi^-$ and $\pi^0\pi^0$ final states have been corrected for. They contribute, however, only 0.6% to σ^{tot} at W = 2 GeV according to the model, a number which is confirmed by a direct mesurement of the exclusive reaction $\gamma\gamma \to \pi^+\pi^-$ [18].

To achieve a good reproduction of their data the PLUTO collaboration had to use a W depending mixture of different topological models (see Table V), with IP topology dominating at low W ($\lesssim$ 4 GeV) and $P_\perp$ limited topologies (LP, QM) dominating at W $\gtrsim$ 7 GeV.

5.5 The unfolding procedure

With the definition of a model for $\gamma\gamma \to$ hadrons, where $\sigma(W)$ is not jet fixed, an unfolding procedure[19] can be applied to search for a smooth function $\sigma(W)$ which reproduces the observed W_{vis} distribution. The unfolding method applied by the PLUTO collaboration has been described in detail at this workshop[20].

By including the experimental luminosity, a full detector simulation for the events and the Q^2 interpolating functions (see Table V) the resulting cross section $\sigma(W)$ is

a) interpolated to $Q^2 = \langle Q^2\rangle$
b) corrected for all event acceptance inefficiencies
c) unfolded for the $W \to W_{vis}$ transition due to resolution and particle losses.

Two types of checks on the reliability of the unfolding step have been performed. The first check concerns the reliability of the estimate of the amount of unobserved or rejected events. This number depends mainly on a correct estimate of events with low observed multiplicity ($N_{CH} \leq 2$) in the simulation model. Fig. 8 (b and c) checks how well the observed charged multiplicity is reproduced, and Fig. 9 checks the charged particles momenta transverse to the e^+e^- beam, since low $p_\perp$ ($p_\perp \lesssim 0.1$ GeV) tracks are not detected with full efficiency. There is no obvious mismatch between the data and their simulation, except for the $P_\perp^2$ distribution of PEP4/9 at $p_\perp^2 > 0.4$ GeV^2 reflecting the exclusive use of a limited $p_\perp$ model. The authors claim, however, that the acceptance efficiency is insensitive to this $p_\perp$ region.

The second check concerns the reliability of the $W_{vis} \rightarrow W$ unfolding. To this end the PLUTO collaboration applied the unfolding procedure to a simulated ee $\rightarrow$ ee + hadrons experiment. Fig. 10 shows the "true" cross section $\sigma(W)$ as a curve, and the reconstructed cross section as a histogram. For $W > 2$ GeV the reconstruction is perfect. Only the steep slope of a $1/W^2$ term below $W = 2$ GeV in Fig. 10b is marginally recovered.

6. Results on σ^{tot} ($\gamma\gamma \rightarrow$ hadrons)

The total cross section results (W, $<Q^2><P^2>$) are shown in Fig. 11. This figure shows the W-dependence of σ from no-tagging, single-tagging and double-tagging measurements, together with their respective average Q^2 and P^2 values. The Q^2 dependence, $\sigma(<W>, Q^2)$, for σ averaged over the W interval $3 < W < 10$ GeV from single-tagging data is shown in Fig. 12. In Fig. 13 the double-tagging cross section in ϕ intervals averaged over the full W(2-20 GeV) and q_i^2 (0.1 - 1.6 GeV^2) regions is given. Fig. 11 - 13 contain all results of the $\gamma\gamma$ total cross section measurements. What do they tell us?

6.1 Analysis of the ϕ dependence

A fit of (9) to Fig. 13 yields the following results

$$\tau_{TT}/\sigma^{eff} = -0.43 \pm 0.23 \tag{19a}$$

$$\tau_{LT}/\sigma^{eff} = -0.01 \pm 0.03 \tag{19b}$$

The latter one is expected because of (13), and the first one indicates $\sigma_{\perp} \gtrsim \sigma_{\parallel}$ (see (12)).

6.2 W-dependence of σ^{tot} at different P^2 and Q^2 values

Here are two points of interest:

1. Does $\sigma(W, Q^2, P^2)$ factorize, as expressed in (17)?
2. Which W^n terms are present in $\sigma(W)$?

By comparing the no-tag and the SAT single-tag ($<Q^2> = 0.44$ GeV) cross sections with the other ones of Fig. 11, it is obvious that W-Q^2-P^2 factorization of σ^{tot} does not work. Even though the SAT and the double-tagging data have about the same level of cross section, i.e. the same amount of suppression as compared to $P^2 = Q^2 = 0$, their W-dependences are clearly different. A potential inconsistency in the data analysis from the different multiplicity parametrizations employed (see section 5.4) might even work into the direction opposite to the observed differences. The other difference between the two experiments is the method to determine W. Whether this difference leads to a systematic difference in σ^{tot} at low W might be checked with the PEP4/9 data which allow to use the single-tagging method - W from unfolding of W_{vis} as calculated from the observed hadrons - in addition to the double-tagging method of eq. (18).

With the data as presented we have to conclude that $\sigma(W, <Q^2> = 0.44, P^2 = 0)$ cannot be converted into $\sigma(W, <Q^2> = <P^2> = 0.3$ GeV$^2)$ by

a factorizing ansatz (Fig. 15). This is even more surprising when looking at Fig. 4. Both experiments demonstrate that for their rather
wide regions of Q^2 - wide as compared to the differences between the
two experiments, where Q^2_{max} from double-tagging and Q^2 from single-
tagging overlap largely - factorization with the GVDM form factor is
consistent with their data, since low and high W data scale alike.

The decomposition of $\sigma(W, <Q^2>)$ into terms of W^0, $1/W$ and $1/W^2$ is
of interest from a t-channel or Regge point of view. The presently
available data suffer from rather large systematic errors (> 20%) at
small W (< 4 GeV) making a meaningful determination of terms with
different powers in W practically impossible. All one can say is that
only the PLUTO SAT data establish a declining cross section with growing
W. It can be fitted (see Fig. 14) to shapes like

$$\sigma(W) = (107 \pm 40) + \frac{933 \pm 112}{W} \quad [GeV, nb] \tag{20a}$$

$$\text{or} \qquad = (0.91 \pm 0.08) \left\{240 + \frac{270}{W}\right\} + \frac{941 \pm 186}{W^2} \quad [GeV, nb] \tag{20b}$$

after GVDM extrapolation to $Q^2 = 0$. The W dependent fraction is stronger
than in the traditional[21] 240 + 270/W parametrization. Because of (20a)
the necessity of a $1/W^2$ term, however, is not established. The no-tag
cross section has too short a leaver arm in W and too large systematic
errors for any conclusion. The double-tag cross section is constant
between 2 and 20 GeV (σ = (380 ± 60) + (70 ± 290)/W) and the single-tag
cross sections at $<Q^2>$ = 5.4 and 45 GeV^2 are consistent with constant,
but have rather large errors.

6.3 Point-like and hadron-like behaviour of the photon

The point we want to investigate here is not whether the point-like
cross section contribution of the photon as predicted by QED and the QPM
exists. It has been clearly observed in the jet analysis of $\gamma\gamma \to$
hadrons[22] and has been discussed in previous talks[2,23] at this workshop.

Nor is the point whether VDM type contribution exist in $\gamma\gamma \to$ hadrons. Again the jet analysis[22,23] demonstrates that - beyond the size of the total $\gamma\gamma$ cross section at low Q^2 - also topological properties of the hadronic final state require VDM type contributions.

The point of concern is how these two pictures of the photon have to be combined in order to describe the facts. For this question the total $\gamma\gamma$ cross section is supplementary to the jet analysis. Firstly because also data at $W \lesssim 5$ GeV can be included and secondly since the VDM picture allows for quantitative predictions of the Q^2- and to some extent also of the W-dependence of σ^{tot} while topological . properties are only implied.

We first look into the Q^2 dependence. Fig. 16 shows (a) the pre-diction of the EVDM[24], of a similar approach by Etim and Masso[25] (EM) who replace the EVDM sum $\sum_n \Gamma_n V_n$ (see eq. (5)) by a different parametri-zation of the measured e^+e^- annihilation cross section, and a prediction by Alexander, Maor and Milstene[26] (AMM) who infer $\sigma(\gamma_v(Q^2), \gamma(P^2=0))$ from data on $\sigma(\gamma_v(Q^2),p)$, $\sigma(\gamma(P^2=0),p)$ and $\sigma(pp)$, basically using the relation $\sigma(\gamma_v(Q^2),\gamma) = \sigma(\gamma_v(Q^2),p) \cdot \sigma(\gamma,p)/\sigma(p,p)$ with appropriate kinematic threshold factors. It is obvious that all these hadronic approaches to the photon fail to describe the Q^2 dependence of $\sigma(\gamma_v(Q^2),\gamma)$ when averaged from $W = 3$ to 10 GeV. In Fig. 16b the same data is shown together with the Q^2 dependence of GVDM[10] and the absolute prediction by the QPM[3]. The GVDM cross section here is normalized to the data where it dominates clearly over the QPM prediction, at $W_{vis} > 4$ GeV, $\cos\theta^*$ $(\gamma, \text{Jet}) > 0.9$, and $Q^2 < 1$ GeV^2. The GVDM cross section takes the form

$$\sigma^{GVDM}(W,Q^2) = \sigma_0 \cdot F^{GVDM}(Q^2) \tag{21}$$

with $\sigma_0 = 232$ nb as average for $3 < W > 10$ GeV

It is obvious that the GVDM prediction falls below the data the more Q^2 increases. On the other hand the QPM prediction fails badly at low Q^2 ($Q^2 \lesssim 15$ GeV^2). We note that the sum of QPM and GVDM happens to coincide with the data at $0.1 < Q^2 < 100$ GeV^2.

The W dependence is investigated in Fig. 17. The AMM[26] prediction, at $Q^2 = 0.44$ GeV^2 (Fig. 17a) may be consistent with the data in the full W range, however at $Q^2 = 5.4$ GeV^2 (Fig. 17b) EVDM is clearly below the data at W > 4 GeV, while the AMM prediction is too low at all W > 2 GeV. In Fig. 18 the sum QPM + GVDM is compared to the PLUTO single-tag data. Since the GVDM does not really predict the W dependence of $\sigma_{\gamma\gamma}$, we chose the ansatz of eq. (21) with $\sigma(W) = 232$nb instead of σ_0. This sum also describes the W dependence at all three Q^2 regions. It thus turns out that the combination of the point-like picture in the QPM approximation with a specific hadronic picture of the photon describes the dependence over a large Q^2 region with a substantial variation of the cross section, and the W dependence of the total cross section.

Fig. 18b,c show also the cross section calculated thru eq. (1) from the following ansatz[27] for F_2 of the photon:

$$F_2(x, Q^2) = F_2^{QCD} (x, Q^2) + F_2^{VDM} (x, Q^2) \qquad (22)$$

where $F_2^{VDM} = 0.2 \cdot \alpha(1-x)$ is the structure function of the hadronic component of the <u>target</u> photon only: Using the - singularity free - leading order QCD (LOQCD) prediction[28] for F_2^{QCD}, with $\Lambda = 200$ MeV, we get from eq. (1) the curves shown in Fig. 18b,c for $Q^2 = 5.4$ and $Q^2 = 45$ GeV^2. They are consistent with the data. For a discussion of other QCD predictions we refer to the talks by Bardeen[1] and Wagner[2].

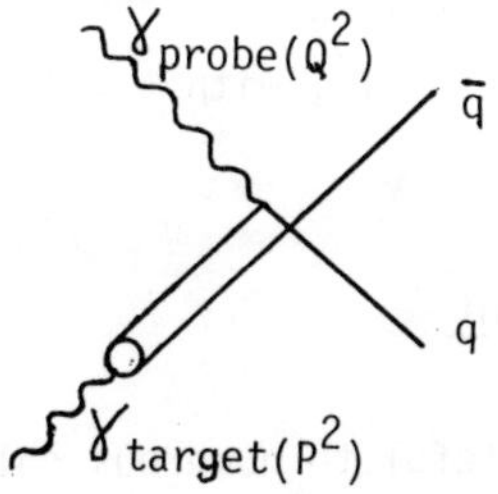

7. Summary and conclusions

- The total cross section, $\sigma^{tot}(\gamma\gamma \to$ hadrons) has been measured in no-tag ($Q^2 = P^2 - 0$), single-tag ($0.1 < Q^2 < 100$ GeV^2) and double-tag ($0.1 \leq P^2$, $Q^2 \leq 1.6$ GeV^2) experiments, in the W range from 2 to 10 (20 for double-tag) GeV.
- A decomposition like $\sigma = A + \dfrac{B}{W} + \dfrac{C}{W^2}$ is unreliable for the present no-tag and single-tag data. There is no need for B, C terms at $Q^2 > 1$ GeV^2, nor in the double-tag data.
- The cross section factorization $\sigma(W,Q^2,P^2) = \sigma(W) \cdot F(Q^2) \cdot F(P^2)$ fails to link single-tag data at low ($<Q^2> = 0.44$) and high Q^2 ($<Q^2> = 5.4$, 45 GeV^2), and also the present single- and double-tag data both at low Q^2 (Q^2, $P^2 \lesssim 1.5$ GeV^2).
- "Hadronic" models which allow to describe $e^+e^- \to$ hadrons and ep scattering fail to describe the full Q^2 dependence in $\gamma\gamma$ scattering.
- The point-like predictions in the QPM approximation which work on average in e^+e^- annihilation at all s fail badly in $\gamma\gamma$ scattering at $Q^2 \lesssim 10$ GeV^2. Gluonic corrections are obviously much more important in $\gamma\gamma \to$ hadrons than in $e^+e^- \to$ hadrons.
- In contrast to $e^+e^- \to$ hadrons, where the sum QPM + GVDM amounts to double counting, the reaction $\gamma\gamma \to$ hadrons is well described by this sum in a large range of Q^2.
- $\sigma(W, Q^2)$ from $F_2^\gamma = F_2^{LOQCD} + F_2^{VDM}$ is also consistent with the data for $Q^2 > 1$ GeV^2, with $\Lambda \sim 200$ MeV.

Acknowledgement

I am grateful to numerous colleagues in the PLUTO and PEP4/9 collaborations for their support in providing results. The organizers, the secretaries and all the UC-Davis crew did a great job to make such a pleasant and fruitful meeting possible.

References

1) W.A. Bardeen, talk given at this workshop

2) W. Wagner, talk given at this workshop

3) V.M. Budnev et al., Phys. Rep. 15C (1975) 181, Yad.Phys. 13 (1971) 353

4) G. Rossi UCSD-10P10-227, Ph.D. thesis and UCSB-TH-83-01

5) PLUTO-Collaboration, Ch. Berger et al., Phys. Lett. 99B (1981) 287

6) J.L. Rosner, Brookhaven report CRISP 7126 (1971)

7) E. Witten, Nucl. Phys. B120 (1977) 189
W.A. Bardeen, A.J. Buras, Phys. Rev. D20 (1979) 166

D.W. Duke, J.F. Owens, Phys. Rev. D22 (1980) 2280

8) J.J. Sakurai, Phys. Lett. 46B (1973) 207

9) J.D. Bjorken and E.A. Paschos, Phys. Rev. 185 (1969) 1975

10) J.J. Sakurai and D. Schildknecht, Phys. Lett. 40B (1972) 121

I.F. Ginzburg and V.G. Serbo, Phys. Lett. 109B (1982) 231

11) M. Greco, Nucl. Phys. B63 (1973) 398

12) Proc. of the Fifth Int. Workshop on Photon Photon Collisions, Aachen 1983, ed. Ch. Berger, Springer Verlag (1983)

13) PEP-9 Two-Photon Collaboration, J.C. Armitage et al., Measurement of the Hadronic Photon-Photon Cross-Section in a Doubly-Tagged Experiment, Contribution to the XXII International Conference on High Energy Physics, Leipzig (1984), and presented at the parallel sessions of this workshop by D. Bintinger.

14) PLUTO-Collaboration, Ch. Berger et al., DESY 84-080, Measurement of the Total Photon-Photon Cross Section for the Production of Hadrons at Small Q^2, and Phys. Lett. B (to be published).

15) PLUTO-Collaboration, Ch. Berger et al., DESY 84-081, A Measurement of the Q^2 and W Dependence of the $\gamma\gamma$ Total Cross Section for Hadron Production, and Z. Phys. C (to be published).

16) PLUTO-Collaboration, in preparation

17) TASSO-Collaboration, R. Brandelik et al., Phys. Lett. 89B (1980) 418

TASSO-Collaboration, M. Althoff et al., DESY 83-130 (1983) and Z. Phys. C - Particles and Fields (to be published).

JADE-Collaboration, W. Bartel et al., Z. Phys. C - Particles and Fields 20 (1983) 187

PLUTO-Collaboration, Ch. Berger et al., Phys. Lett. 95B (1980) 313

C. Bacci et al., Phys. Lett. 86B (1979) 234 (ADONE)

J.L. Siegrist, Ph. D. Thesis SLAC-225 (1980) (MARK II)

M.S. Alam et al., Phys. Rev. Lett. 49 (1982) 357 (CLEO)

LENA-Collaboration, B. Niczyporuk et al., Z. Phys. C -
Particle and Fields 9 (1981) 1

18) PLUTO-Collaboration, Ch. Berger et al., Z. Phys. C 26 (1984) 199

19) V. Blobel, Proceedings of the 1984 CERN school of computing,
Aiguablava, September 1984,(to be published)

20) A. Bäcker, Talk given at this workshop

21) J.L. Rosner, BNL report 17552 (1972) 316
T.F. Walsh, J. Physique C2 Suppl. 3 (1974) 77

22) JADE-Collaboration, W. Bartel et al., Phys. Lett. 107B (1981) 163
PLUTO-Collaboration, Ch. Berger et al., Z. Phys. C26 (1984) 191

23) F. Foster   talk given at this workshop
PLUTO-Collaboration, presented at the parallel sessions of this
workshop by D. Schmidt

24) U. Maor and E. Gotsman, Phys. Rev. D28 (1983) 2149
and U. Maor, private communication

25) E. Etim and E. Masso, Z. Phys. C18 (1983) 117

26) G. Alexander, U. Maor and C. Milstene, Phys. Lett. 131B (1983) 224.

27) C. Peterson, T.F. Walsh, P.M. Zerwas, Nucl. Phys. B174 (1980) 424

28) E. Witten, ref. 7.

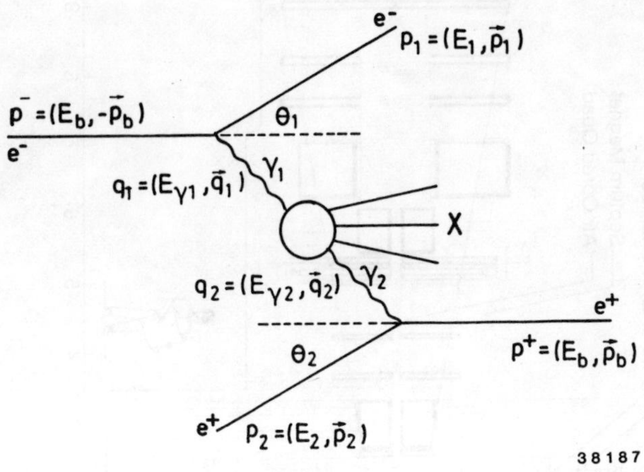

Fig. 1: Definition of variables for the reaction $e^+e^- \rightarrow e^+e^- + X$

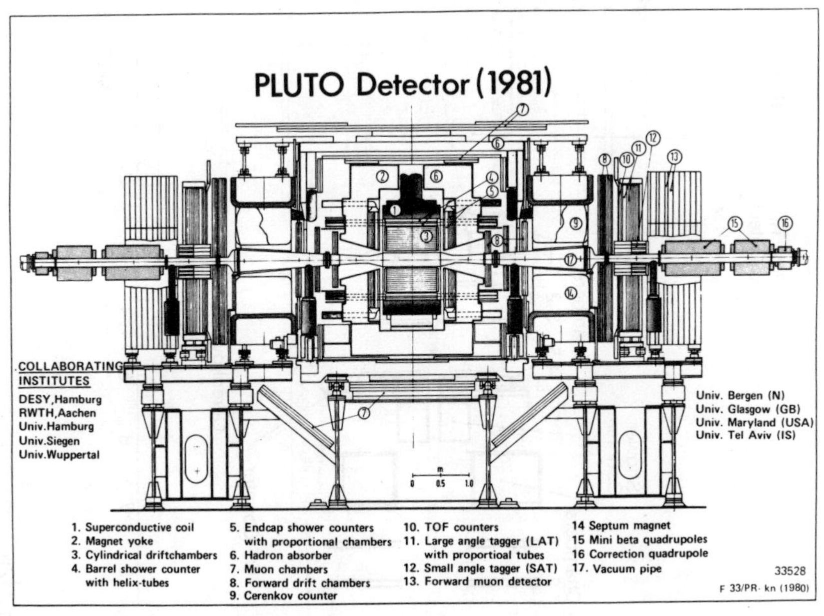

Fig. 2a

Pole Tip
Muon Detectors
3 2 1
Pole Tip
Calorimeter
Muon Detectors
Hexagonal Calorimeter
Outer Drift Chamber
Time Projection Chamber
Inner Drift Chamber
Septum Magnet
Air Cored Quad
Shower
Counter
Na I
Drift
Chambers
Muon Chambers
3 2 1
4-5 3 2 1
180 mrads
TOF Counter
Cerenkov
9 8 7 6 5 4 3 2 1 0 1 2 3 4 5 6 7 8 9
METERS

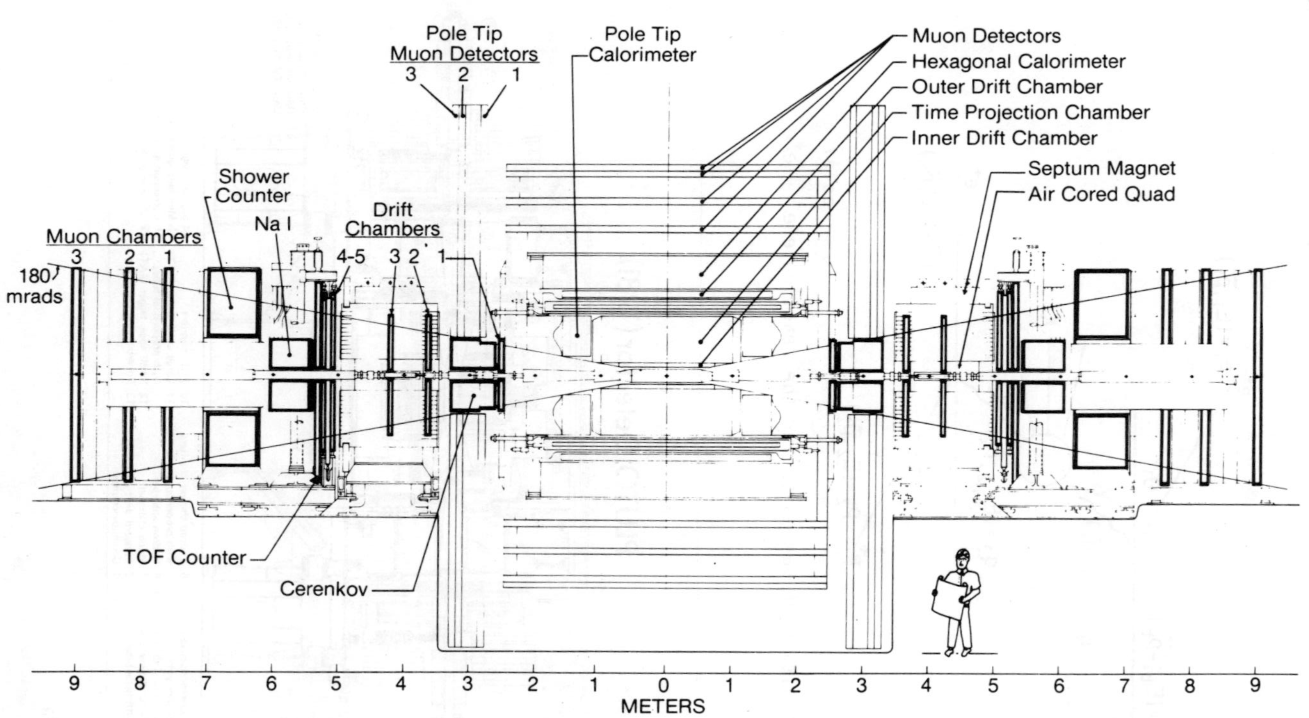

Fig. 2b

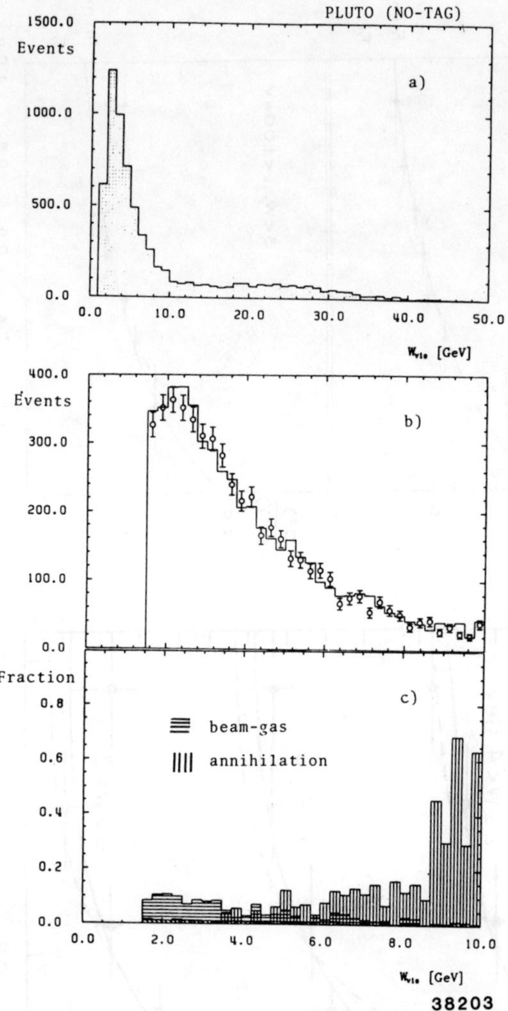

Fig. 3: No-tag event distributions: (a) all W_{vis}, data, (b) $W_{vis} <$ 10 GeV, data as circles, MC simulation including background as histogram, (c) estimated background as fraction of the observed sample, from e^+e^- annihilation (vertical lines), from beam gas (horizontal lines), and from $\gamma\gamma \to \tau\tau$ (skew lines).

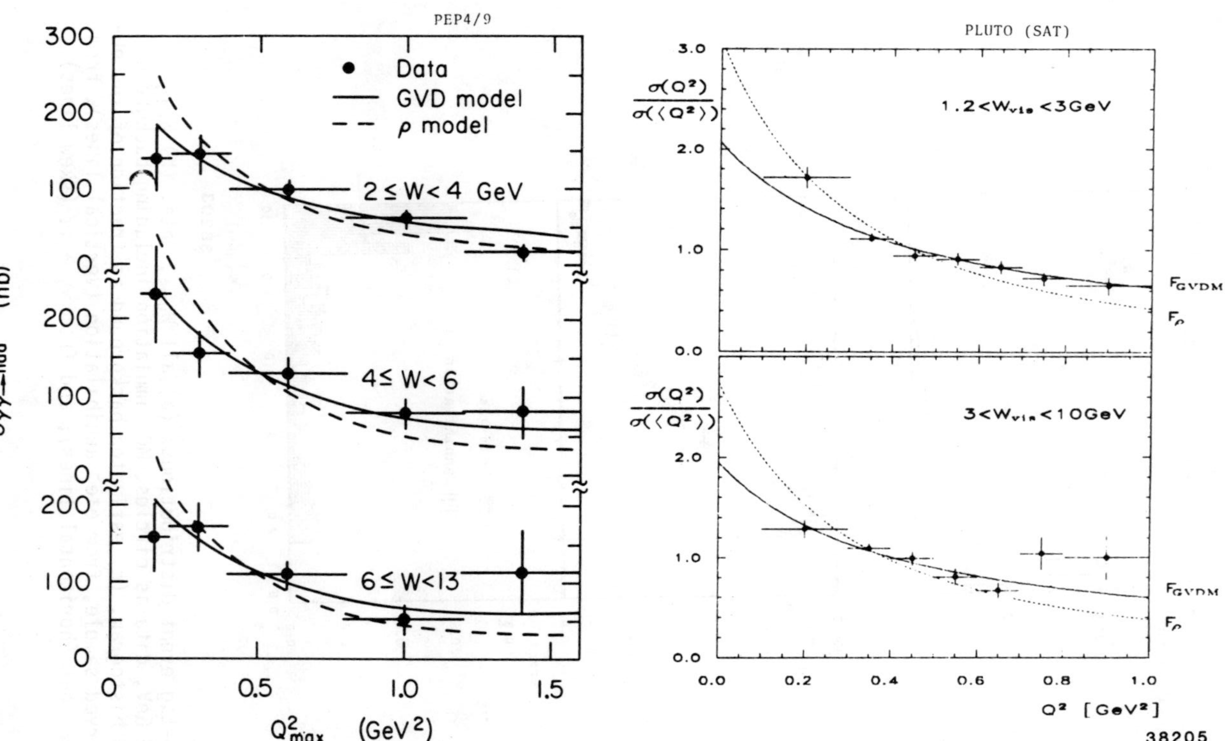

Fig. 4: Q^2 distributions from double-tagging (the larger Q^2 value) and from single-tagging data, for several W_{vis} intervals, in comparison with GVDM and ρ-pol form factor.

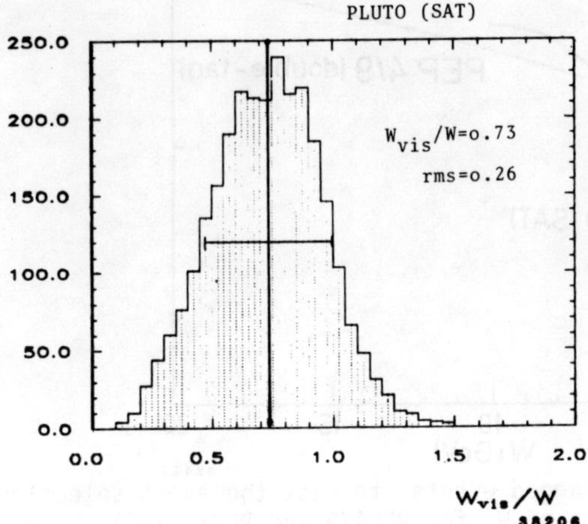

Fig. 5:

The W_{vis} resolution (W_{vis}/W) for PLUTO SAT from MC simulation.

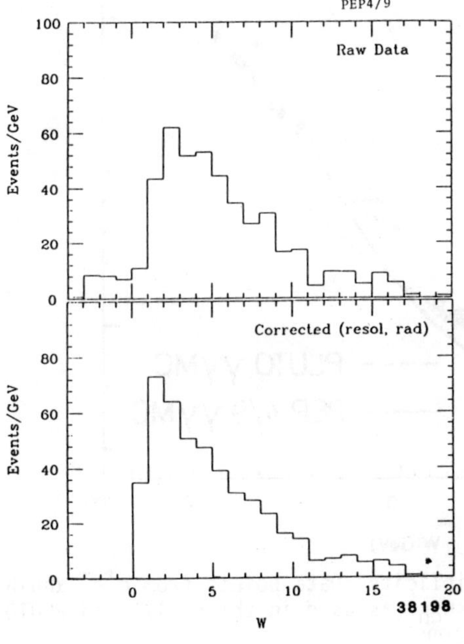

Fig. 6:

The W distribution from PEP4/9 double-tagging data: as measured, and after correction for resolution and radiative effects.

308

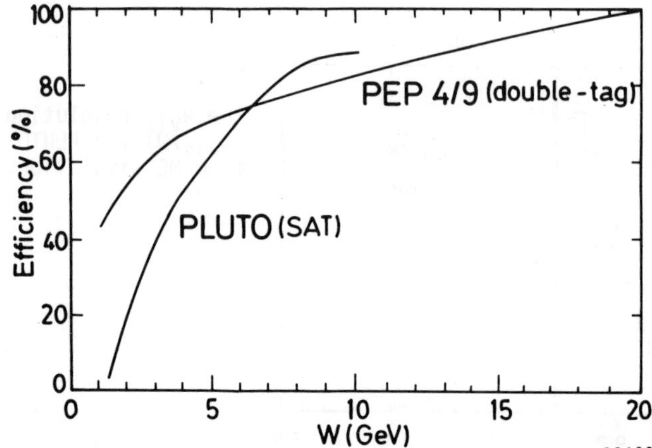

Fig. 7: The efficiency for tagged events to pass the event selection criteria, as function of W, for PEP4/9 and PLUTO(SAT).

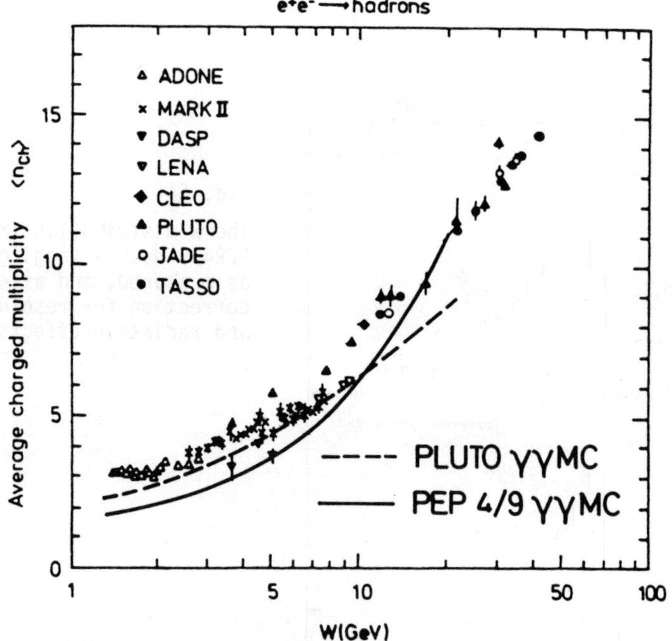

Fig. 8a: Average charged multiplicity. Data points from e^+e^- annihilation. Curves show $\langle n_{ch} \rangle$ as used in the PEP4/9 and PLUTO $\gamma\gamma \to$ hadrons simulations.

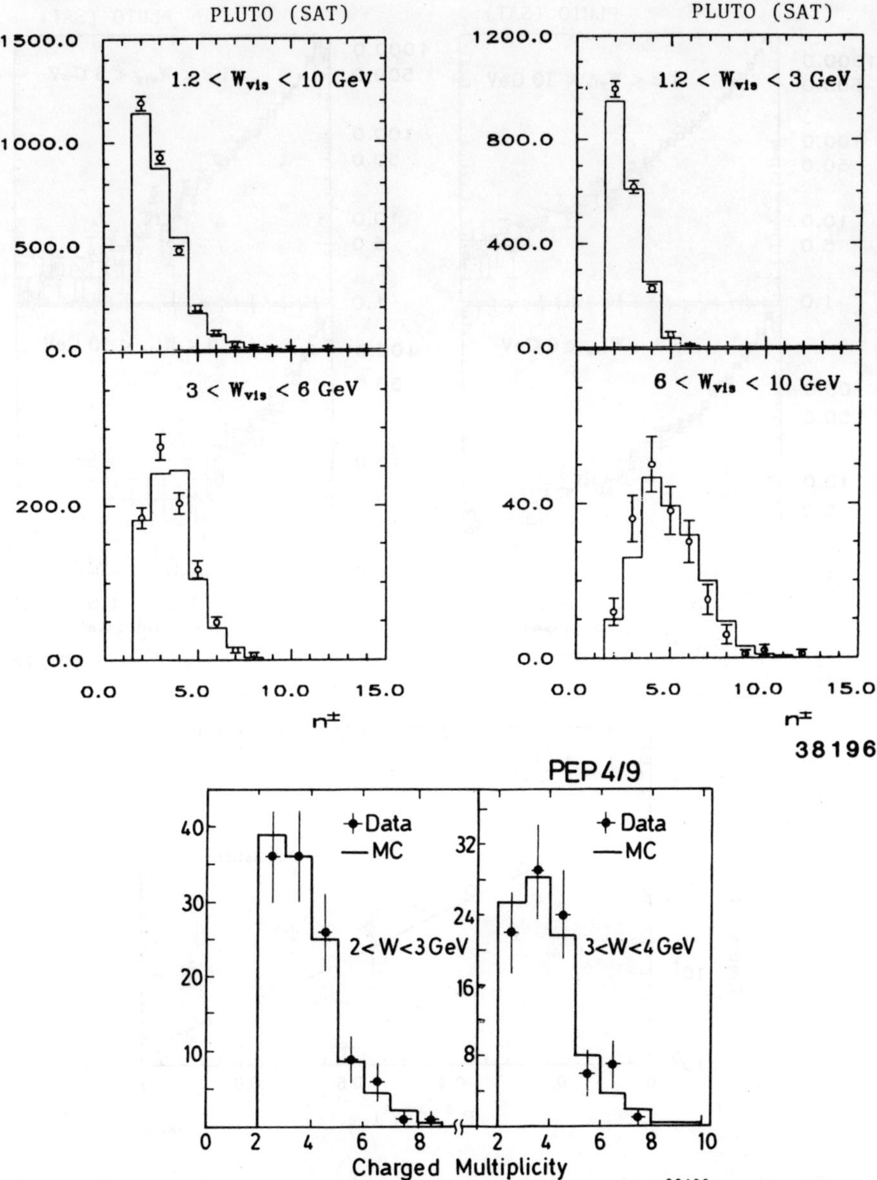

Fig. 8b: Charged particle multiplicity distribution in selected W_{vis} regions, from PLUTO (SAT) and PEP4/9 double-tagging, data (circles) and MC simulation (histogram).

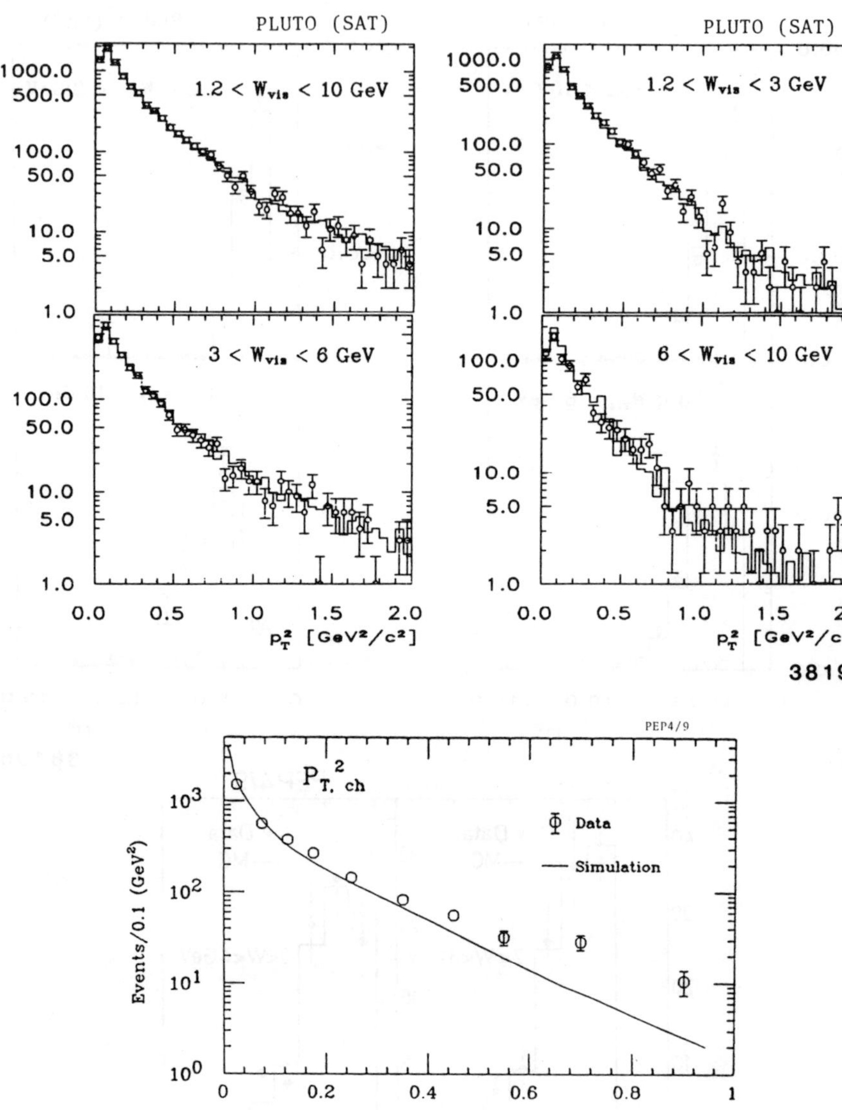

Fig. 9: Charged particle $P_\perp^2$ (w.r.t. beam) distribution from PLUTO (SAT) and PEP 4/9 double-tagging, data (circles) and MC simulation (histogram, curve).

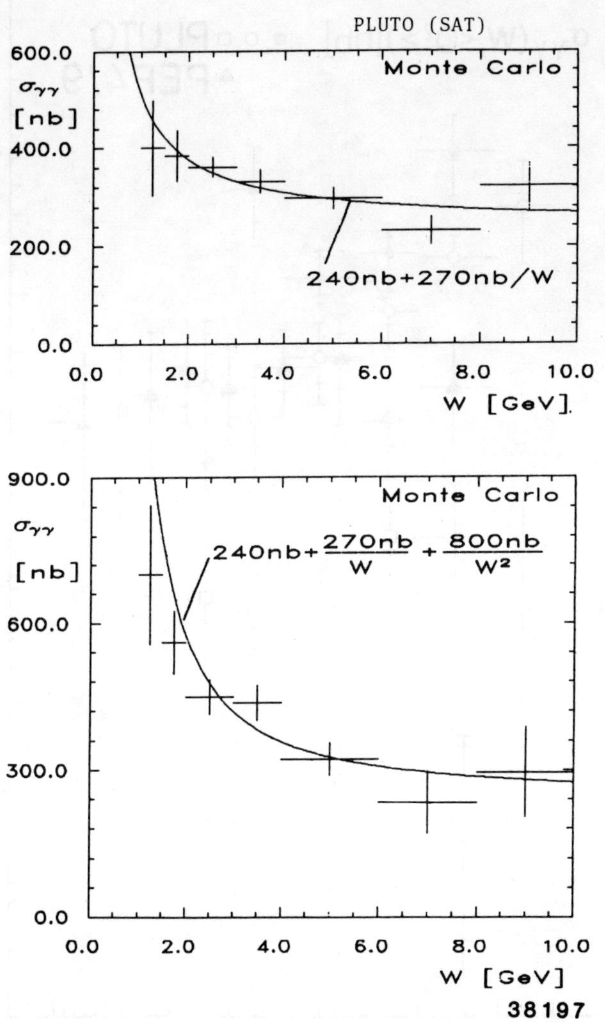

Fig. 10: Test of the unfolding procedure. The curves are the input cross section for simulated experiments (PLUTO-SAT), and the crosses are the results from the analysis.

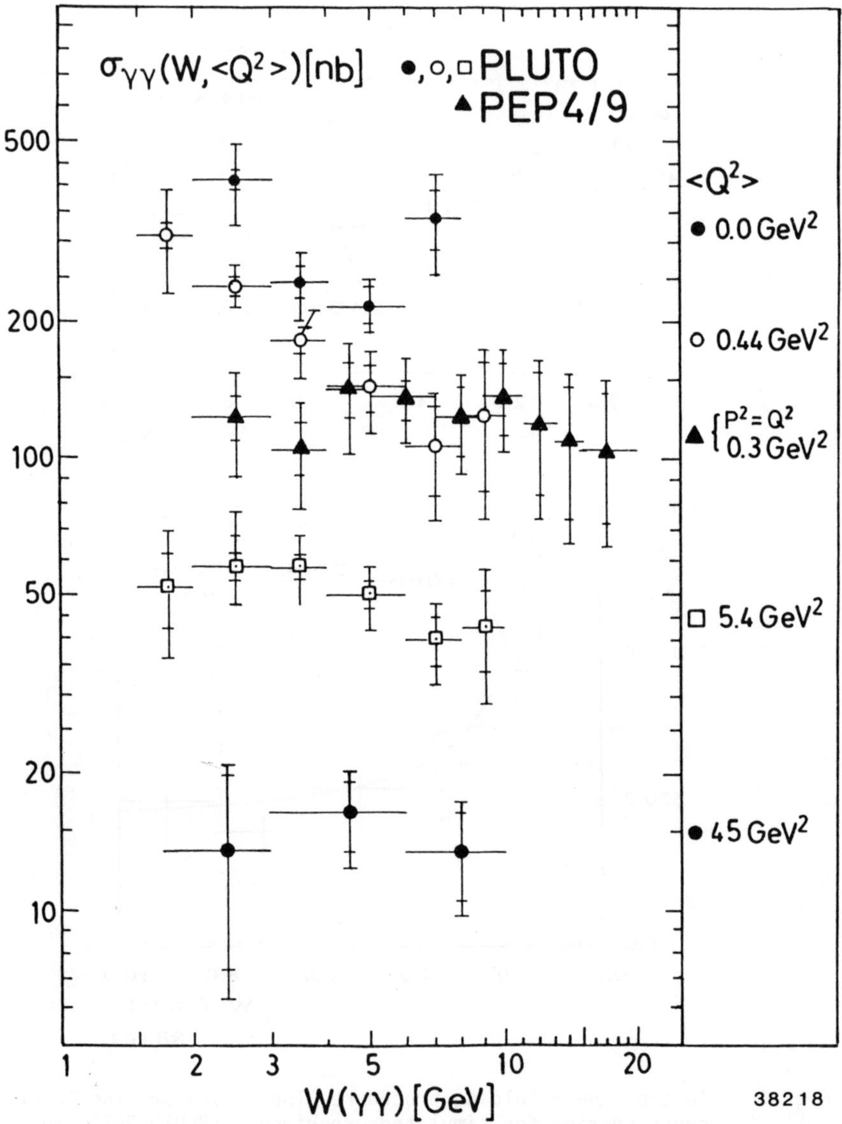

Fig. 11: The cross section $\sigma_{\gamma\gamma}$ (W, $<Q^2>$, $<P^2>$) for no-tagging, single-tagging and double-tagging measurements. All PLUTO data are at $<P^2>$ = 0.01 GeV^2. The double-tag data in the region 5<W<11 GeV, originally presented in 1 GeV bins (see Fig. 15), have been rebinned here by the author into 2 GeV bins.

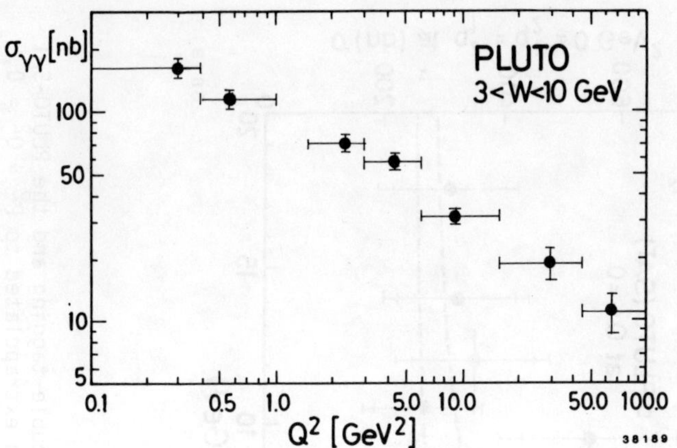

Fig. 12: The Q^2 dependence of $\sigma_{\gamma\gamma}$, averaged over $3 < W < 10$ GeV, from the PLUTO single-tagging data.

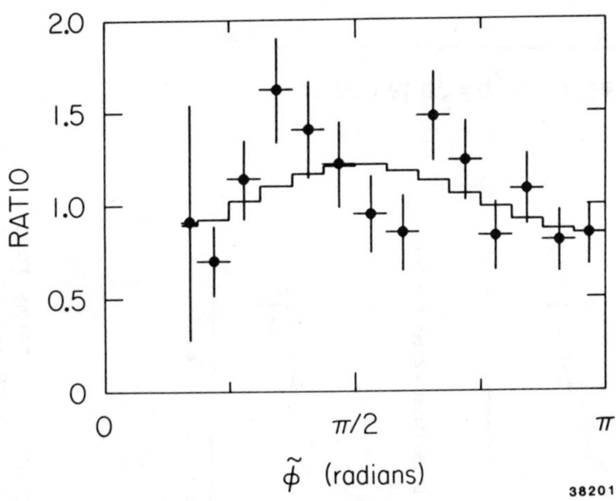

Fig. 13: Ratio of σ_{eff} plus interference terms (τ) to σ_{eff} vs. aco-planarity $\tilde{\phi}$ in $\gamma^*\gamma^*$ center of mass, for $W > 2$ GeV. The histogram is the best fit for σ_{eff}, τ_{TT} and τ_{TL} in eq. (9).

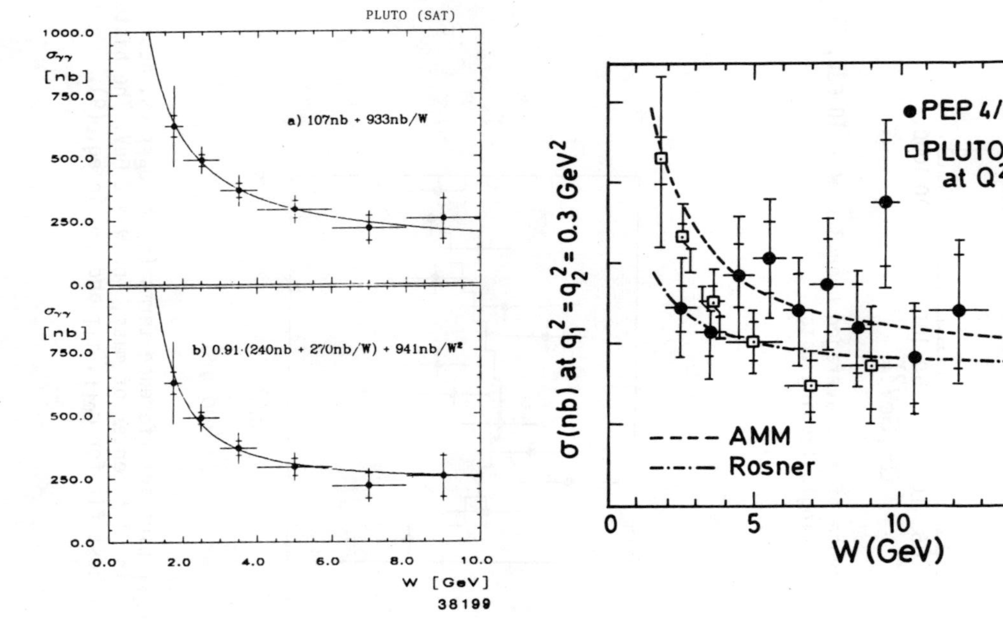

Fig. 14: Fits of power terms to $\sigma(W)$: (a) with a W^0 and W^{-1}, (b) the traditional (ref. 21) VDM estimate and a W^{-2} term.

Fig. 15: The PEP4/9 double-tagging and the PLUTO-SAT cross section extrapolated to $P^2 = Q^2 = 0$, using GVDM form factors, in comparison with the old (Rosner, ref. 21) and the improved (AMM, ref. 26) VDM predictions.

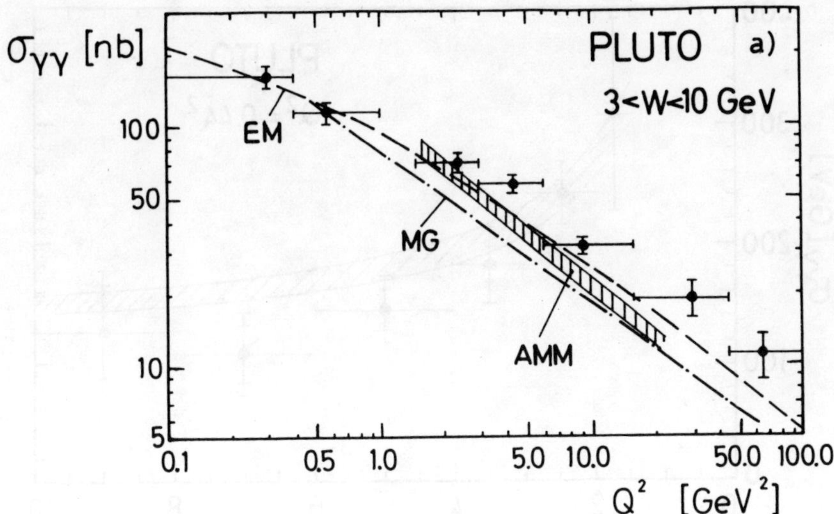

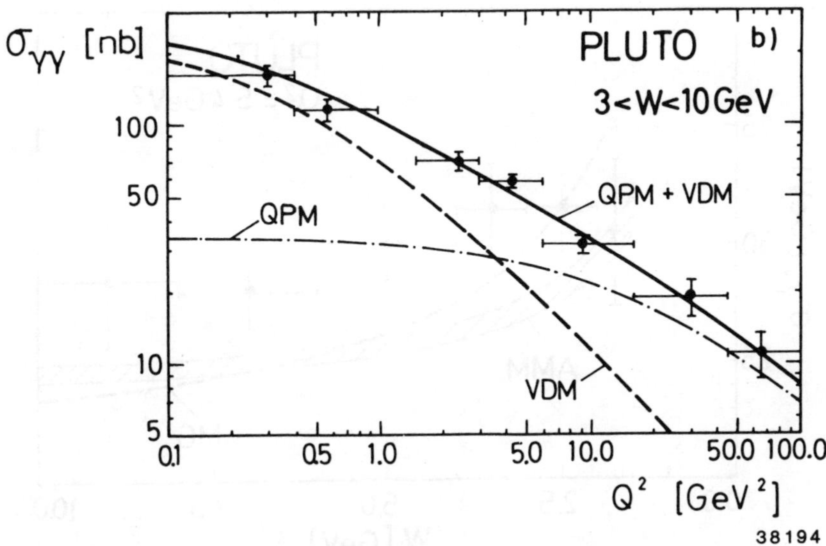

Fig. 16: The Q^2 dependence of $\sigma_{\gamma\gamma}^{tot}$, for $3 < W < 10$. The PLUTO data are compared (a) to two versions of EVDM (by Etim and Masso [25] (EM) and by Maor and Gotsman [24] (MG)) and to the factorization approach by Alexander et al. [26] (AMM); and (b) to a GVDM prediction (VDM, see text eq. (19)) and the QPM prediction.

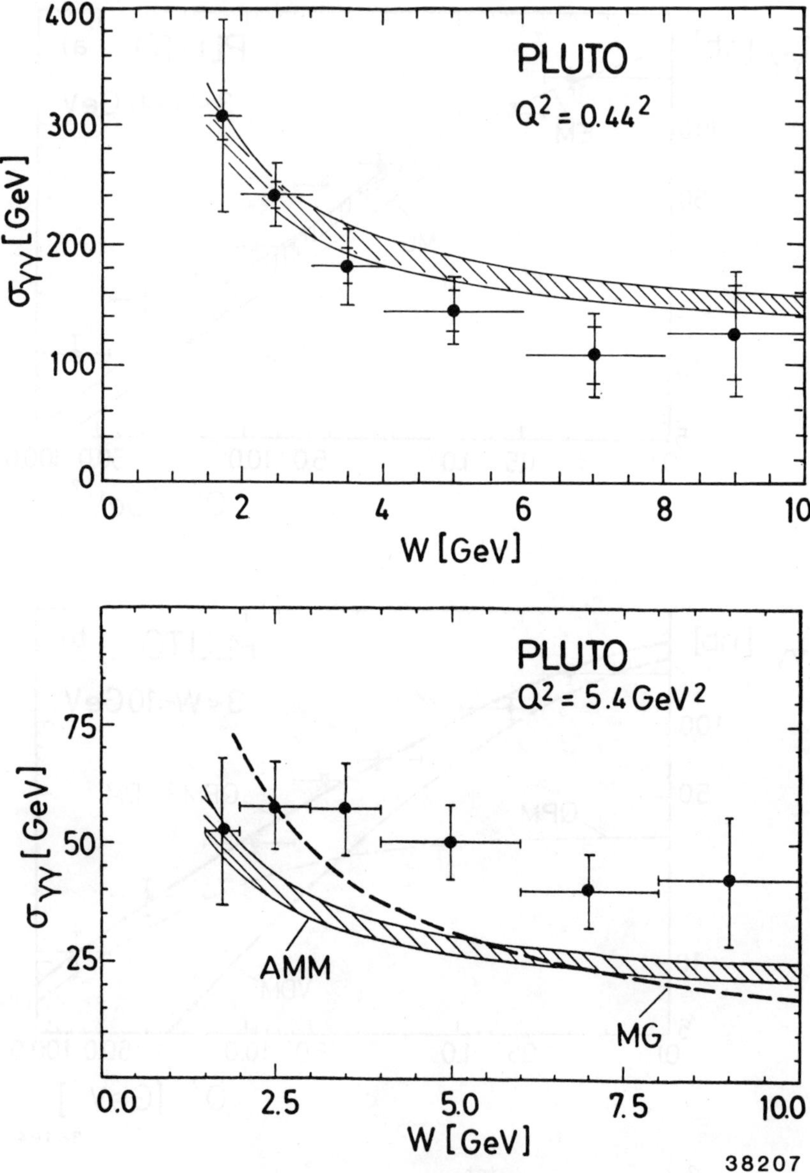

Fig. 17: The W dependence of $\sigma_{\gamma\gamma}^{tot}$, (a) at $\langle Q^2 \rangle = 0.44$ GeV^2 and (b) at $\langle Q^2 \rangle = 5.4$ GeV^2. The curves are explained in Fig. 16.

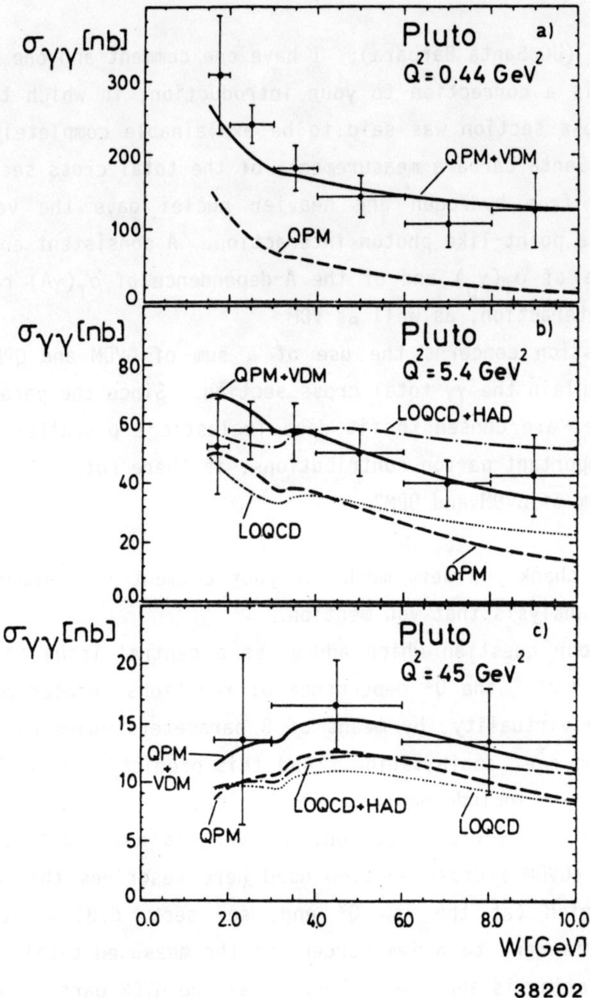

Fig. 18: The W dependence of $\sigma_{\gamma\gamma}^{tot}$ at $\langle Q^2 \rangle$ = 0.44, 5.4 and 45 GeV². The curves labeled QPM and QPM + VDM are defined in Fig. 16, the one labeled LOQCD is from the leading order QCD prediction[27] for F_2^γ, with Λ = 200 MeV, and LOQCD + HAD includes a hadronic component, $F_2^{had} = \alpha \cdot 0.2(1-x)$.

Questions and Discussion

D.O. CALDWELL (UC Santa Barbara): I have one comment and one question. The comment is a correction to your introduction, in which the $Q^2 = 0$ γ-p total cross section was said to be explainable completely by VDM. In fact, the Santa Barbara measurements of the total cross sections for real photons from hydrogen and heavier nuclei gave the very first evidence for a point-like photon interaction. A consistent analysis of the magnitude of $\sigma_T(\gamma_p)$ and of the A-dependence of $\sigma_T(\gamma A)$ required a point-like interaction, as well as VDM.

The question concerns the use of a sum of GVDM and QPM contributions to explain the $\gamma\gamma$ total cross section. Since the parameters of the GVDM model are chosen to fit deep inelastic e-p scattering, which surely has important parton contributions, is there not double counting in using a sum of GVDM and QPM?

G. KNIES: I thank you very much for your comment - I simply was not aware of the analysis that you mention.

As to your question which addresses a central issue of my talk. The GVDM "predicts" the Q^2 dependence of reactions induced by photons with variable virtuality, by means of 8 parameters adjusted to fit ep scattering and e^+e^- annihilation. And this prediction should - as you state correctly - include parton contributions.

The absolute $\gamma\gamma$ cross section, however, is not really predicted. The absolute (GVDM-) cross section used here describes this data in a kinematic region (at the low Q^2 end, see sect. 6.3) where the QPM contribution is down to a few percent of the measured total. Here the GVDM normalization is anchored. Therefore, the GVDM part is not underestimated. With this correct or consistent normalization the GVDM part is lower than the data at large Q^2 (see Fig. 16). The excessive cross section at large Q^2 is - that is my understanding - due to the point-like (QED) coupling of the target photon to partons, as calculated by the QPM. The GVDM part is responsible for the hadron-like coupling (thru ρ, ω, ϕ...) of the target photon to partons, which of course interact with the probe photon (at large Q^2) as incoherent partons. At low Q^2, the probe photon acts as vector meson, mainly. To the extent that QPM does not 'contain' VDM-like γ-hadron transitions for the target photon, I think there is no double counting in the sum of GVDM and QPM. The data seem to say: the sum of QPM and GVDM is neither too low, i.e. is sufficient, nor too high, i.e. no double-counting.

A. JANAH (Kansas State Univ.): I would like to make a comment. The no-tag and low Q^2 single-tag data, where VDM effects are being invoked, are also precisely where integer charge quark (gauge ICQ) effects may be important.

The color-octet charge suppression factor $m_g^2/(Q^2+mg^2)$ itself, integrated over Q^2, will affect the W dependence of the effective point like contribution to the $\gamma\gamma$ cross section $\sigma_{\gamma\gamma}$ (W). Moreover, the presence of charged gluon productions to the pointlike $\gamma\gamma$ cross-section may drastically alter the expected W-dependence of this quantity. Naively, one expects a slower W fall-off for the gluon contribution compared to the quark contribution. This needs to be checked carefully.

To test this pointlike contribution, however, one should, to be safe, put in some sort of transverse momentum cut $p_T > p_{TC}$, and attempt to measure $\sigma_{\gamma\gamma}$ (w, p_{TC}). For jet-like events, p_T can be taken to be p_T (jet). P_{TC}, the p_T cut-off, should be 2 GeV at least. This will, of course, limit the data to high W.

RELATED PROCESSES

Paul Kessler

Laboratoire de Physique Corpusculaire, Collège de France, Paris
FRANCE

ABSTRACT

Two types of processes, related to photon-photon collisions,
are considered: Deep inelastic Compton scattering, and photon pair production. The relevant theoretical and experimental literature is reviewed.

1. Introduction

As is well known, high-energy, high-p_T reactions involving photons allow an easier analysis in perturbative QCD, as compared with
purely hadronic processes, due to the direct coupling between photon
and quark; indeed such reactions tend to involve a smaller number of
parton subprocesses; in addition their calculation is less arbitrary,
as there are less (theoretically unknown) distribution and/or fragmentation functions. One may thus conclude that processes involving
photons are better suited for investigating the structure of hadrons
and the dynamics of their constituents.

If that conclusion is true for processes involving one photon
(photoproduction, direct-photon production), it should be true a
fortiori for processes with two photons: (i) Photon-photon collisions; (ii) deep-inelastic Compton scattering; (iii) photon pair
production (see fig. 1). Here we shall discuss reactions of type
(ii) and (iii). It is to be noticed that, to some extent, they were
already discussed in previous Workshops on Photon-Photon Collisions
1-3).

2. Deep inelastic Compton scattering (DICS)

That reaction was first considered by Bjorken and Paschos [4] on the basis of Feynman's parton model. The authors emphasized that the reactions $\gamma p \longrightarrow \gamma X$ and $e p \longrightarrow e X$ are equivalent tools for investigating the internal structure of the proton. In both cases the incident particle is assumed to be scattered incoherently from the pointlike constituents (partons). The main difference is that, in deep inelastic Compton scattering, the cross section should be proportional to the fourth power of the parton charge; while in deep inelastic electron scattering, the square of that charge is involved.

These considerations remain basically valid in the quark-parton model, i. e. in lowest-order perturbative QCD (see fig.2,(i)).

The subsequent theoretical literature has been essentially devoted to going beyond the lowest order, i. e. computing various other contributions (see fig. 2, (ii)-(vi)); showing, however, that under certain conditions those contributions may be kept small with respect to the Born term (i). (Since the latter allows one to check the quark charge and propagator, it is indeed important to define the conditions where that term predominates).

One notices that all terms (i), (ii), (iii), (iv) are of order $\alpha^2 \alpha_s^0$, since all structure functions and fragmentation functions involving photons ($G_{q/\gamma}$, $G_{g/\gamma}$, $D_{\gamma/q}$, $D_{\gamma/g}$) contain (using the leading-log approximation) a factor $\ln (Q/\Lambda) \sim 1/\alpha_s$. Terms (vi) are of order $\alpha^2 \alpha_s$. Term (v) is of order $\alpha^2 \alpha_s^2$, but - as shown by Combridge [5] - may become significant in specific kinematic configurations.

There are differences among various authors, in the sense that they include or neglect various types of graphs, and use different theoretical ingredients (distribution and fragmentation functions, definition of Q^2, value of Λ ...). Nevertheless, qualitatively at least, their conclusions are all compatible.

Tu and Wu [6] consider only contributions (i), (ii), (iii); moreover they neglect those diagrams in (ii) and (iii) which involve γg coupling.

They also make the following cut-offs: In (ii) the interacting quark emanating from the photon should take over at least 80% of its energy; in (iii) the photon radiated from the quark produced should take over at least 80% of the latter's energy. They assume E_γ^{lab} = 150 Gev.

Calling x the scaling parameter used for characterizing the parton content of the photon, conclusions are given for fixed $x \simeq 0.1$:

$$(ii)/(i) \simeq (iii)/(i) \simeq 0.1$$

where we call (i), (ii), (iii) the respective yields of diagrams (i), (ii) (total), (iii) (total); when x is increased, both above ratios steadily decrease.

Thus, according to Tu and Wu, for $x \gtrsim 0.1$ the Born term should provide at least 80% of the total yield.

Igushi, Niégawa and Uranishi [7] compute diagrams (i), (ii), (iii), (iv). In the distribution and fragmentation functions involving photons, they include a VDM part. No cut-offs are applied. Yields are given as functions of p_T of the outgoing photon, at fixed values of the incident photon's lab energy E_γ^{lab}, as well as of the outgoing photon's lab energy E or of the photon's c. m. emission angle θ_{cm} (fig. 3). One here notices that the Born term tends to dominate for $p_T \gtrsim 2$ GeV/c at E_γ^{lab} = 150 GeV, and $p_T \gtrsim 10$ GeV/c at E_γ^{lab} = 5000 GeV.

The next paper by Igushi and Niégawa [8] contains the same calculations, but this time taking account of smearing effects, affecting both the initial partons and the outgoing particles resulting from parton fragmentation; they do this in an approximate way, i. e. introducing only average transverse momenta. The yields modified by smearing are shown in fig. 4. One notices that smearing effects appear significant, and tend to increase the yield, at low p_T. However the Born term still tends to dominate for $p_T \gtrsim 3$ GeV/c at an incident energy of 150 GeV and $p_T \gtrsim 10$ GeV/c at E_γ^{lab} = 5000 GeV.

Duke and Owens [9] as well compute the terms (i)-(iv), and compare their total contribution with that of (i) alone. It is obvious from fig. 5-7 that the Born term dominates for x_T ($\equiv 2 p_T / s$) $\gtrsim 0.4$.

Those authors also compute the total contribution of terms (vi); they show (fig. 8) that $R = \left[(i) + (vi)\right]/(i)$ remains very close to 1, even at small x_T ($\gtrsim 0,2$), for $y = 0$; and at somewhat higher x_T ($\gtrsim 0,4$) for finite y.

Here again one concludes that the Born term should always dominate in the high-x_T range.

A paper by Fontannaz and Schiff [10] contains a computation of the box diagram (v). As fig. 9 shows, its contribution can become significant, as compared to that of the Born term, in the very forward direction (corresponding to small values of the scaling parameter x, i. e. to a large gluon content of the proton).

Recent work by Aurenche, Douiri, Baier, Fontannaz and Schiff [11] represents the most complete theoretical study of DICS. Fig. 10 shows a comparison - computed by the authors for various values of E_γ^{lab} and various p_T values, with y varying continuously - between the Born term (i), the sum of leading terms (i)-(iv), and the box diagram (v). The Born term, here again, appears to saturate the total contribution of leading terms when x_T is large enough ($\gtrsim 0.3$-0.4). The box diagram can become significant, again, at large positive values of y.

Fig. 11, 12 show a comparison, as given by the same authors, between the sum of all contributions (i)-(vi) and the Born term alone, for various incident energies, with p_T varying continuously (in fig. 11 $y = 0$, while in fig. 12 positive values of y were integrated over). In addition fig. 14 shows the ratio $\left[(i) + (vi)\right]/(i)$. Again one notices that the total correction to the Born term tends to become small at sufficiently large values of the outgoing photon's transverse momentum.

Finally Fig. 13 shows a fit, presented by the same authors, of their predictions with preliminary experimental data from the NA 14 Collaboration at CERN; account is taken of the contamination from indirect photons due to π^0, η production.

One concludes that deep inelastic Compton scattering seems to provide a good check for perturbative QCD at large p_T, as the perturbative series appears to converge quite well. A good agreement is obtained, as one notices, between theoretical predictions and recent experimental data. In particular the standard fractional-charge model for quarks thus appears to be confirmed.

3. Photon pair production

That process was first considered in the pioneering paper of Berman, Bjorken and Kogut [12], where $q \bar{q} \rightarrow \gamma\gamma$ was mentioned among the elementary hard-collision processes. Again, as for DICS, continuous treatment in the theoretical literature started only in recent years.

Here again the fundamental contribution is provided by the Born term (fig. 14, (i)), while additional contributions (fig. 14, (ii)-(v)) are to be considered as well. One notices that terms (i) and (iii) are all of order $\alpha^2 \alpha_s^0$ (except for the last diagram of (iii), which is $\sim \alpha^2 \alpha_s^2$), while terms (iv) are of order $\alpha^2 \alpha_s$ and term (ii) $\sim \alpha^2 \alpha_s^2$.

Considering only the Born term (i), Soh, Pac, Lee and Choi [13] suggest that it would be interesting to compare it with $g\gamma$ production in hadronic collisions (assumed to be dominated by $q \bar{q} \rightarrow g\gamma$) in order to derive the ratio α_s/α.

Krawczyk and Ochs [14] compare reactions $\pi^\pm p \rightarrow \gamma\gamma X$, $\pi^\pm p \rightarrow e^+ e^- X$ and $\pi^\pm p \rightarrow \pi^0 \pi^0 X$, all proceeding via $q \bar{q}$ annihilation (lowest-order diagrams only). They show in particular a characteristic difference between π^+ and π^- production of photon pairs (even more striking than in the case of lepton pair production): While the ratio

$$R \equiv \frac{\sigma(\pi^- p \rightarrow \gamma\gamma X)}{(\pi^+ p \rightarrow \gamma\gamma X)} \simeq \frac{e_u^4 \, u(x) + e_d^4 \, \bar{d}(x)}{e_d^4 \, d(x) + e_u^4 \, \bar{u}(x)}$$

(where u, d are distribution functions of the quark in the proton) goes to 1 in the limit of the sea region ($x \rightarrow 0$), it can become as large as $\simeq 32$ in the valence region.

Hemmi [15] suggested that, on the basis of diagram (i), it might be possible to obtain a hint on the quark mass by measuring the production cross section and the angular distribution of photon pairs.

Combridge [5], and independently Carimalo, Crozon, Kessler and Parisi [16], showed that the box diagram (ii) might compete with the Born term (i) when $W_{\gamma\gamma}/\sqrt{s}$ becomes small (see fig. 15; notice that the contribution from (ii) is quite model-dependent).

Contogouris, Marleau and Pire [17] compare $p\,p \rightarrow \gamma\gamma X$ and $p\,p \rightarrow \pi^0\pi^0 X$. In the calculation of the former process, they include terms (i), (ii), and in addition the first diagram of (iii). The comparison, shown in fig. 16, depends on the parameter Δz_e, defining $z_e = p_{T2}/p_{T1}$ (p_{T2}, p_{T1} being the transverse momenta of both photons or pions produced) and $z_e^{min} = 1 - \Delta z_e$. One notices that, when Δz_e is kept sufficiently small, the ratio of yields $(\gamma\gamma/\pi^0\pi^0)$ is rising very fast with increasing transverse momentum.

The same authors also provide a guess for the $O(\alpha_s)$ corrections to the Born term, i. e. the set of diagrams (iv), by using the "soft-gluon approximation". They obtain an enhancement factor (which happens to be the same for photon pair and for lepton pair production) of

$$1 + \left[\alpha_s/(2\pi)\right] C_F \pi^2 \quad \text{with } C_F = 4/3,$$

i. e. $\simeq 50\%$ with $\alpha_s = 0.25$.

Berger, Braaten and Field [18] compute the contributions of (i), (ii) and most of the bremsstrahlung diagrams (iii); see fig. 17, where the corresponding yields are shown, for $\pi^- p \rightarrow \gamma\gamma X$, as functions of z_e (one notices that, among the separate contributions of diagrams (iii), that of the first one by far predominates).

In a recent paper of Carimalo, Crozon, Kessler and Parisi [19], account is taken of a possible resonant contribution (v). The value such a contribution can take is shown in fig. 18 (for two different machine energies, and considering the case where both γ's come out at 90° in the overall c. m. frame, so that $W_{\gamma\gamma} = 2\ p_T$), under various assumptions: arbitrary mass M_R (the authors let it vary continuously); $J_R = 0$; $\Gamma(R \to \gamma\gamma) = 1$ keV; in the expression of the production cross section, $\delta(M_R - 2\ p_T)$ is replaced by $(2\Delta p_T)^{-1}$, where Δp_T is the experimental momentum resolution of either photon, for which a realistic value is taken. It can be seen that under the conditions assumed, at $E = 540$ GeV, a resonant contribution would predominate over the $\gamma\gamma$ continuum for $M_R \lesssim 12$ GeV.

Till now, only one experimental measurement of photon pair production was reported in the literature, namely the one performed at the CERN ISR by Kourkoumelis et al. [20]. In that experiment both photons were measured at $\simeq 90°$ in the overall c. m. frame (here the same as the lab frame) with opposite p_T; photons unaccompanied by hadrons were selected, thus reducing the background from π^0 and other neutral mesons simulating single photons, and also suppressing the contribution of bremsstrahlung diagrams.

Selecting events with both $p_T > 3$ GeV/c, 31 ± 16 $\gamma\gamma$ pairs were obtained after background subtraction, together with 25 ± 6 $e^+ e^-$ pairs (measured under the same conditions). 7 ± 4 of those electron pairs were attributed to resonant contributions due to various bottonium states, so that finally the following ratio of yields was obtained:

$$\gamma\gamma / e^+ e^- \text{ (continuum)} = 1.7 \pm 1$$

while theoretically (taking account of contributions (i) + (ii) to $\gamma\gamma$ production, as compared to Drell-Yan lepton pair production), one expects that ratio to be ≈ 1.

In spite of its crudeness, that result appears - according to a recent study by Jyamaran, Rajasekaran and Rindani [21] - to provide a decisive test against the gauged integer-charge model which has been patronized by some theorists in the last years; that model, indeed, predicts considerably higher values of the ratio $\gamma\gamma / e^+ e^-$ (fig. 19).

Finally, in a very recent study [22], Aurenche, Baier, Douiri, Fontannaz and Schiff are showing preliminary results of a calculation including the $O(\alpha_s)$ corrections to the Born term, i.e. diagrams (iv). The latter appear to give a large contribution (see fig. 20), but in fact that contribution crucially depends on the choice of $z_e^{min} \equiv (p_{T2}/p_{T1})^{min}$; when z_e^{min} comes closer to 1, it sharply decreases.

We here conclude that, so far, $\gamma\gamma$ production has allowed one to check the fractionnaly-charged quark model. In order to see some of the finer effects (contribution of the box diagram, of resonances ...), more precise measurements will have to be performed in the future.

Acknowledgments

The author is grateful to D. Schiff and M. Fontannaz for useful discussions on the subject treated in this report.

He also wishes to express his gratitude to R. Lander and his staff for the kind hospitality extended to him at Tahoe.

References

1) W. J. Willis, Proceedings of the Fourth International Colloquium on Photon-Photon Interactions, Paris, ed. G. London (World Scientific, Singapore, 1981), p. 331.

2) P. M. Zerwas, Proceedings of the Fourth International Colloquium on Photon-Photon Interactions, Paris, ed. G. London (World Scientific, Singapore, 1981), p. 361.

3) R. D. Field, Proceedings of the Fifth International Workshop on Photon-Photon Collisions, Aachen, ed. Ch. Berger (Springer-Verlag, Berlin, 1983), p. 1.

4) J. D. Bjorken and E. A. Paschos, Phys. Rev. 185, 1975 (1969).

5) B. L. Combridge, Nucl. Phys. B174, 243 (1980).

6) T. S. Tu and C. M. Wu, Nucl. Phys. $\underline{B156}$, 493 (1979).

7) K. Iguchi, A. Niēgawa and Y. Uranishi, Prog. Theor. Phys. $\underline{63}$, 1711 (1980).

8) K. Iguchi and A. Niēgawa, Z. Phys. $\underline{C9}$, 135 (1981).

9) D. W. Duke and J. F. Owens, Phys. Rev. $\underline{D26}$, 1600 (1982).

10) M. Fontannaz and D. Schiff, Z. Phys. $\underline{C14}$, 151 (1982).

11) P. Aurenche, A. Douiri, R. Baier, M. Fontannaz and D. Schiff, Z. Phys. $\underline{C24}$, 309 (1984).

12) S. M. Berman, J. D. Bjorken and J. B. Kogut, Phys. Rev. $\underline{D4}$, 3388 (1971).

13) K. S. Soh, P. Y. Pac, H. W. Lee and J. B. Choi, Phys. Rev. $\underline{D18}$, 751 (1978).

14) M. Krawczyk and W. Ochs, Phys. Lett. $\underline{79B}$, 119 (1978).

15) S. Hemmi, Prog. Theor. Phys. $\underline{63}$, 1073 (1980).

16) C. Carimalo, M. Crozon, P. Kessler and J. Parisi, Phys. Lett. $\underline{98B}$, 105 (1981).

17) A. P. Contogouris, L. Marleau and B. Pire, Phys. Rev. $\underline{D25}$, 2459 (1982).

18) E. L. Berger, E. Braaten and R. D. Field, Nucl. Phys. $\underline{B239}$, 52 (1984) (see also 3)).

19) C. Carimalo, M. Crozon, P. Kessler and J. Parisi, Phys. Rev. $\underline{D30}$, 576 (1984). That paper also contains a background study in $\gamma\gamma$ production, which is being reported in a parallel session of this Workshop.

20) C. Kourkoumelis et al., Z. Phys. $\underline{C16}$, 101 (1982).

21) T. Jayaraman, G. Rajasekaran and S. D. Rindani, Univ. of Madras Preprint (1984).

22) D. Schiff, Invited Talk given at the XVth Symposium on Multiparticle Dynamics, Lund, 1984. Preprint LPTHE Orsay 84/36.

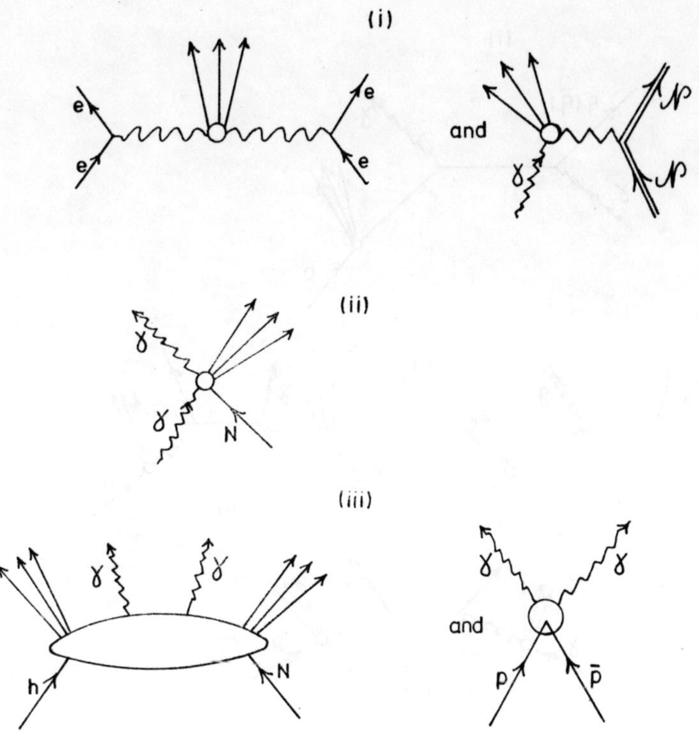

Fig. 1. Various types of two-photon reactions.

(i) Photon-photon collisions in ee storage rings or colliders, and Primakoff effect.

(ii) Deep inelastic Compton scattering from a nucleon target.

(iii) Photon pair production in hard collisions and in p$\bar{\text{p}}$ annihilation.

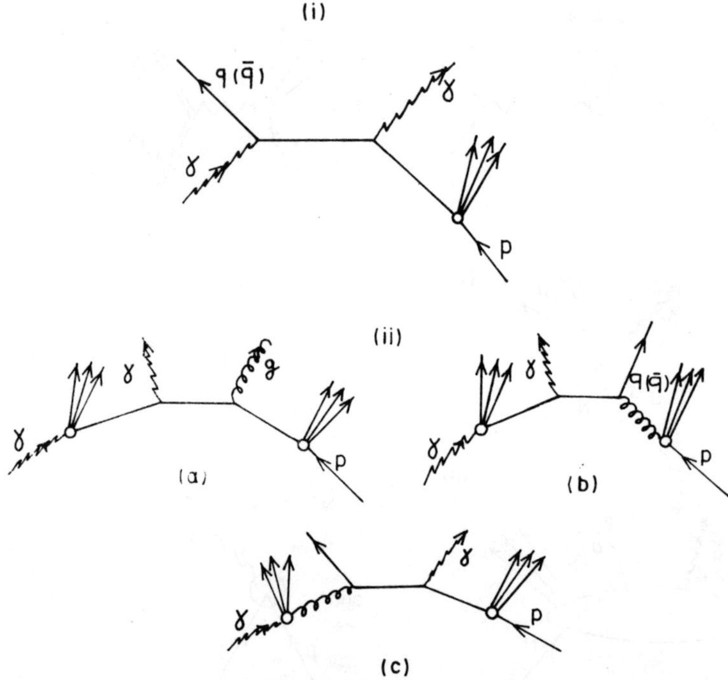

Fig. 2. Contributions to deep inelastic Compton scattering from a
proton target.

(i) Born term.

(ii) Contributions involving the photon's structure function.

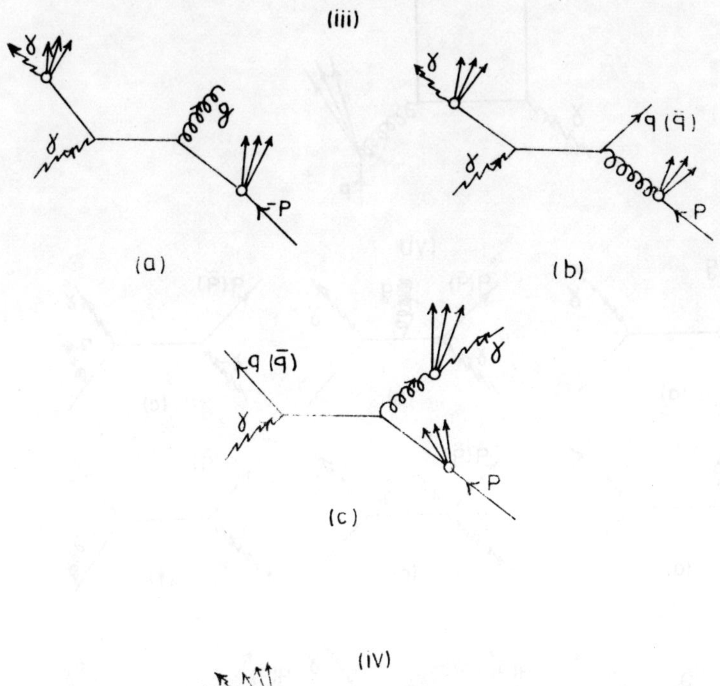

Fig. 2 (continued)

    (iii) Contributions involving quark (or gluon) bremsstrahlung.

    (iv) Contributions involving both the photon's structure
        function and quark (or gluon) bremsstrahlung.

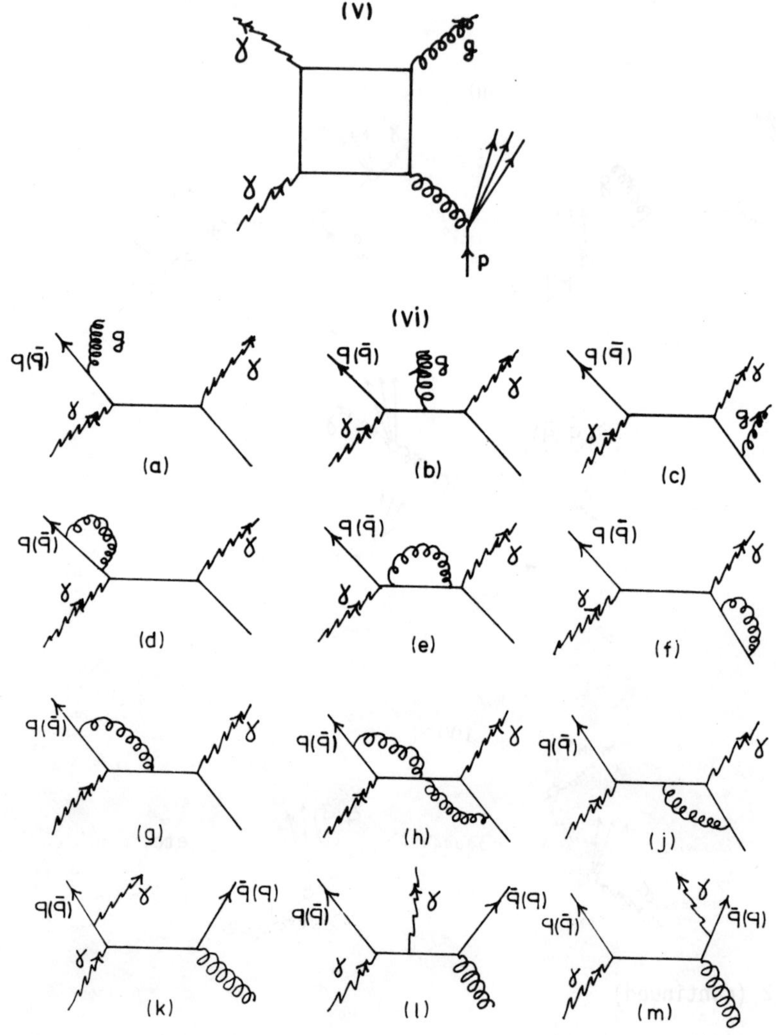

Fig. 2 (continued)

(v) Box diagram.

(vi) $O(\alpha_s)$ corrections to the Born term (for simplicity, the proton vertex has been omitted).

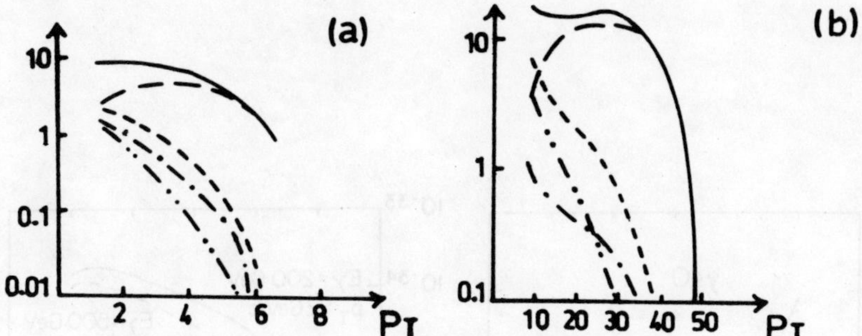

Fig. 3. p_T^5 times $E\, d\sigma/d^3p$ (in units nb GeV^3) vs. p_T (GeV/c) for DICS. Curves are characterized as follows: ——— total; ————— contribution of (i); ——————— (ii); —·—·— (iii); —··—··— (iv). (From ref. 7)

(a) E^{lab} = 150 GeV, E = 60 GeV. (b) E^{lab} = 5000 GeV, θ_{cm} = 90°

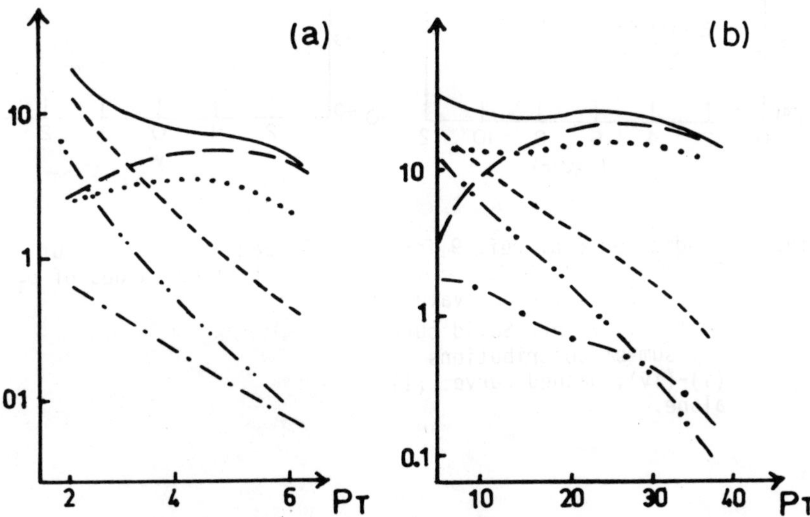

Fig. 4. p_T^5 times $E\, d\sigma/d^3p$ (in units nb GeV^3) vs. p_T (GeV/c) for DICS, including smearing effects. Curves are characterized as in fig. 3; in addition: ········ total without smearing. (From ref. 8.)

(a) E^{lab} = 150 GeV, θ_{cm} = 90°. (b) E^{lab} = 5000 GeV, θ_{cm} = 90°.

334

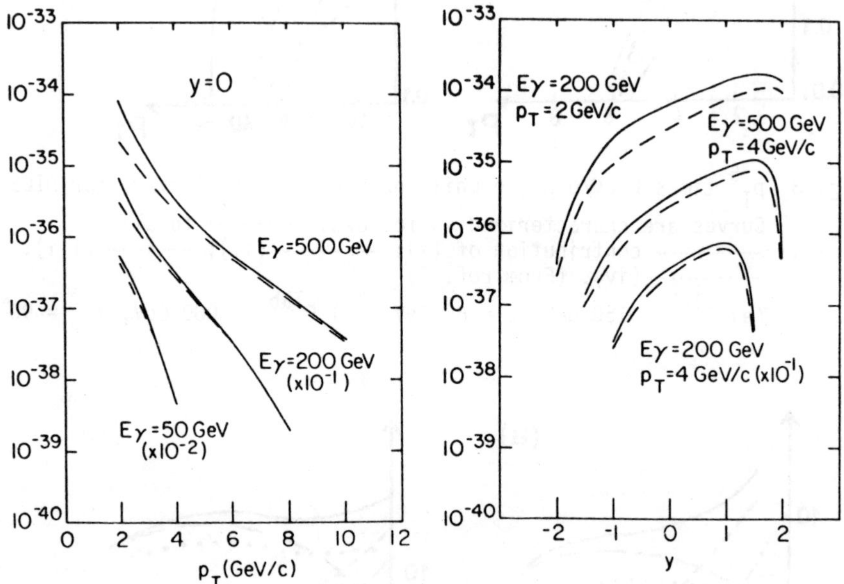

Fig. 5. Prediction from ref. 9 for $E\,d^3\sigma/dp^3$ ($cm^2\ GeV^{-2}\ c^3$), at $y = 0$ and various values of E_γ^{lab}, vs. p_T. Solid curve: sum of contributions (i)-(iv); dashed curve: (i) alone.

Fig. 6. Same as fig. 5, but vs. y at fixed values of p_T.

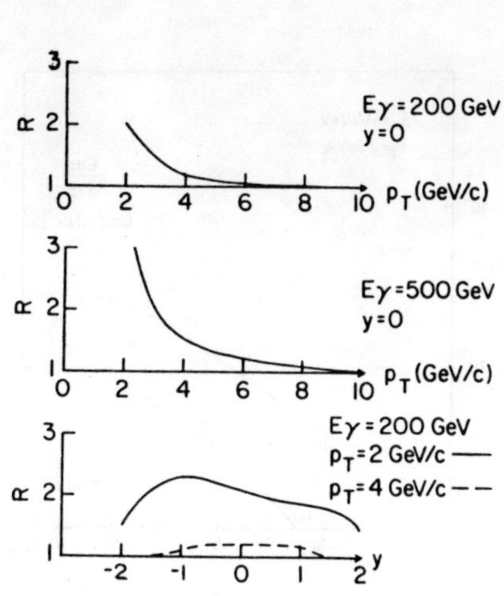

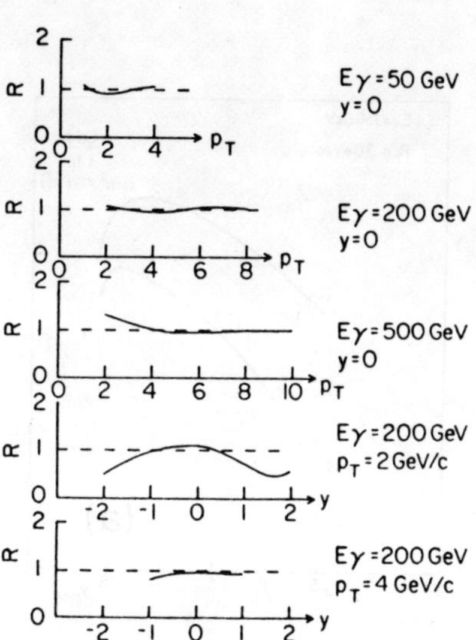

Fig. 7. Ratio R of the sum of contributions (i)-(iv) to the contribution of (i) alone, vs. p_T, at fixed values of E_{lab} and y (or vs. y, at fixed values of E_{lab} and p_T). (From ref. 9.)

Fig. 8. Ratio R of $[(i) + (vi)]/(i)$ vs. p_T at fixed values of E_{lab} and y (or vs. y at fixed values of E_{lab} and p_T). (From ref. 9.)

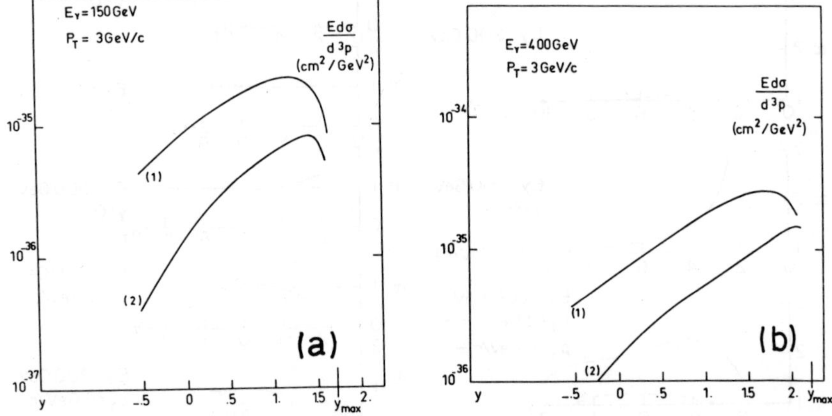

Fig. 9. Single inclusive cross section vs. y at (a) E_{lab} = 150 GeV, p_T = 3 GeV/c, (b) E_{lab} = 400 GeV, p_T = 3 GeV/c. Curve (1) shows the contribution of (i), curve (2) that of (v). (From ref. 10.)

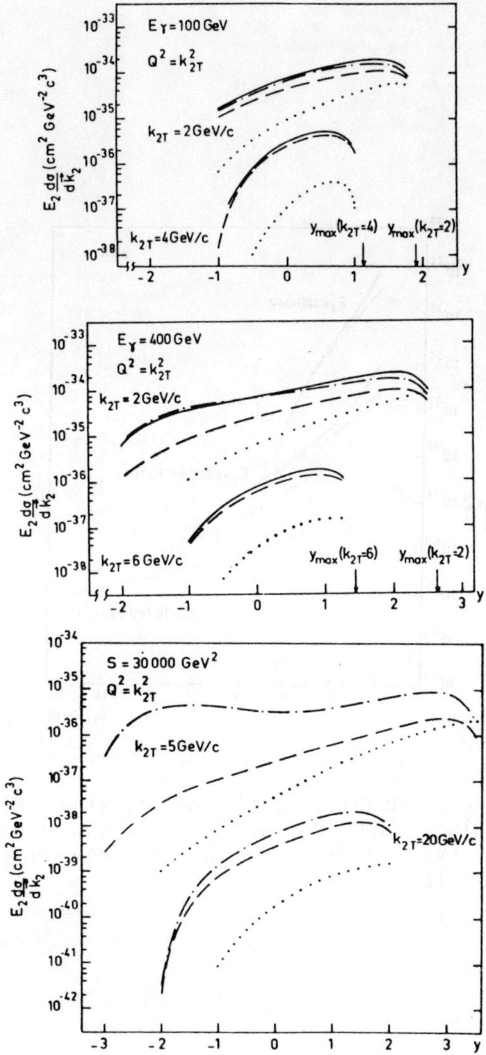

Fig. 10. Single inclusive cross section vs. y at fixed p_T values for three different energies. The solid (dash-dotted) curve represents the total contribution of (i)-(iv) with distribution and fragmentation functions, involving the photon, given by leading-log (quark-parton model) expressions. The dashed curve represents (i) alone, while the dotted one shows the contribution of (v). (From ref. 11.)

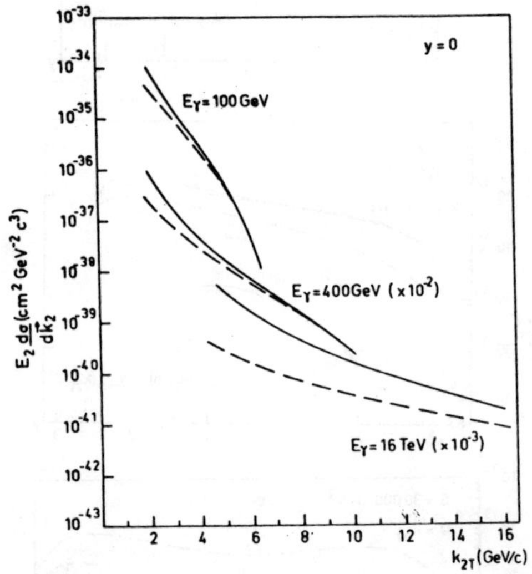

Fig. 11. Single inclusive cross section vs. p_T at y = 0 and for three
energy values. The solid curve is the sum of all contributions (i)-(vi), while the dashed one represents (i) alone.
(From ref. 11.)

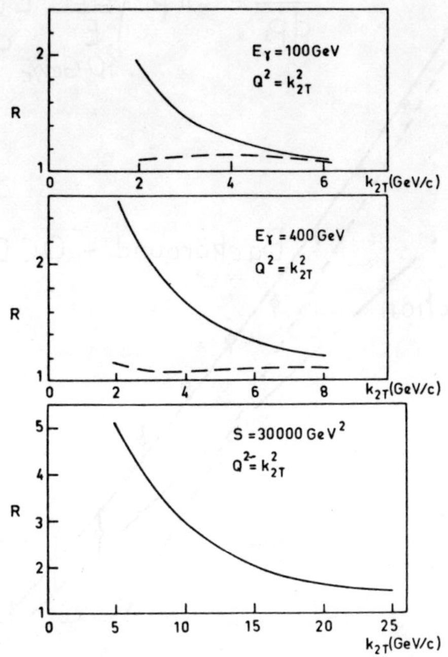

Fig. 12. Ratio R vs. p_T for three different energies, y being integrated over between 0 and y_{max}, of the sum of all terms (i)–(vi) to (i) alone (solid curve), and of (i) + (vi) to (i) (dashed curve). In (c) the dashed curve practically coincides with the abscissa. (From ref. 11.)

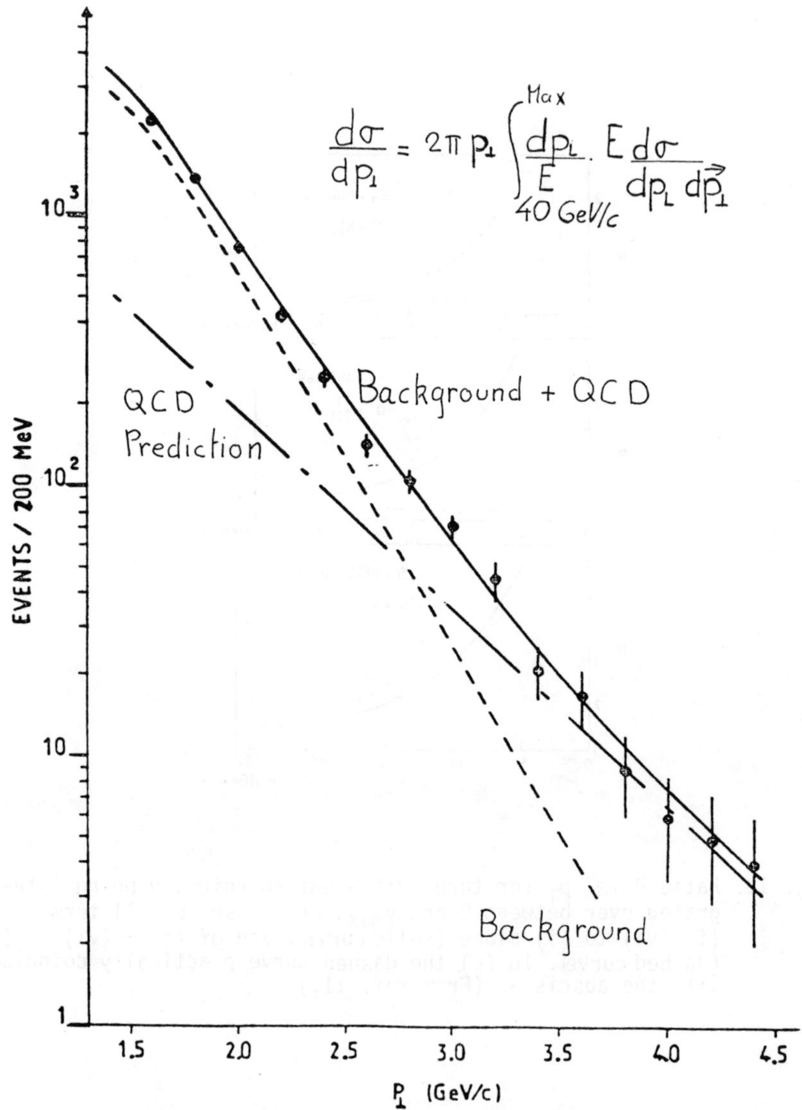

Fig. 13. Fit of preliminary data on DICS from the NA 14 Collaboration at CERN. The theoretical prediction includes the indirect-photon background due to π^0, η, ... (Privately communicated by the authors of ref.11.)

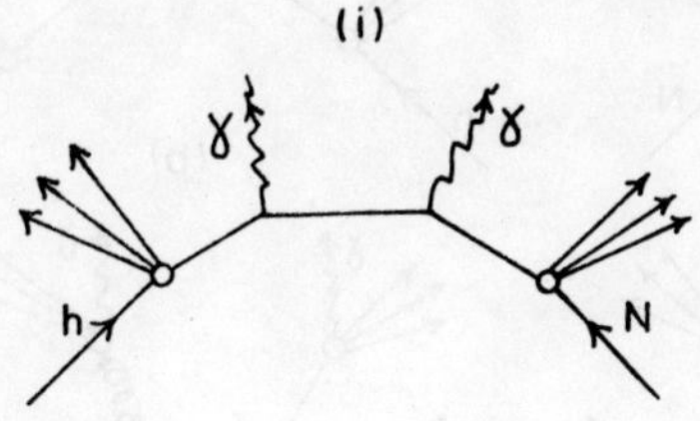

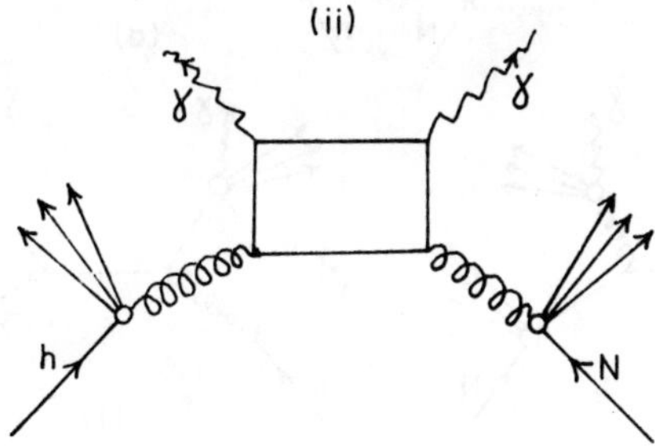

Fig. 14. Contributions to photon pair production in hadron-nucleon collisions.

(i) Born term.
(ii) Box diagram.

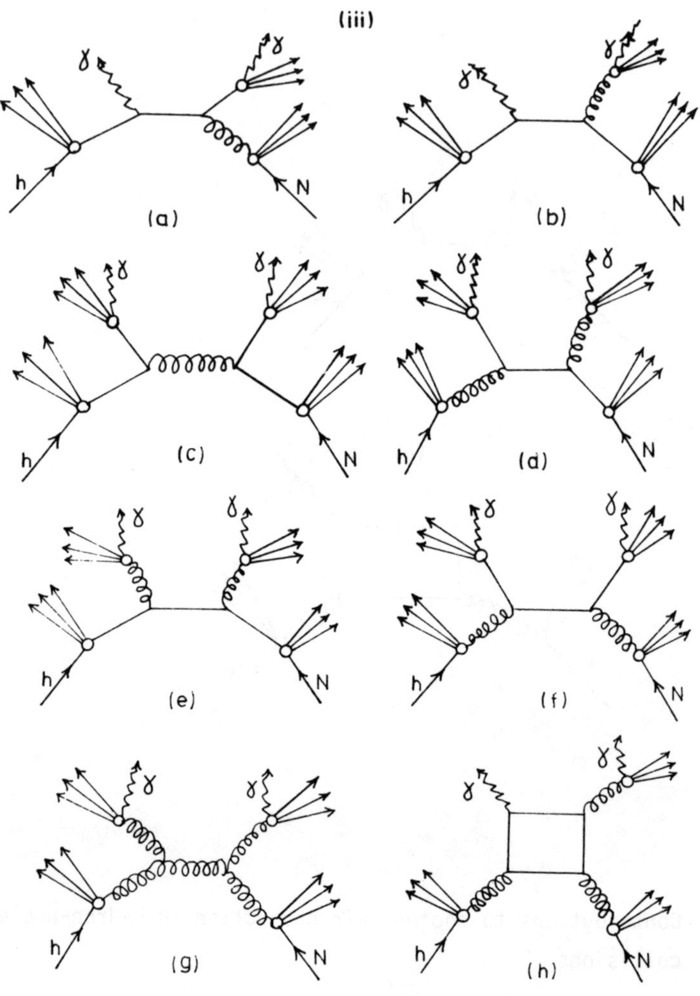

Fig. 14 (continued)

(iii) Contributions involving quark (or gluon) bremsstrahlung.

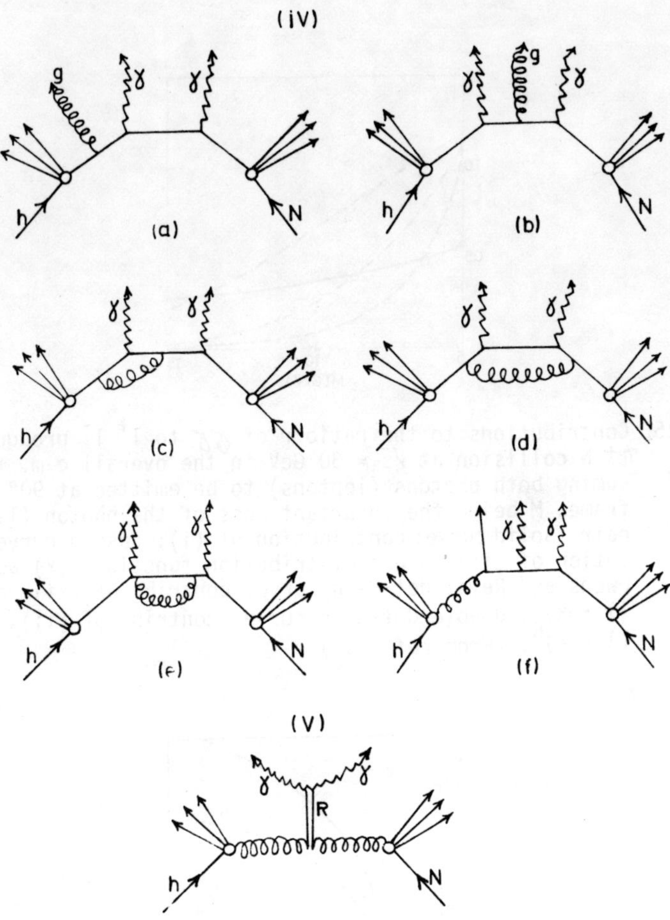

Fig. 14 (continued)

(iv) $O(\alpha_s)$ corrections to the Born term.
(v) Possible resonant contributions.

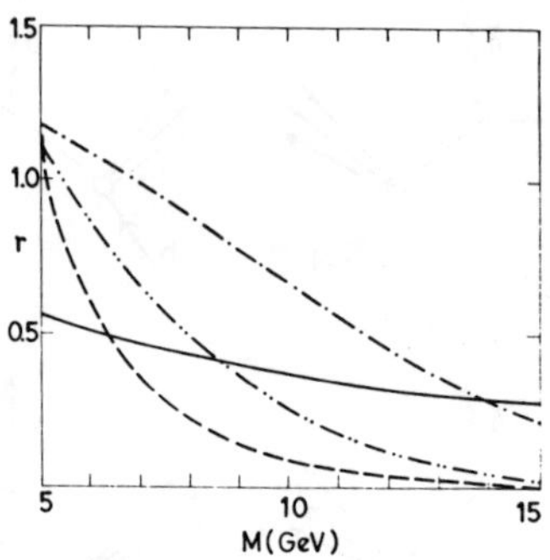

Fig. 15. Contributions to the ratio r of $\gamma\gamma$ to $1^+ 1^-$ production in a
π^+ N collision at $\sqrt{s}$ = 30 GeV in the overall c.m. frame, assuming both photons (leptons) to be emitted at 90° in that
frame, M being the invariant mass of the photon (lepton)
pair. Solid curve: contribution of (i); dashed curve: contribution of (ii), gluon distribution function g(x) according to
Owens and Reya; dot-dash curve: contrib. of (ii), g(x) $\sim$
$(1 - x)^4$; double dot-dash curve: contrib. of (ii), g(x) $\sim$
$(1 - x)^6$. (From ref. 16.)

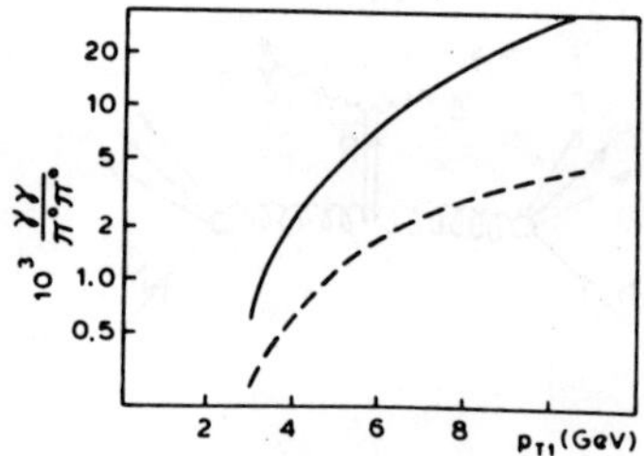

Fig. 16. Ratio of $\gamma\gamma$ to $\pi^0\pi^0$ yield, times 10^3, vs. p_{T1}, at $\sqrt{s}$ = 63
GeV, assuming both photons (pions) to be emitted at 90° in the
overall c.m. frame. Solid curve: Δz_e = 1 GeV/p_{T1}; dashed curve: Δz_e = 1/2. (From ref. 17.)

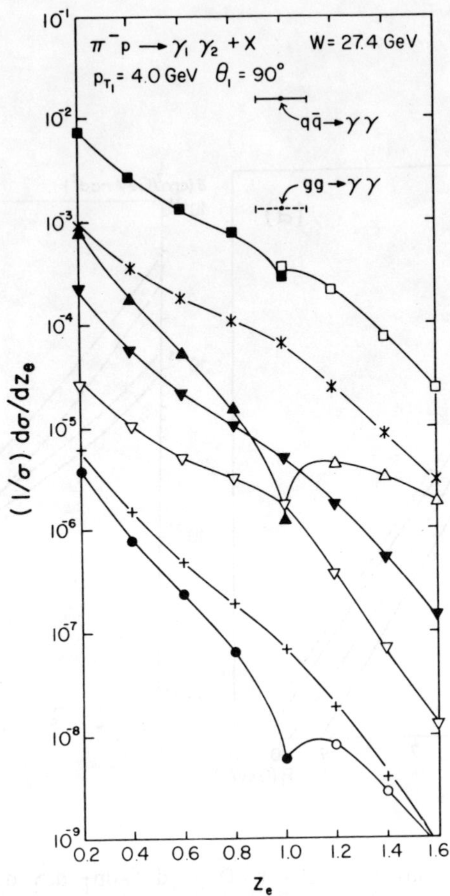

Fig. 17. Contributions, from various bremsstrahlung terms (iii), to the z_e distribution of the "away-side" photon γ_2, for a photon trigger γ_1, under the conditions shown in the figure (θ_2 being integrated over). The individual subprocesses are labeled as follows: $gq \rightarrow \gamma_1(q \rightarrow \gamma_2)$ (solid squares); $gq \rightarrow \gamma_2(q \rightarrow \gamma_1)$ (open squares); $qq \rightarrow (q \rightarrow \gamma)(q \rightarrow \gamma)$ (asterisk); $q\bar{q} \rightarrow \gamma_1(g \rightarrow \gamma_2)$ (up pointing solid triangles); $q\bar{q} \rightarrow \gamma_2(g \rightarrow \gamma_1)$ (up pointing open triangles); $gq \rightarrow (g \rightarrow \gamma)(q \rightarrow \gamma)$ (down pointing solid triangles); $gg \rightarrow (q \rightarrow \gamma)(\bar{q} \rightarrow \gamma)$ (down pointing open triangles); $gg \rightarrow (g \rightarrow \gamma)(g \rightarrow \gamma)$ (plus signs); $gg \rightarrow \gamma_1(g \rightarrow \gamma_2)$ (solid dots); $gg \rightarrow \gamma_2(g \rightarrow \gamma_1)$ (open dots). The $z_e = 1$ delta function contributions from (i) and (ii) have arbitrarily been spread over 0.2 units of z_e. (From ref. 18.)

346

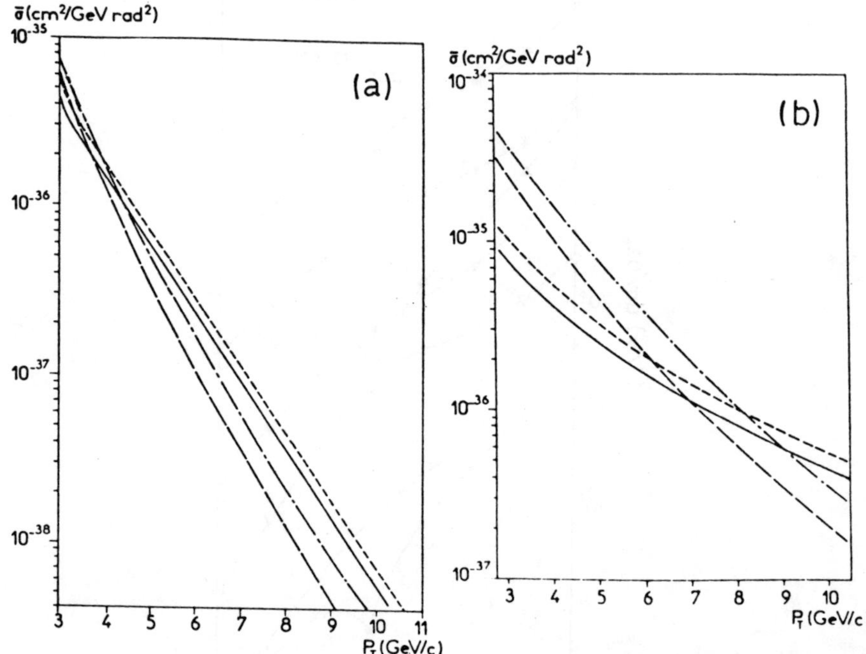

Fig. 18 . Various contributions to $\bar{\sigma} \equiv (d^3\bar{\sigma}/dp_T\, d\theta_1\, d\theta_2)$ ($\theta_1 = \theta_2 = 90°$ in the overall c.m. frame) for (a) $pp \longrightarrow \gamma\gamma X$ at $\sqrt{s}$ = 60 GeV; (b) $p\bar{p} \longrightarrow \gamma\gamma X$ at $\sqrt{s}$ = 540 GeV. Solid curve: contribution of (i); long-dashed curve: contribution of (ii); dot-dash curve: contribution of (v), i.e. of a hypothetical resonant structure of arbitrary mass $M_R = 2\, p_T$ (we let that mass vary continuously. For comparison, the analogous differential cross section for lepton pair production (Drell-Yan effect) is also shown (short-dashed curve). (From ref. 19.)

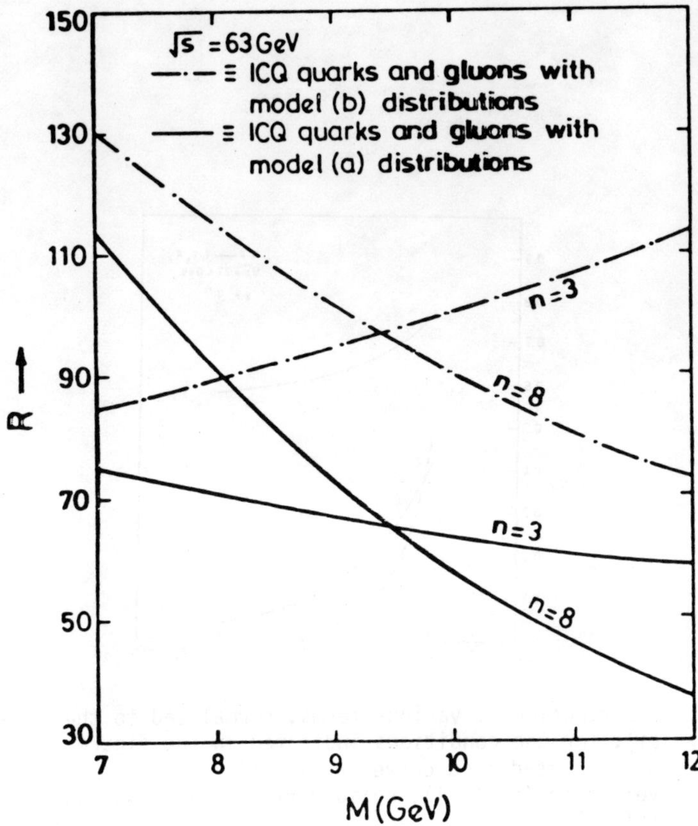

Fig. 19. Ratio R, as predicted in ref. 21, of photon pair to lepton pair production in pp collisions at $\sqrt{s}$ = 63 GeV, vs. M (invariant mass of the photon or lepton pair), according to the gauged integer-charged quark model (ICQ). Model (a) for quark distributions is taken from Pakvasa et al., Phys. Rev. D10, 2124 (1974); model (b) is taken from Peierls et al., Phys. Rev. D15, 1397 (1977). n = 3 means: gluon distribution function $\overline{g(x)} \sim (1 - x)^3$; n = 8 means: $g(x) \sim (1 - x)^8$.

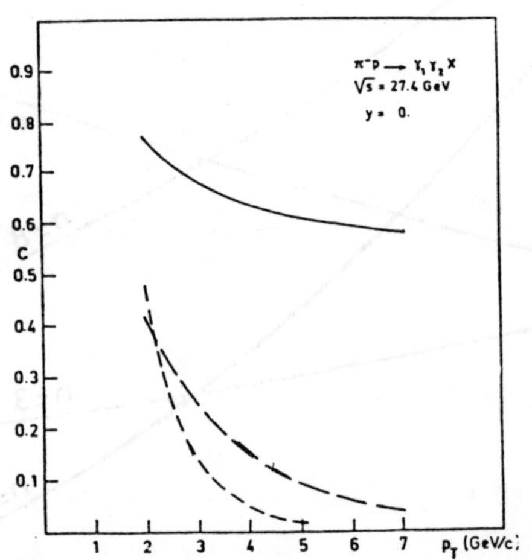

Fig. 20. Contribution of various terms, normalized to the Born term (i), for the conditions indicated in the figure and z_e^{min} = 0.5. Short-dashed curve: ratio (ii)/(i); long-dashed curve: ratio (iii)/(i); solid curve: ratio (iv)/(i). (From ref. 22.)

SUMMARY OF THEORETICAL CONTRIBUTIONS
=======================================

P.M. Zerwas

Inst. f. Theor. Physik, RWTH Aachen

INTRODUCTION

The theoretical contributions to this $\gamma\gamma$ workshop, presented in two parallel sessions, can be divided into five different classes:

1. <u>Leptons and radiative corrections</u>

F. Berends:	Complete lowest order calculation of $ee \to 4$ leptons
F. Berends:	Radiative corrections to multiperipheral processes
W. v. Neerven:	Radiative corrections to multiperipheral processes
P. Kessler:	Radiative corrections to deep inelastic $e\gamma$ scattering

2. <u>Integer charge quark models</u>

T. Jayaraman:	Integer-charged quarks and jet production
A. Janah:	e tagging and integer-charged quark effects in $\gamma\gamma$-mediated jets
J. Field:	Integer charge quarks and resonances
T. Maruyama:	Direct γ in e^+e^- annihilation [exp.]

3. QCD and resonances

K. Liu: $\gamma\gamma \rightarrow$ vector mesons

4. QCD at short distances

I. Antoniadis: 2nd order QCD predictions for structure
                              functions

M. Drees: Structure functions and QCD

P. Aurenche: Higher order QCD corrections to $\gamma\gamma \rightarrow$ hadrons

K. Tsokos: Gluon contributions to exclusive $\gamma\gamma$

P. Kessler: Direct $\gamma\gamma$ in hadronic collisions

5. New particles

K. Grassie: Structure functions and SUSY at HERA

Y. Srivastava: Excited fermion signatures

The material on the photon structure function has been extensively covered in a plenary session on this subject /1/, likewise $\gamma\gamma$ production in hadronic interactions /2/. Consequently, these points won't be taken up again. Theoretical problems of resonance production, section (3), are discussed in detail within the accompanying experimental summary talk /3/. Instead, the results of a measurement of direct γ's in e^+e^- annihilation testing integer-charged quark models, will be summarized here in the appropriate context.

1. LEPTONS AND RADIATIVE CORRECTIONS

Increasing statistics in four-lepton final states of e^+e^- collisions, ee $\rightarrow$ eeee, ee $\rightarrow$ ee$\mu\mu$ etc., and in hadron production, ee $\rightarrow$ ee + hadrons, required a more detailed theoretical analysis of

the QED part in these channels than has been available up to now.
These processes have been
calculated in the past by
evaluating the multiperipheral
diagrams either analytically
in the equivalent particle
approximation or numerically
by Monte Carlo methods /4/.
Extensive theoretical work has
now been finished on (i)
inclusion of non-multiperipheral contributions and (ii)
radiative corrections to
multiperipheral processes.

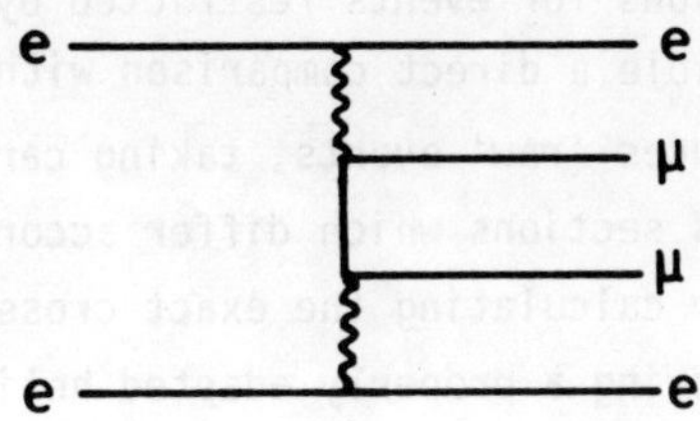

(i) <u>Non-multiperipheral diagrams /4/</u>

The complete evaluation of non-multiperipheral diagrams is

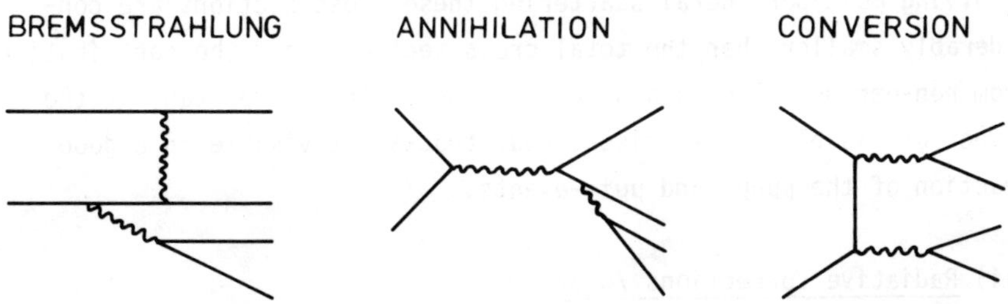

not only required by the increasing accuracy of experimental data on
conventional final states as eeμμ, but it is also very important for
estimating the QED background to CELLO type events ee → μμ + 2 jets
and to possible Higgs signals in ee → Z + H, Z → μ⁺μ⁻ and H → 2 jets,

after extending the program into the weak sector.

For this purpose a Monte Carlo event generator has been constructed which produces unweighted events for all possible four-lepton
final states /5/. The program evaluates total cross sections and cross
sections for events restricted by kinematical cuts,and thus makes
possible a direct comparison with experimental data. The generator first
produces 'raw' events, taking carefully account of the peaks in the
cross sections which differ according to the class of diagrams chosen.
After calculating the exact cross section on the amplitude level by
employing a properly adapted helicity method, a rejection algorithm
selects the final event sample.

Some of the results are displayed in the first two tables. The
total cross sections, table 1, for processes involving multiperipheral
scattering are very large, but the vast majority of these events
remains hidden in the beam pipe. The visible cross sections in table 2
are defined by demanding one or more tracks at an angle of at least 25
degrees with respect to the beams and, for ee $\rightarrow$ 4e, a cut of 1 GeV on
the invariant mass of every outgoing e^+e^- combination. For processes
involving multiperipheral scattering these cross sections are considerably smaller than the total cross sections, and the contribution
from non-peripheral diagrams becomes increasingly important if the
number of visible tracks rises. Four tracks are visible in a good
fraction of the $\mu\mu\mu\mu$- and $\mu\mu\tau\tau$-events.

(ii) <u>Radiative corrections /6-8/</u>

A precise measurement of the photon structure function requires
the calculation of QED radiative corrections to the cross section for
ee $\rightarrow$ ee + hadrons. After subtraction of the contributions from
inelastic Compton scattering, the radiative corrections in a narrow
sense fall into four categories /6/: (i) peripheral corrections to the
e^- and e^+ currents; (ii) γ exchange between e^+ and e^-; (iii) γ exchange
between leptons and hadrons; and (iv) corrections inside the hadronic

system. If for fixed invariant energy all hadronic degrees of freedom
are summed up, class (iii) does not interfere with the Born term, and
QED corrections to the hadronic system should remain small. The most
important corrections are expected to be those which affect the multiperipheral scattering mechanism locally. They consist of three
different parts:

VERTEX   VACUUM   BREMSSTRAHLUNG
CORRECTION   POLARIZATION   (SOFT + HARD)

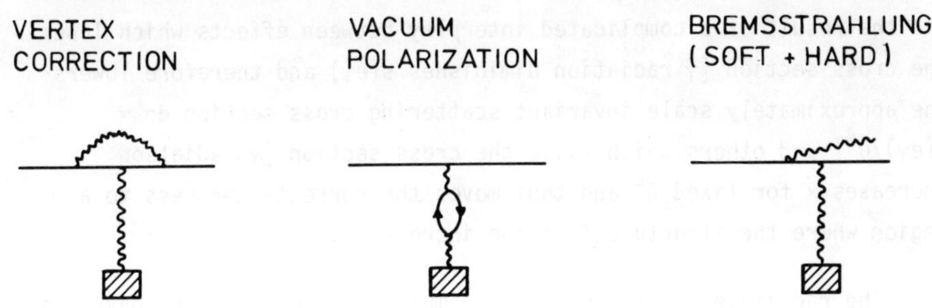

Hard γ bremsstrahlung makes the application of unfolding procedures
necessary in experimental analyses since the momentum transfer to the
hadronic blob is altered; otherwise, a model for the structure function
must be employed which then determines the size of the radiative
correction.

For these radiative corrections to μ pair production ee $\rightarrow$ ee$\mu\mu(\gamma)$,
a Monte Carlo event generator has been constructed /7/, which can be
applied to any experimental situation. (It can also be used to study
radiative corrections to deep inelastic eγ scattering, the structure
function is then approximated by the parton model expression without
radiative gluon corrections.) Corrections to the total cross section
are very small, see table 3. For large angle e tagging, however,
differential cross sections are in general substantially affected by
radiative corrections. This is demonstrated in fig. 1 which shows the
Q^2 variation of the cross section for $.085 \leq \Theta_+ \leq .8$ (and opposite

cone) $\left[E_b = 17.5 \text{ GeV}, W_{\mu\mu} \geq 1 \text{ GeV}, E_+ \geq 8 \text{ GeV}, \Theta_- \leq .03, \cos\Theta(\mu^{\pm}) \leq .996\right]$.

An analytic calculation of radiative corrections to deep inelastic $e\gamma$ scattering comes to a similar conclusion /8/. The photon structure function is used in the leading log approximation of QCD. The change of the Bjorken x distribution by γ radiative corrections is displayed in fig. 2 for momentum transfer $Q^2 = 5$ and 20 GeV^2. The final numbers are the result of a complicated interplay between effects which lower the cross section [γ radiation diminishes $s(e\gamma)$ and therefore lowers the approximately scale invariant scattering cross section $d\sigma \propto s(e\gamma)/Q^4$] and others which raise the cross section [γ radiation increases x for fixed Q^2 and thus moves the current-γ process to a region where the structure function increases].

The radiative corrections of the multiperipheral graphs have also been carried out by the authors of ref. /6/. In addition they have analyzed photon exchange between electron and positron currents. This has been done for the production of a pointlike pseudoscalar meson, resulting in a very small correction of order 10^{-5}. From this we can infer that such corrections are negligible for other $\gamma\gamma$ final states too. Two-photon processes appear to be factorizable to a high degree of accuracy /6/, an important result for precision measurements of the photon structure function.

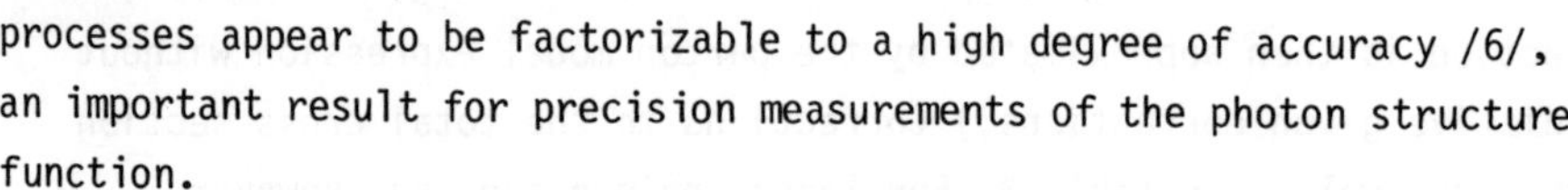

2. INTEGER CHARGE QUARK MODELS /12,13/

Soon after the introduction of quarks as basic building blocks
of hadrons, attempts have been made to assign integer charges to
quarks /9/. This requires enlarging the number of quarks by postulat-
ing color degrees of freedom which interpret fractional electric
charges as averages over integer electric charges in color triplets.
The charge matrix of up and down quarks decomposes into a color
singlet component, carrying the standard fractional quark charges,
and the eighth member of a color octet, $Q = Q_0 + Q_8$,

$$
\begin{matrix} \text{up} \\ \text{down} \end{matrix}
\begin{bmatrix} 0 & 1 & 1 \\ -1 & 0 & 0 \end{bmatrix}
=
\begin{bmatrix} 2/3 & 2/3 & 2/3 \\ -1/3 & -1/3 & -1/3 \end{bmatrix}
+
\begin{bmatrix} -2/3 & 1/3 & 1/3 \\ -2/3 & 1/3 & 1/3 \end{bmatrix}
$$

$$
\text{color} \rightarrow \qquad \text{COLOR-SINGLET} \qquad \text{COLOR-OCTET}
$$

Gauging this model to introduce forces, the standard $SU(3)_C$ x
$SU(2)_L$ x $U(1)$ symmetry at energies $0(TeV)$ is spontaneously broken in
two steps /10/. At a scale $0(100\ GeV)$ it breaks down to $SU(3)_C$ x $U(1)_f$
and finally at $<< 0(1\ GeV)$ to
$U(1)_{em}$. In contrast to the
standard model, 4 gluons are
electrically charged, and they
are massive, $m_g << 1\ GeV$. One
member U^0 of the gluon octet
mixes with the $U(1)$ gauge field
A_f [the same way as W^3 and B do
in the standard model], creating
a massless photon state A and a

$$
\begin{array}{c}
SU(3)_C \text{ x } SU(2)_L \text{ x } U(1) \\
\downarrow \quad \sim 0(100\ GeV) \\
SU(3)_C \text{ x } U(1)_f \\
\downarrow \quad <<0(1\ GeV) \\
U(1)_{em}
\end{array}
$$

massive gluon state $\tilde{U}$, the mixing angle being small $e_f\cos\omega = g_s\sin\omega = e$,

$$
\text{physical photon} \qquad A = A_f\cos\omega + U^0 \sin\omega \qquad\qquad m_A = 0
$$

$$
\text{physical gluon} \qquad \tilde{U} = -A_f\sin\omega + U^0 \cos\omega \qquad\qquad m_{\tilde{U}} > 0
$$

Both these vector bosons contribute to lepton-quark scattering

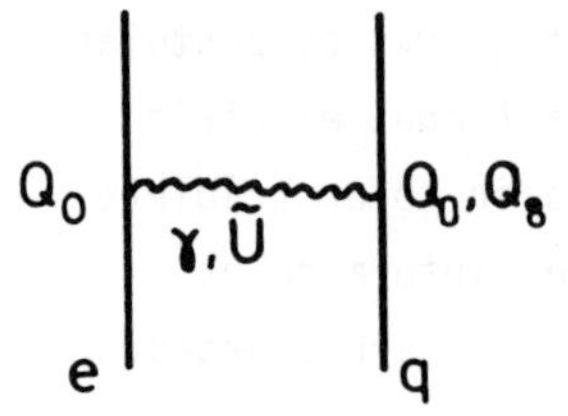

processes. For the color octet component the interference pattern is destructive,

$$\propto \frac{1}{q^2} Q_0 + \left\{ \frac{1}{q^2} - \frac{1}{q^2 - m_g^2} \right\} Q_8$$

$$\equiv \frac{1}{q^2} Q_{eff}(q^2)$$

suggesting the definition of an effective quark charge

$$Q_{eff}(q^2) = Q_0 + \Delta(q^2, m_g^2)\, Q_8$$

$$\Delta(q^2, m_g^2) = \frac{m_g^2}{m_g^2 - q^2} \quad \begin{cases} 1 & \text{for } q^2 = 0 \\ 0 & \text{for } q^2 \to \infty \end{cases}$$

While the octet component is fully effective for $q^2 = 0$, it is screened for $|q^2| \gg m_g^2$ in which limit the charge matrix is reduced to the same form as in the standard model. Experiments testing the quark sector of gauged ICQ models must therefore be carried out with on-shell photons. The contributions of charged (longitudinal) gluon partons, on the other hand, survive in the asymptotic region.

Below color threshold, only 2γ processes can be sensitive to integer quark charges since $\langle 1|8|0 \rangle = 0$ for color singlet hadrons, but $\langle 1|8 \times 8|0 \rangle \sim \langle 1|1|0 \rangle$ can be finite. For very high color thresholds, however, virtual color state excitation might be suppressed so strongly that even 2γ processes cannot test integer quark charges in general; the parton language in which those processes are calculated, could then be inappropriate. This caveat in mind, the following 2γ processes have been explored as ICQ tests.

(i) $\gamma\gamma \to$ 2 jets

Normalizing this cross section by the μ pair cross section

$$R_{\gamma\gamma} = \frac{d\sigma(ee \to ee + 2\ jets)}{d\sigma(ee \to ee + \mu\mu)} \quad ,$$

the ratio is 34/27 for two generations of fractionally charged quarks. The much larger value 10/3 for non-gauged integer-charged Han-Nambu quarks is experimentally ruled out. The situation is more subtle in the gauge version of the theory. Hadron production is not only mediated by quark production but also by gluon pair production.

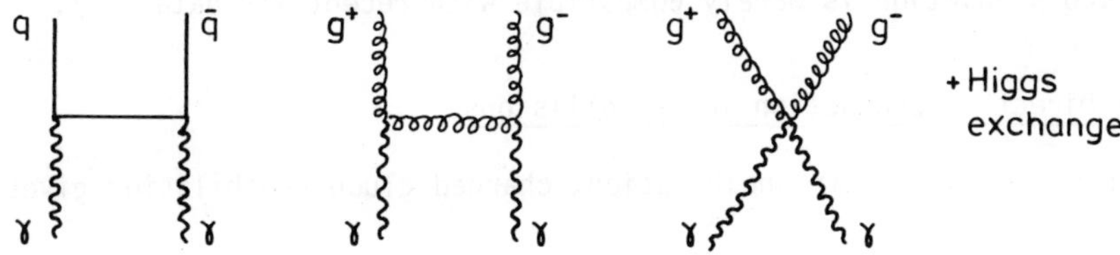

The quark contribution is proportional to

$$\sum_{q,i} Q^4 = \frac{1}{3} \sum_q \Big| \sum_i Q_{eff}(q_1^2)\, Q_{eff}(q_2^2) \Big|^2$$

$$= 3 \sum_q Q_0^4 + \Big[\frac{4}{3} \sum_q Q_0^2 \Big] \frac{m_g^4}{(m_g^2 + q_1^2)(m_g^2 + q_2^2)} + \frac{4}{27} N_q \frac{m_g^8}{(m_g^2 + q_1^2)^2 (m_g^2 + q_2^2)^2}$$

The q^2 dependent terms have to be handled with care in EPA - and even more so for gluon pair production /11-14/. Table 4 shows the sensitivity of ICQ tests in three domains of the gluon mass /12/.

Much attention has been paid to the PLUTO single tag data /15/. An excess of events over the prediction of the fractionally charged quark model has been observed at low q^2 and transverse momenta up to

358

the 6 GeV range. The excess decreases with increasing q^2. Using ICQ to explain the excess above $p_\perp \gtrsim 3$ GeV [below this limit deviations from the parton model must do the rest] /13,14/, a gluon mass in the range between 175 MeV and 350 MeV is compatible with the data, fig. 3 a,b. However, a better understanding of higher orders and non-perturbative effects is required before positive support can be claimed.

(ii) <u>Deep inelastic Compton scattering</u>

This process is microscopically built up by Compton scattering on quark and charged gluon targets /13,16/. The cross section turns out to be 2.5 to 5 times bigger than the standard QCD cross section. Even though higher order corrections have not been worked out yet, this ICQ prediction is barely compatible with recent ISR data /17/.

(iii) <u>Direct $\gamma\gamma$ production in pp collisions</u>

Besides quark pair annihilation, charged gluon annihilation gives

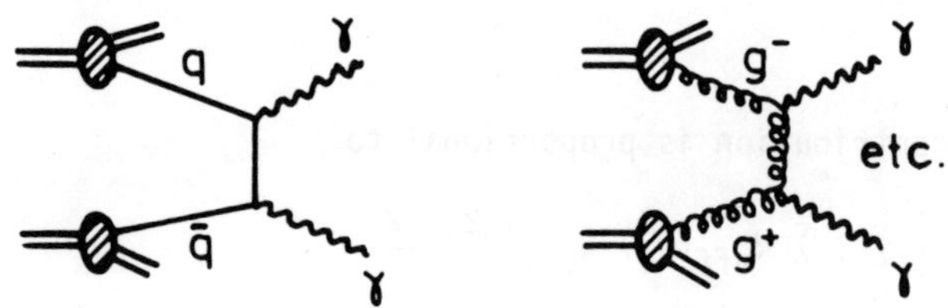

a very large yield of $\gamma\gamma$ pairs /13,18/. Measured in units of the Drell-Yan cross section,

$$R = \left. \frac{d\sigma(pp \rightarrow \gamma\gamma X)/dy d\cos\Theta}{d\sigma(pp \rightarrow e^+e^- X)/dy d\cos\Theta} \right|_{\substack{y = 0 \\ \Theta = \pi/2}}$$

the ratio is ≈ 1 for fractionally charged quarks, but shoots up to a value of ≈ 70 below and ≈ 100 above color threshold in the gauged ICQ model. In view of the experimental result /19/ R = 1.7 ± 1, also this

gauged version of the ICQ model appears strongly disfavored, keeping
the aforementioned caveat in mind though.

(iv) <u>Direct γ's in e^+e^- annihilation</u>

Photons carry the physical information from the femto-universe
($d \lesssim 10^{-15}$ cm) to large distances without distorting interactions.
They are therefore a powerful probe of quark properties /20-22/ at

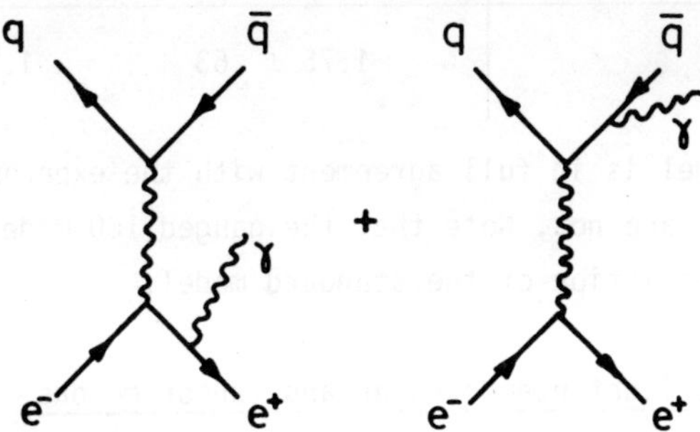

short distances. Hard photons emitted at large angles to final state
$q,\bar{q}$ jets, indeed come from this perturbative QCD regime. The cross
section splits into an ordinary QED part, with the γ emitted from the
lepton current, a part with γ emission from the quark current /20/
and an interference term between both /21/,

$$\frac{d\sigma}{d\cos\Theta} = \sum_q e_q^2 \left(\frac{d\sigma}{d\cos\Theta}\right)_{ini} + \sum_q e_q^3 \left(\frac{d\sigma}{d\cos\Theta}\right)_{intf} + \sum_q e_q^4 \left(\frac{d\sigma}{d\cos\Theta}\right)_{fin}$$

Θ being the angle between the γ and the e^- axis, the sum running over
all flavors and colors, and the indexed cross sections defined for unit
charges.

These cross sections have recently been measured, and the charge
sums are compared with the data /23/, for fractionally charged quarks
and Han-Nambu quarks

	$\sum e_q^4$	$\sum e_q^3$
fractionally charged quarks	$\dfrac{35}{27} = 1.30$	$\dfrac{19}{4} = 2.11$
Han-Nambu quarks	$\dfrac{11}{3} = 3.67$	$\dfrac{11}{3} = 3.67$
data	$1.75 \pm .63$	$1.97 \pm .61$

The standard model is in full agreement with the experimental analysis, Han-Nambu quarks are not. Note that the gauged ICQ model coincides here with the prediction of the standard model.

(v) $\gamma\gamma$ widths of light pseudoscalar and tensor mesons

As an alternative to the debated use of current algebra techniques for $\eta' \to \gamma\gamma$ /24/, one might be tempted to analyze $\gamma\gamma$ decays of pseudo-scalars (and tensors alike) in simple constituent $q\bar{q}$ bound state models though such models do not explicitly incorporate the approximate chiral symmetry of QCD. Large couplings of the SU(3) singlet to gluons require mixing with glueball states. A model of this type has been formulated in ref. /25/. Pseudoscalar and tensor mesons are considered as a nonrelativistic quarkonium system with SU(3) symmetry breaking and binding corrections for transition amplitudes. A linear quarkonium-gluonium mixing matrix is built in with mass dependent SU(3) breaking.

Fitting masses, $\gamma\gamma$ widths of η,η', 1γ transitions between pseudo-scalars and vectors, and OZI rule constraints, the fractionally charged quark model provides an adequate description of the data whereas ICQ models provide a very poor fit (prob $\ll 10^{-4}$). Mixing to gluonium is needed in the standard model, with a substantial glue content in η', $\langle gg|\eta'\rangle \approx -.65$, and a small one in η, $\langle gg|\eta\rangle \approx .14$. The model makes interesting predictions for ι decay channels. As anticipated, it is difficult to find a convincing picture of the π^0.

Applying the model to tensor mesons offers an overall satisfactory description in standard QCD, with almost ideally mixed pure quarkonia f and f'. Gauged ICQ models, on the other hand, appear to be excluded, not providing a consistent approach to the data.

4. QCD AT SHORT DISTANCES

(i) Large $p_\perp$ hadron spectra in $\gamma\gamma$ collisions /26/

Besides the pointlike γ Born term, hadron production in $\gamma\gamma$ collisions involves corrections from gluon bremsstrahlung (including loops) and anomalous γ components which yield a contribution formally of the same size as the Born term. These higher order QCD terms give

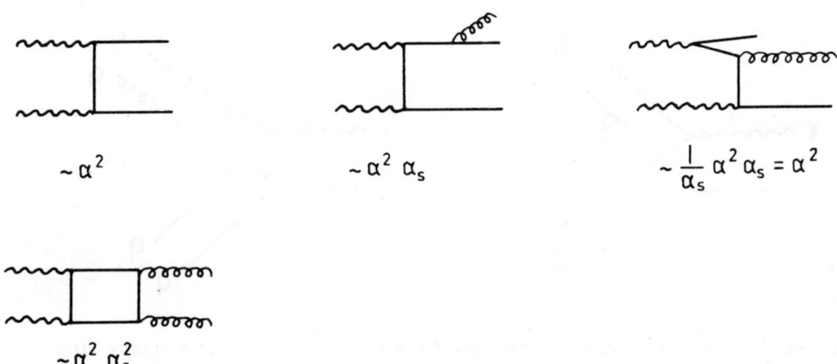

a rather modest correction to the pointlike Born term /26, preliminary/. They do not resolve the discrepancy between the Born term calculations and the TASSO and CELLO data /27/. Naturally, it might be speculated that the VDM part develops a (perturbative, power behaved?) high $p_\perp$ tail. Estimates from two photo-production

$p_\perp$	$\dfrac{d\sigma^{full}/dp_\perp^2}{d\sigma^{Born}/dp_\perp^2}$
2 GeV	~ 1.37
3 GeV	~ 1.26
4 GeV	~ 1.19

experiments are shown in fig. 4. This channel deserves further detailed discussion.

(ii) <u>Exclusive $\gamma\gamma$: rôle of gluons /29/</u>

Convincing arguments have been presented that exclusive 2-particle production in $\gamma\gamma$ collisions at large energy but fixed scattering angle is the convolution of a perturbatively calculable hard scattering amplitude and a quark distribution amplitude /28/. A large gluon component in η' requires extending the theory from "quark final states" to "gluon final states". In the example below this is illustrated for the process $\gamma\gamma \to \pi^0\eta'$ /29/. Taking quark and gluon distributions

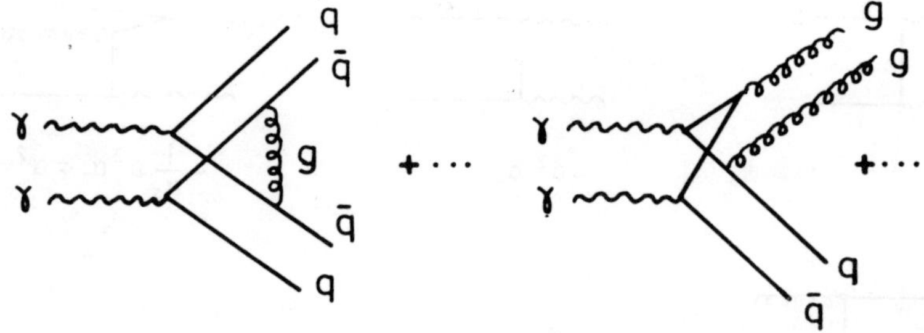

$\sim [x(1-x)]^{const}$, the normalization fixed by $f_{\eta'}$, f_g, the quark and gluon η' decay constant, respectively, the cross section depends sensitively on the ratio $f = f_g/f_{\eta'}$,

$$\frac{d\sigma}{dt} = 32\ \pi\alpha^2\ <e_q^2>^2\ \frac{F_\pi^2(t)}{s^2}\ f_{\eta'}^2\ \left|I_q + \frac{f_g}{f_{\eta'}}\ I_g\right|^2$$

as shown in fig. 5 for various sets of exponents in the wave function. The process is therefore well suited to measure the amount of gluons in the η' in a rather direct way.

5. NEW PARTICLES

(i) <u>SUSY particle production at HERA /30/</u>

A host of supersymmetry particles can be produced in HERA,

$$ep \rightarrow \tilde{q}\, \overset{(-)}{\tilde{q}}$$

$$ep \rightarrow \tilde{q}\, \tilde{g}$$

$$ep \rightarrow \tilde{g}\, \tilde{g}$$

for a mass range of $10 < m_{\tilde{g}} < 60$ GeV, $20 < m_{\tilde{q}} < 80$ GeV. These particles are either produced directly or through the anomalous γ component.

The contributions of the various mechanisms are displayed in fig. 6. The anomalous γ components contribute little to $\tilde{q}\, \bar{\tilde{q}}$ pairs, but they are very important for $\tilde{q}\, \tilde{g}$ final states, as a result of color factors etc. $\tilde{g}\, \tilde{g}$ production is only possible through the anomalous γ component. A comparison of the various channels reveals that $\tilde{q}\, \bar{\tilde{q}}$ dominates for $m_{\tilde{q}} < m_{\tilde{g}}$, $\tilde{q}\, \tilde{g}$ production dominates for $m_{\tilde{g}} < m_{\tilde{q}} < 2m_{\tilde{g}}$, and $\tilde{g}\, \tilde{g}$ becomes dominant if $2\, m_{\tilde{g}} < m_{\tilde{q}}$. The envisaged rates are of the order of 20 events/day.

(ii) <u>Excited leptons and quarks /31/</u>

Since the minimum mass scale of excited leptons and quarks is presumably set at a level of $O(100$ GeV$)$, $\gamma\gamma$ collisions at energies

available in the near future don't provide an experimentally easy
access to this exciting new domain. Under these conditions the virtual
t-channel exchange of new particles looks the most promising approach.
Two examples shall follow. A heavy neutral lepton E^0 could mediate
$\gamma\gamma \to \nu\bar{\nu}$ production, with a cross
section

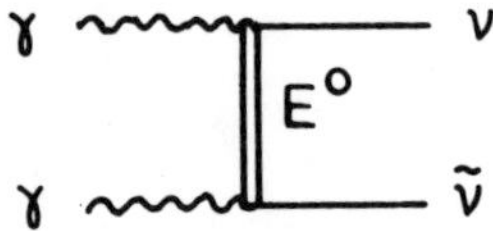

$$\left.\frac{\sigma(\gamma\gamma \to \nu\bar{\nu})}{\sigma(e^+e^- \to \mu^+\mu^-)}\right|_{\sqrt{s}_{ee} \sim m_E}$$

$$\sim \frac{1}{2}\left(\frac{s_{\gamma\gamma}}{m_E^2}\right)^2 \quad .$$

Interference effects can occur between diagrams of normal quark and
excited quark exchange, resulting in

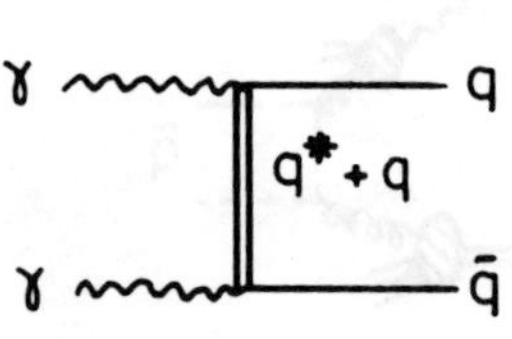

$$s^2\frac{d\sigma}{dp_\perp^2} \propto 1 + \frac{f_q^2}{4}\left(\frac{s_{\gamma\gamma}}{m_*^2}\right)^2 \sin^2\Theta$$

Even though the violation of scale
invariance, quadratic in $s_{\gamma\gamma}$, is
dramatic, it needs large $\gamma\gamma$ energies
to approach a physically interesting
area.

ACKNOWLEDGEMENTS

R. Lander and his colleagues deserve many thanks for organizing
such an enjoyable meeting. This report has been written during a stay
at SLAC, and I would like to thank S. Brodsky and S. Drell for their
kind hospitality.

REFERENCES

/1/ W.A. Bardeen, these proceedings.

/2/ P. Kessler, these proceedings.

/3/ W. Ko, these proceedings.

/4/ J.A.M. Vermaseren, Nucl. Phys. B229 (1983) 347.

/5/ F.A. Berends, P.H. Daverveldt and R. Kleiss, Complete lowest
order calculations for four-lepton final states in electron-
positron collisions, Leiden preprint, 1984.
F.A. Berends, P.H. Daverveldt and R. Kleiss, Total and visible
cross sections for multilepton events in e^+e^- collisions,
Leiden preprint, 1984.

/6/ W.L. van Neerven and J.A.M. Vermaseren, Phys. Lett. 142B (1984) 80.

/7/ F.A. Berends, P.H. Daverveldt and R. Kleiss, Radiative corrections
to the process $e^+e^- \rightarrow e^+e^-\mu^+\mu^-$, Leiden preprint, 1984.

/8/ S. Ong, C. Carimalo and P. Kessler, Phys. Lett. 142B (1984) 429.

/9/ M.Y. Han and Y. Nambu, Phys. Rev. 139 (1965) B1006.

/10/ J.C. Pati and A. Salam, Phys. Rev. D8 (1973) 1240; Phys. Rev. D10
(1974) 275; Phys. Rev. Lett. 36 (1976) 355;
R.N. Mohapatra, J.C. Pati and A. Salam, Phys. Rev. D13 (1976) 1733;
G. Rajasekaran and P. Roy, Pramana 5 (1975) 303.

/11/ A. Janah and M. Özer, Can two-photon jets in e^+e^- colliding beams
be used to measure the quark charges? Maryland preprint MdDP-PP-
81-221, 1981.

/12/ A. Janah, this workshop.

/13/ T. Jayaraman, this workshop.

/14/ R.M. Godbole, T. Jayaraman, J.C. Pati, S.D. Rindani and
G. Rajasekaran, Phys. Lett. 142B (1984) 91.

/15/ J. Dainton, PLUTO Collaboration, Proc. Brighton Conf., 1983.

/16/ A.V. Efremov, S.V. Jvanov and S.V. Mikhailov, JETP Lett. 32 (1980) 669.

/17/ T.S. Virdee, Proc. Brighton Conference, 1983.

/18/ T. Jayaraman, Contribution, Leipzig Conf., 1984.

/19/ C. Kourkoumelis et al., Z. Physik C16 (1982).

/20/ T.F. Walsh and P.M. Zerwas, Phys. Lett. 44B (1973) 195.

/21/ S.J. Brodsky, C.E. Carlson and R. Suaya, Phys. Rev. D14 (1976) 2264.

/22/ E. Laermann, I. Schmitt, T.F. Walsh and P.M. Zerwas, Nucl. Phys. B207 (1982) 205.

/23/ T. Mayurama, MAC Collaboration, this workshop.

/24/ M. Chanowitz, these proceedings.

/25/ J. Field, this workshop; paper in preparation.

/26/ P. Aurenche, this workshop;
see also W.J. Stirling, $\gamma\gamma$ Workshop, Aachen 1983.

/27/ R. Brandelik et al., TASSO Collaboration, Phys. Lett. 107B (1981) 290;
CELLO Collaboration, s. N. Wermes, $\gamma\gamma$ Workshop, Aachen 1983.

/28/ S.J. Brodsky and G. Peter Lepage, Phys. Rev. D24 (1981) 1808;
G. Farrar, these proceedings.

/29/ G.W. Atkinson, J. Sucher and K. Tsokos, Phys. Lett. 137B (1984) 407.

/30/ M. Drees and K. Grassie, Parametrizations of the photon structure function and applications to supersymmetric particle production at HERA, Dortmund preprint DO-TH 84/13, 1984.

/31/ Y. Srivastava, this workshop.

Table 1: Total cross sections ee → 4 leptons /5/.

final state	beam energy (GeV)			
	10	17.5	25	50
eeee	$0.920 \ 10^7$ $\pm \ 0.011 \ 10^7$	$1.070 \ 10^7$ $\pm \ 0.015 \ 10^7$	$1.233 \ 10^7$ $\pm \ 0.018 \ 10^7$	$1.459 \ 10^7$ $\pm \ 0.025 \ 10^7$
ee$\mu\mu$	98.9 $\pm \ 0.6$	131.4 $\pm \ 2.2$	154.0 $\pm \ 0.9$	205.9 $\pm \ 1.2$
$\mu\mu\mu\mu$	$162.2 \ 10^{-6}$ $\pm \ 1.5 \ 10^{-6}$	$77.3 \ 10^{-6}$ $\pm \ 1.6 \ 10^{-6}$	$48.7 \ 10^{-6}$ $\pm \ 0.4 \ 10^{-6}$	$18.06 \ 10^{-6}$ $\pm \ 0.16 \ 10^{-6}$
$\mu\mu\tau\tau$	$96.8 \ 10^{-6}$ $\pm \ 1.0 \ 10^{-6}$	$55.2 \ 10^{-6}$ $\pm \ 1.0 \ 10^{-6}$	$36.7 \ 10^{-6}$ $\pm \ 0.4 \ 10^{-6}$	$15.87 \ 10^{-6}$ $\pm \ 0.14 \ 10^{-6}$
eeee*	$0.920 \ 10^7$ $\pm \ 0.010 \ 10^7$	$1.070 \ 10^7$ $\pm \ 0.013 \ 10^7$	$1.229 \ 10^7$ $\pm \ 0.016 \ 10^7$	$1.459 \ 10^7$ $\pm \ 0.025 \ 10^7$
ee$\mu\mu$*	97.2 $\pm \ 0.5$	129.6 $\pm \ 2.1$	152.1 $\pm \ 0.8$	203.8 $\pm \ 1.1$

* Contribution of the multiperipheral Feynman diagrams only.

Table 2: Visible cross sections ee $\rightarrow$ 4 leptons $\left[E_{beam} = 17.5 \text{ GeV}\right]$ /5/.

final state	tracks visible	cross section (nb)		σ for multiperipheral graphs only (nb)	
eeee	ee	1.65 ± 0.27		1.48 ± 0.02	
	eee	1.76 ± 0.95	10^{-3}	1.93 ± 0.20	10^{-3}
	eeee	1.2 ± 0.7	10^{-4}	2.01 ± 0.34	10^{-5}
ee$\mu\mu$	e	0.65 ± 0.04		4.0 ± 0.9	10^{-3}
	μ	43.0 ± 2.4		44.4 ± 0.4	
	ee	2.8 ± 0.4	10^{-5}	6.0 ± 5.6	10^{-9}
	$\mu\mu$	29.9 ± 1.6		31.6 ± 0.3	
	eμ	1.4 ± 1.2	10^{-2}	0.71 ± 0.27	10^{-2}
	eeμ	1.17 ± 1.16	10^{-3}	5.5 ± 4.4	10^{-7}
	$\mu\mu$e	5.4 ± 3.8	10^{-4}	1.4 ± 0.4	10^{-3}
	ee$\mu\mu$	2.30 ± 0.18	10^{-4}	7.4 ± 5.9	10^{-6}
$\mu\mu\mu\mu$	μ	14.3 ± 1.3	10^{-6}		
	$\mu\mu$	10.43 ± 1.3	10^{-6}		
	$\mu\mu\mu$	7.65 ± 1.1	10^{-6}		
	$\mu\mu\mu\mu$	25.5 ± 1.5	10^{-6}		
$\mu\mu\tau\tau$	τ	8.20 ± 0.53	10^{-6}		
	μ	0.63 ± 0.12	10^{-6}		
	$\tau\tau$	8.32 ± 0.43	10^{-6}		
	$\mu\mu$	1.30 ± 0.29	10^{-6}		
	$\mu\tau$	2.79 ± 0.37	10^{-6}		
	$\tau\tau\mu$	4.95 ± 0.42	10^{-6}		
	$\mu\mu\tau$	2.23 ± 0.46	10^{-6}		
	$\mu\mu\tau\tau$	23.93 ± 1.13	10^{-6}		

Table 3: Peripheral radiative corrections to the total cross section
ee → eeμμ(γ) /7/.

E_b (GeV)	σ_o (nb)	σ_{rad} (nb)
10	97.1 ± 0.3	97.3 ± 0.5
20	137.2 ± 0.5	138.7 ± 0.8
50	202.4 ± 0.8	202.3 ± 1.2
100	261.4 ± 1.0	262.8 ± 1.6

Table 4: Experimental sensitivity to integer charge quarks /12/.

	$m_g < 10$ keV	$m_g \sim 1 - 100$ MeV			$m_g > 1$ GeV
		double tag	single tag	untagged, complete	
	Δ_1, Δ_2 << 1	Δ_1, Δ_2 << 1	Δ_1 << 1 $\Delta_2 \lesssim 1$	$\Delta_1, \Delta_2 \lesssim 1$	$\Delta_1, \Delta_2 \sim 1$
quark color	drops out	drops out	small	large	full
gluon	shd. survive	shd. survive	shd. survive	large	full
expt'l verdict	possible	?			ruled out

Fig. 1 Q^2 distribution in tagged ee → eeµµ(γ) /7/.

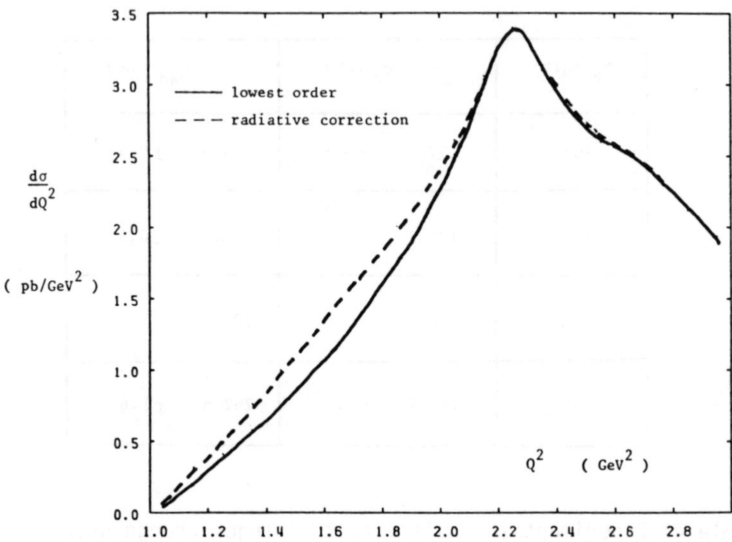

Fig. 2 Radiative corrections
to deep inelastic
eγ scattering /8/.

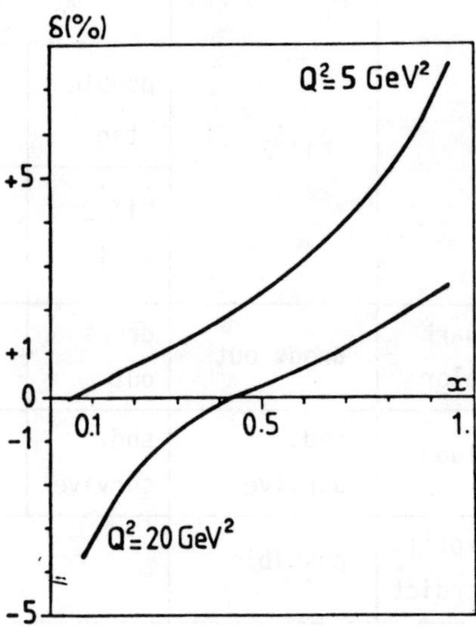

Fig. 3   Fits of the ICQ gauge model to $e\gamma \to e + \text{high } p_\perp \text{ jet} + X$ /14/.

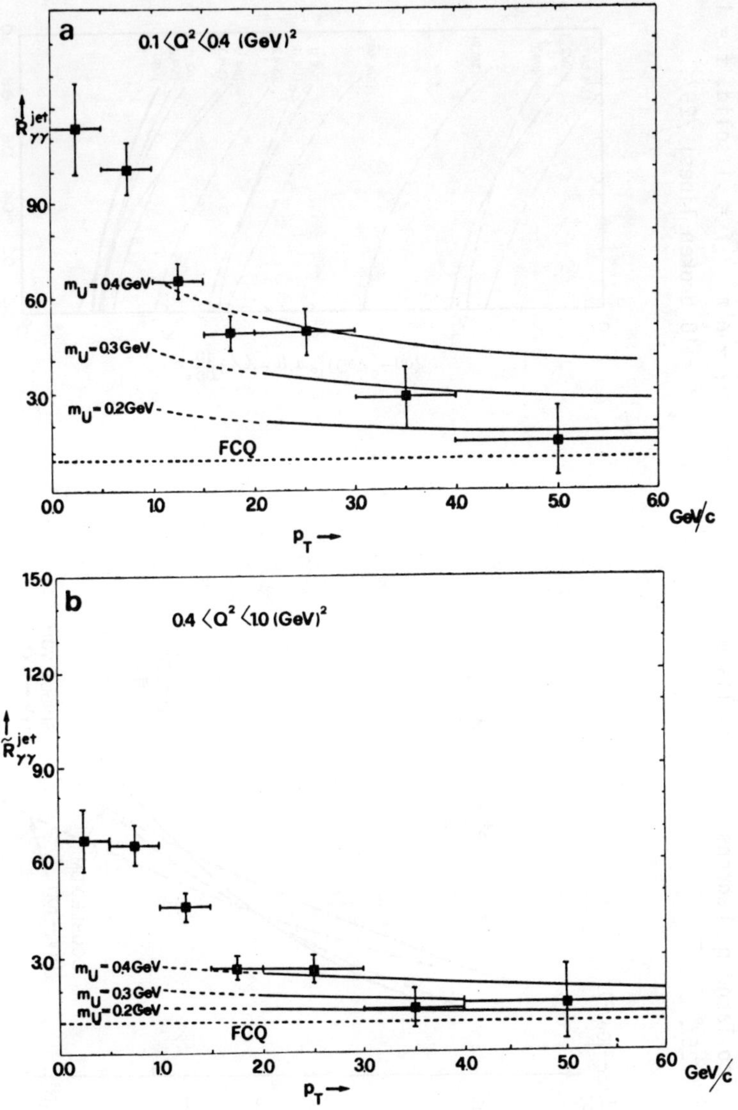

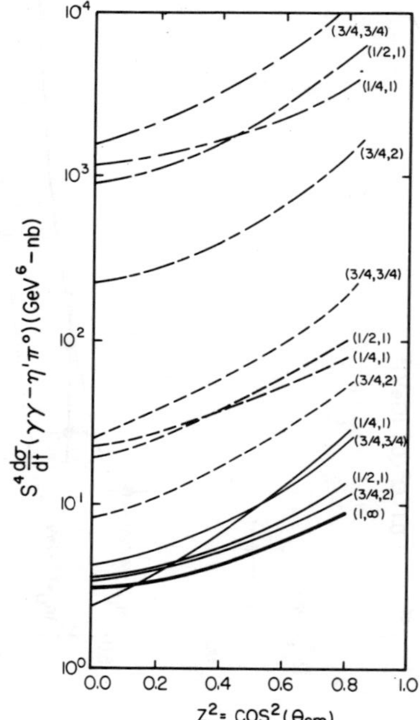

Fig. 4 Estimates of various contributions to large $p_\perp$ hadrons in $\gamma\gamma$ collisions /26/.

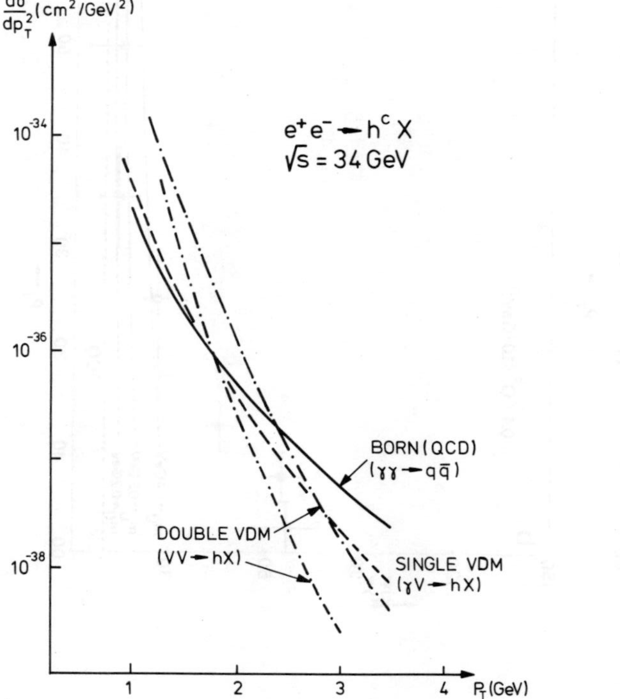

Fig. 5 Leading order QCD predictions for $\gamma\gamma \to \eta'\pi^0$ [f = .1 solid, f = 1 dashed, f = 10 broken lines] /29/.

Fig. 6   Supersymmetric particle production at HERA /30/.

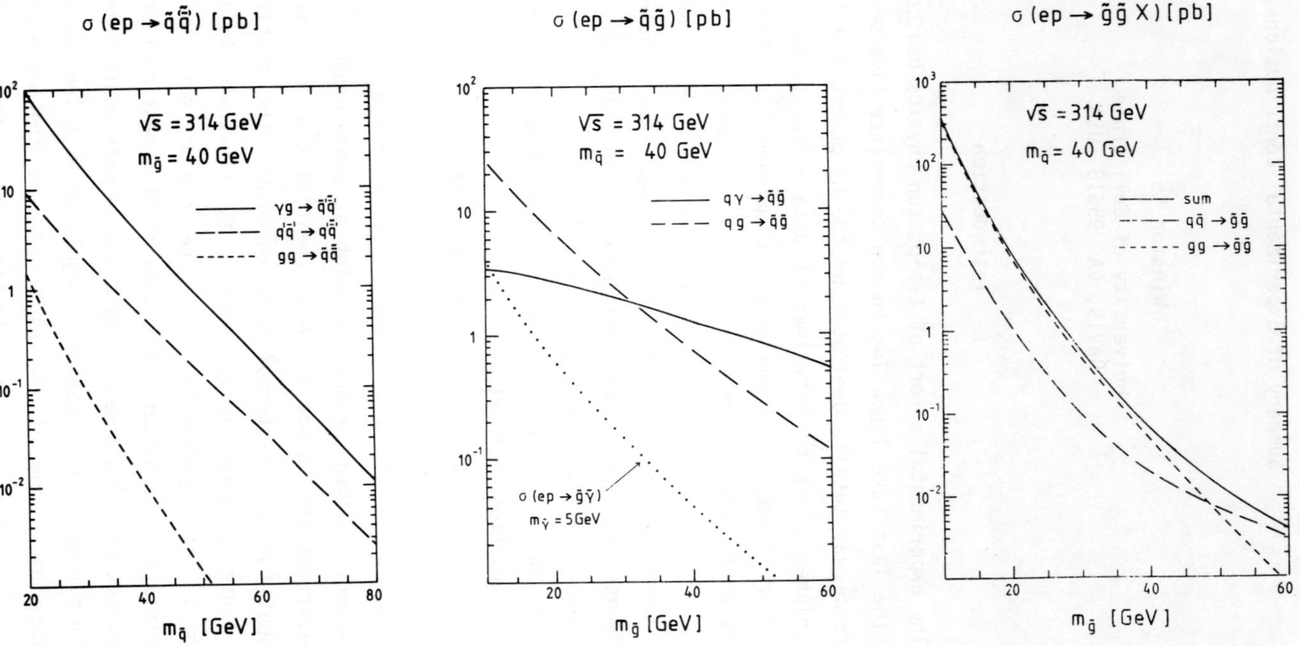

Summary of Experimental Contributions

Winston Ko
University of California
Davis, CA 95616 USA

I. Introduction

The experimental effort of two-photon physics has come a long way since the last Lake Tahoe Two-Photon Conference five years ago. That conference was totally dominated by theory papers and there was only one finished piece of experimental data - eta prime measurement by Mark II at SPEAR - and two very preliminary "illustrative" results. In this conference, there are in total 35 experimental papers. Five of the eight parallel sessions are devoted to experimental results. While there was only one (PEP9 QED) paper contributed by the PEP experiments at the Fifth International Workshop on Photon-Photon Collisions in Aachen last year, now there are 12. The experimental study of photon-photon collisions is now in a mature stage and is a truly international effort.

Photon-photon experiments at e^+e^- colliders cover a wide range of photon-photon center-of-mass energies. This can be illustrated by the PEP9 measurement of total cross sections in Fig. 1. The center-of-mass energy reached 20 GeV, as high as photo-production experiments have ever gotten. Since one experiment at e^+e^- colliders covers all the photon-photon center-of-mass energies, the statistics at each given center-of-mass energy is poor, as illustrated by the error bars in Fig. 1. All experiments, therefore, can use more luminosity. Lack of luminosity also hinders the study of high mass pair production, one of the most exciting areas brought out in this conference.

This summary is organized roughly as were the parallel sessions. It plays back some of the highlights of the experimental contributions. Coherent reviews and references of the individual subjects are found in the excellent papers in this proceeding by H. Kück, F. Foster, A. Cordier, F. Erné, W. Wagner, and G. Knies. We will deal with Inclusive Reactions first, from soft to hard, then Exclusive Reactions, from soft to hard.

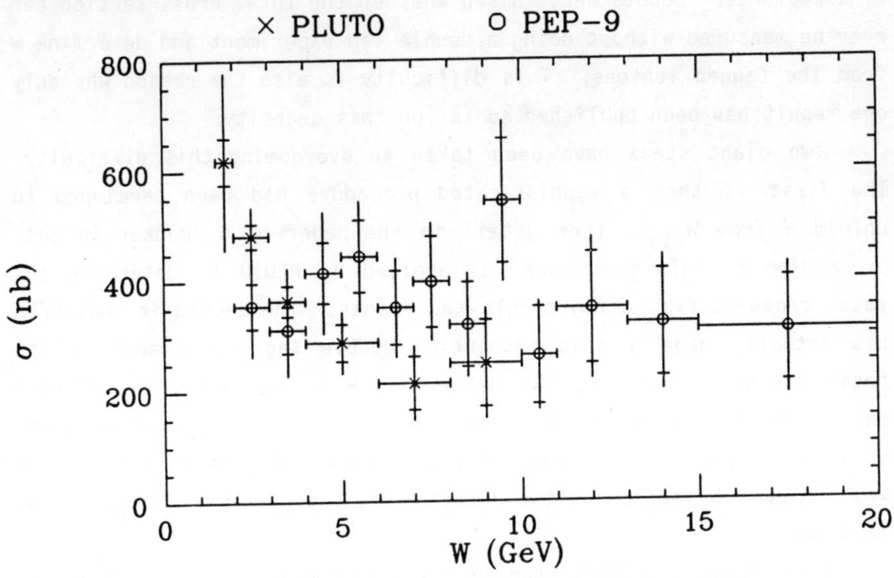

Figure 1.   Total γγ cross-section are measured up to center of mass energy W = 20 GeV.

II.   Total Cross-section

The simple and fundamental quantity of total cross-section for photon-photon interaction was a point of contention in several previous two photon conferences. Two much improved measurements of total cross-section are reported in this conference. We are witnessing the greatest progress of this topic since PLUTO's first measurement in 1981.

The previous difficulty in measuring the total cross-section centers around the determination of W, the photon-photon center of mass energy. The total cross-section had to be expressed as function of W. However, unlike e^+e^- annihilation and hadronic interactions, W cannot be controlled by varying beam energies. In single tag and no tag events, it has to be inferred from W_{vis}, the total visible hadron energies. Since the fraction of hadron energy visible depends on acceptance, the determination of W is entangled with the determination of acceptance. Doubts were raised whether the total cross-section can ever be measured without doing a double tag experiment and determine W from the tagged leptons. This difficulty is also the reason why only one result has been published so far on this quantity.

Two giant steps have been taken in overcoming this difficulty. The first is that a sophisticated procedure had been developed to unfold W from W_{vis}. (See detail in the paper by A. Bäcker in this proceeding.) This procedure was applied by PLUTO to determine the total cross-section using single tag events. The second is that PEP9 has actually made a high resolution double tag measurement of the total cross-section, and has determined W by computing the missing mass from the leptons in the reaction $e^+e^- \rightarrow e^+e^- X$. The requirement of very accurate measurement of the tagging energies in this method was achieved by employing two arrays of NaI crystals with $\sigma_E/E=1\%$ at 14.5 GeV.

Both PLUTO and PEP9 studied the dependence of the total cross-section as function of Q^2, the mass square of the virtual photon (Fig. 2). If the photon behaves merely as the hadron ρ (in Vector Dominance Model), then the Q^2 dependence of the cross-section will behave like a ρ-pole form factor, $f_\rho(Q^2) = 1/(1+Q^2/m_\rho^2)^2$.

Rather than that, both experiments find the Generalized Vector Dominance (GVD) form factors of Sakarai and Schildknecht fit the data very well (solid curves). The GVD form factors for transverse and scalar photons can be written as

$$f_T(Q^2) = \sum_V \frac{r_V}{(1+Q^2/m_V^2)^2} + \frac{1-\sum_V r_V}{1-Q^2/m_o^2} \quad \text{and} \quad f_S(Q^2) = \tfrac{1}{2} \sum_V \frac{r_V}{(1-Q^2/m_V^2)^2} \frac{Q^2}{m_V^2}.$$

The sum is over the contributions from ρ, ω and ϕ vector mesons, (with r_ρ, r_ω and r_ϕ at 0.65, 0.08 and 0.05 respectively). A continuum term with a point-like behavior is included for transverse photons and

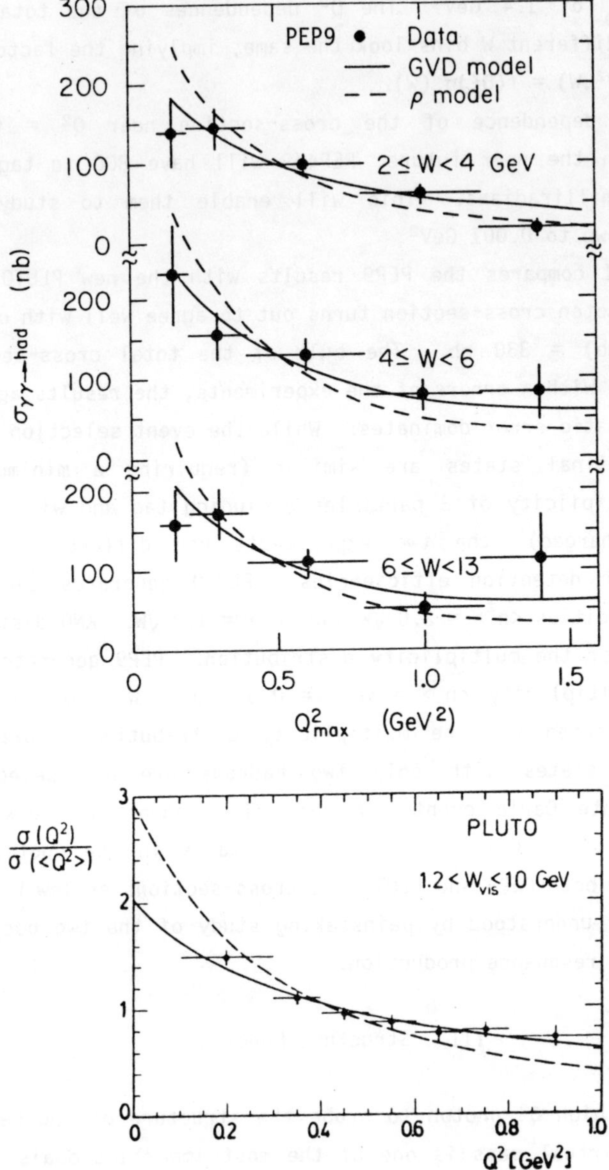

Figure 2. Q^2 dependence of the total cross-sections. Q^2_{max} is the larger of two Q^2 in the double tag experiment.   Broken lines are for ρ form factor and solid lines are for GVD model.

uses mass m_o of 1.4 GeV. The Q^2 dependences of the total cross-section for different W bins look the same, implying the factorization relations $\sigma(Q^2,W) = f(Q^2)\sigma_o(W)$.

The Q^2 dependence of the cross-section near $Q^2 = 0$ is of interest. In the near future, PEP4/9 will have BGO to tag leptons down to 5 milliradians. This will enable them to study the Q^2 dependence down to 0.001 GeV^2.

Figure 1 compares the PEP9 results with the new PLUTO result. The photon-photon cross-section turns out to agree well with $\sigma^2_{\gamma p}/\sigma_{pp} \sim$ $(115\mu b)^2/(40mb) = 330$ nb. The bulk of the total cross-section is hadron-like. Within errors of the experiments, the results agree. At low W, systematic error dominates. While the event selection criteria for hadron final states are similar (requiring a minimum total detected multiplicity of 3 particles excluding tag and with at least two being charged), the two experiments use different models to correct their detection efficiencies. PLUTO generates events with mean multiplicities $<n^{\pm}> = 2.0 \sqrt{W}$ and $<n^o> = 1.3 \sqrt{W}$. KNO distribution is assumed for the multiplicity distribution. PEP9 generates events with mean multiplicity $<n^{\pm}> = <n^o> = 0.5 \pm 0.26$ W. Poisson distribution is assumed for the multiplicity distribution. Furthermore, since final states with only two hadrons are not selected and therefore Monte Carlo events of this kind cannot be checked, PEP9 chooses not to include two body final states in the quoted cross-section. For both PEP9 and PLUTO the cross-sections at low W can only be completely understood by painstaking study of the two body events including the resonance production.

III. Structure Functions

Using a high Q^2 photon to probe the structure of another photon with small virtual mass is one of the most important goals for two-photon physics. There is a concerted international effort to achieve this goal. The papers on this subject contributed to this conference all use very similar methods of analysis. This is particularly important in the study of the Q^2 evolution of the structure function, in which different experiments at different Q^2 have to be compared.

The photon structure function, if originated purely from a point-like interaction, is particularly of interest since $F_2(x,Q^2)=f(x)\log\frac{Q^2}{\Lambda^2}$. with the QCD parameter Λ showing in the leading term. The Q^2 evolution therefore gives a sensitive measure of Λ. Because of the log dependence, however, quite a large lever-arm in Q^2 is required in order to make an accurate measurement of Λ. A compilation of F_2 measurements in Fig. 3 gives a Λ between 100 MeV and 500 MeV with best fit at 200 MeV. An alternative measurement of Λ is to compare the absolutely normalized $F_2(x)$, at a given Q^2, with the theory. This is done by PLUTO and TASSO collaborations (Fig. 4). Similar Λ values are obtained.

The simplicity of the two photon process's relation to the QCD scale is very exciting. The sensitivity of measuring Λ by photon-structure function is quite impressive.

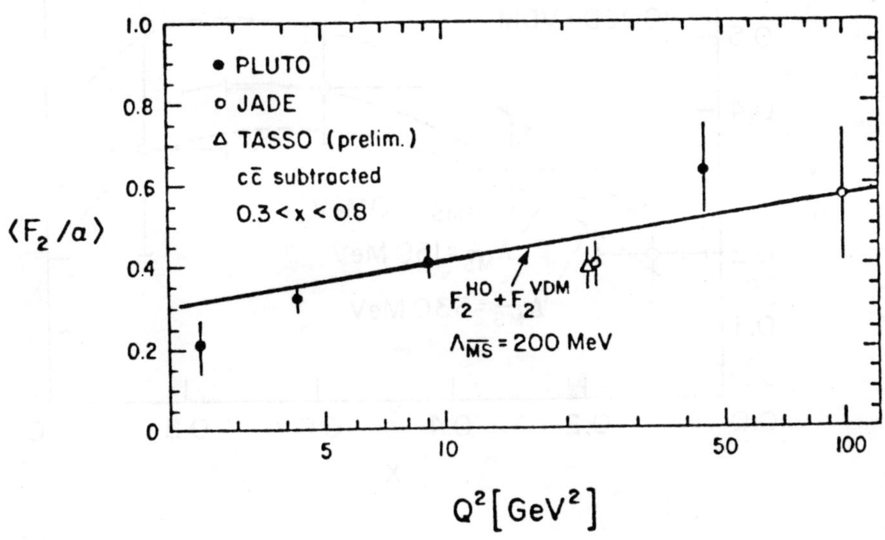

Figure 3. The Q^2 evolution of the structure function.

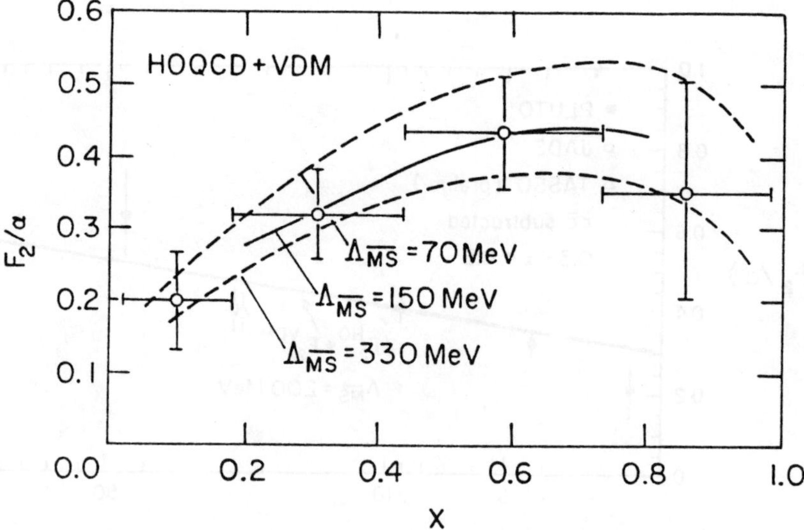

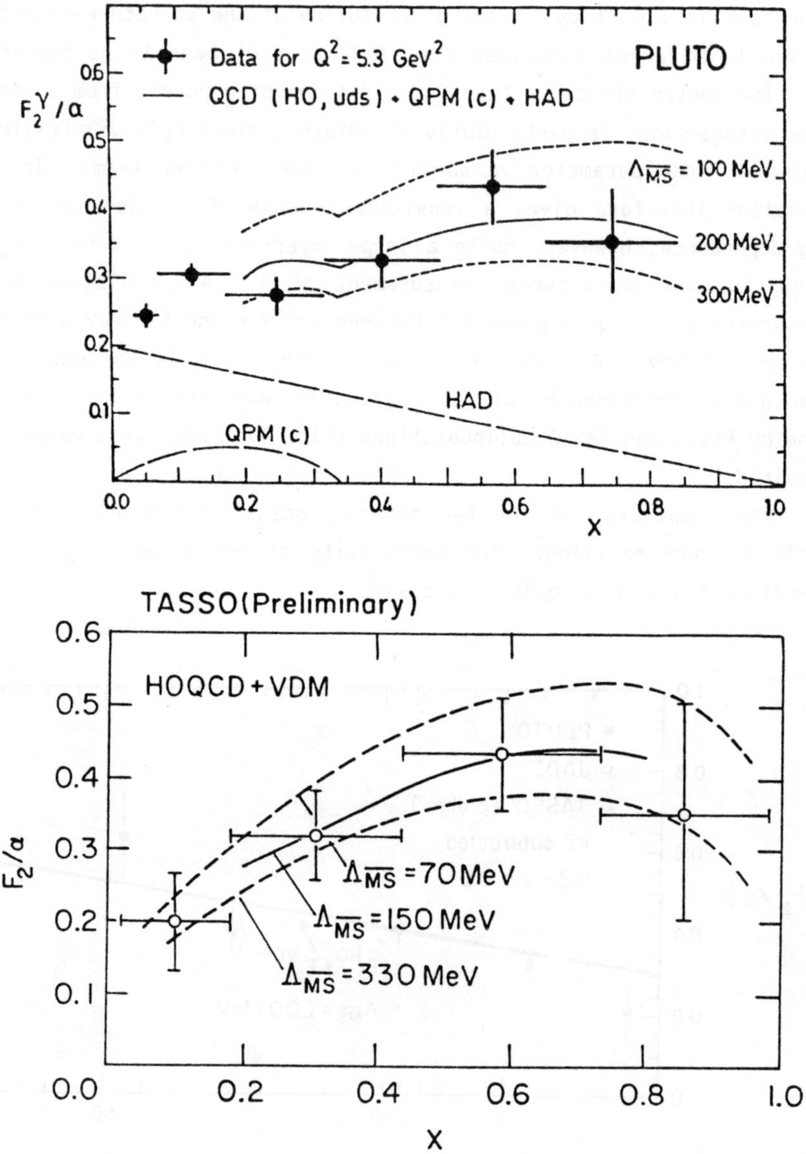

Figure 4.   Structure functions $F_2(x)$ as measured by PLUTO and TASSO. Comparison is made with absolute normalization for different Λ values.

There are, however, several complications in extracting Λ. First, the factor $\log \frac{Q^2}{\Lambda^2}$ is an approximation with zero target mass. When either the target photon virtue mass or the constituent mass (e.g. for the charm quark) is comparable or larger than Λ, the Q^2 dependence is really $\log \frac{Q^2}{\Lambda^2 + 0(m^2)}$ and clearly the target mass has a direct effect on the Λ value extracted. TASSO finds that there is at least a factor of two difference in Λ value with and without the target mass correction. In the near future the effect of target mass can be studied directly. Double tag experiments can measure the effect of the target photon directly and the experiments with particle identification can study the effect of charm mass.

The second complication is that in addition to the point-like scattering, the hadron-like VDM part also contributes to the photon structure function. Although VDM events dominate only at small Q^2, their fraction is still substantial at $Q^2 \sim 5$ GeV^2. The VDM background would affect the slope in Fig. 3 since at present we cannot afford to throw away events below $Q^2 = 10$ GeV. Of course, the absolute normalization in Fig. 4 is also affected by the VDM background. Theoretical models take account of the VDM part. As an indication that this procedure is reasonable, PEP4/9 shows in Fig. 5 that for Q^2

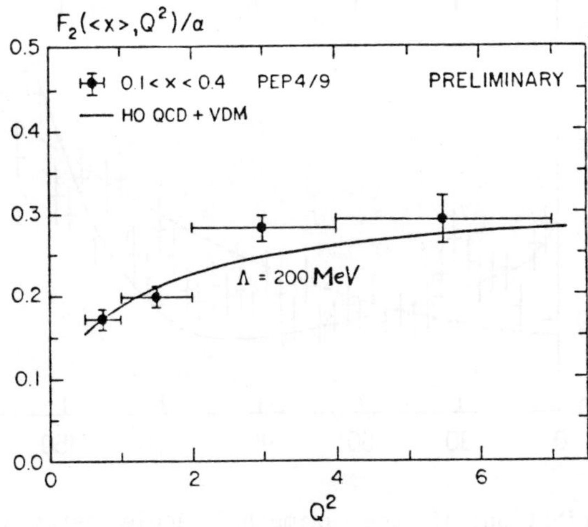

Figure 5. Q^2 evolution of F_2 at low Q^2 as measured by PEP4/9.

between 1 and 7 GeV, where VDM is a large part of the cross-section, the data still gives a Λ of around 200 MeV. It would be very interesting to study this transition region and find out the VDM contribution empirically.

IV. High p_T Jets

The explicit separation of the point-like photon-photon inter-action from VDM is important; attempts are made to do so from the different event topologies. A common technique is the jet analysis, as point constituents manifest themselves in the form of jets. In photon-photon interactions, two jets are expected from quark-pair production. However, the VDM events are also expected to have two jets, one from each vector meson. There are differences between point-like jets and VDM jets. For example, taking the track with the highest p_T and observing the distribution of the azimuthal angle ϕ between it and other tracks, weighted by their p_T, the point-like jets point more at the forward and backward directions. TASSO shows (Fig. 6) that the amount of point-like events is indeed quite sizeable.

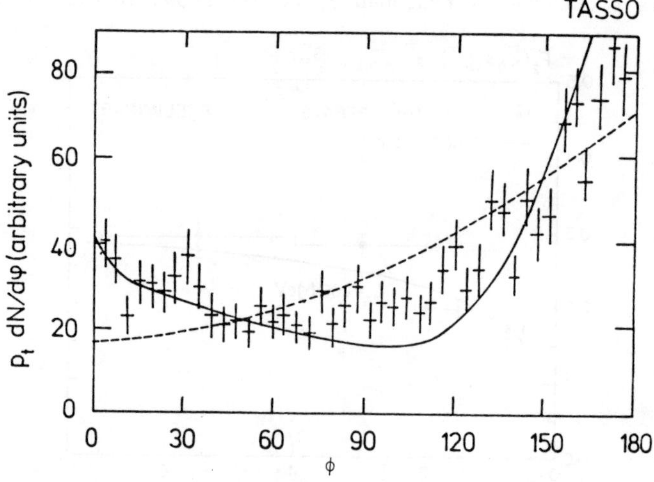

Figure 6. Distribution of the azimuthal angle between track with highest P_T and others weighted by their P_T. Solid line is expected from point-like jets; dashed lines VDM.

The most prominent difference between VDM and point-like jets lies in the transverse momentum, p_T, with respect to photon-photon axis. VDM jets have a limited p_T, characteristic of hadronic inter-action. Point-like jets do not have such an explicit limit and have a p_T distribution similar to QED events like $\gamma\gamma \to \mu\mu$. This is very nicely illustrated by CELLO's result on the p_T^2 distribution in Fig. 7. PLUTO uses the p_T distribution to study the fraction of point-like events as a function of Q^2 (Fig. 8). Using thrust analysis and forcing all events into two jets, the p_T^2 distribution is compared with the VDM prediction. At low Q^2, VDM dominates for p_T^2 up to 5 GeV^2. As Q^2 increases, a larger and larger fraction of the events is point-like. At the highest Q^2 value, $\langle Q^2 \rangle = 49$ GeV^2, almost all events are point-like.

Using high P_T jets to find the fraction of point-like events at different Q^2 can give one more confidence in extracting Λ from the Q^2 evolution of the structure functions. The same procedure applied to events at different x enables one to pinpoint the contribution of point-like and VDM contributions to the structure function.

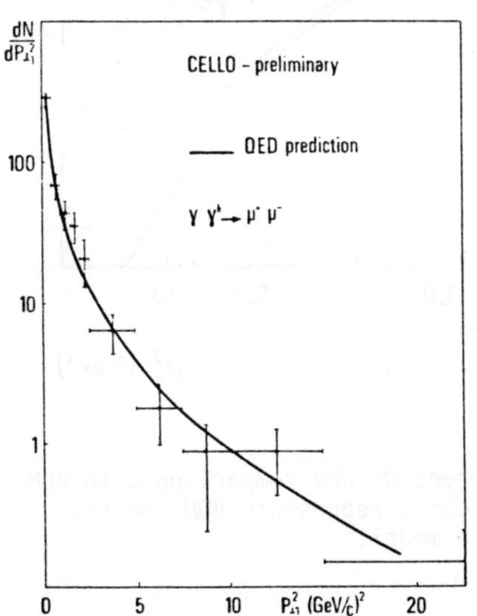

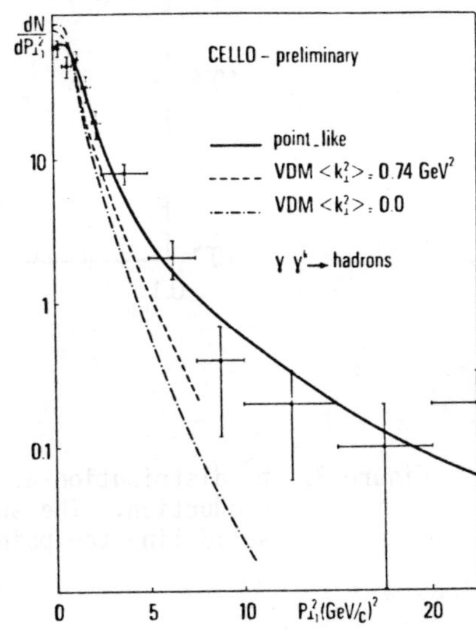

Figure 7.   Comparison between the P_T^2 distribution of muons and hadrons produced in $\gamma\gamma$ interaction.

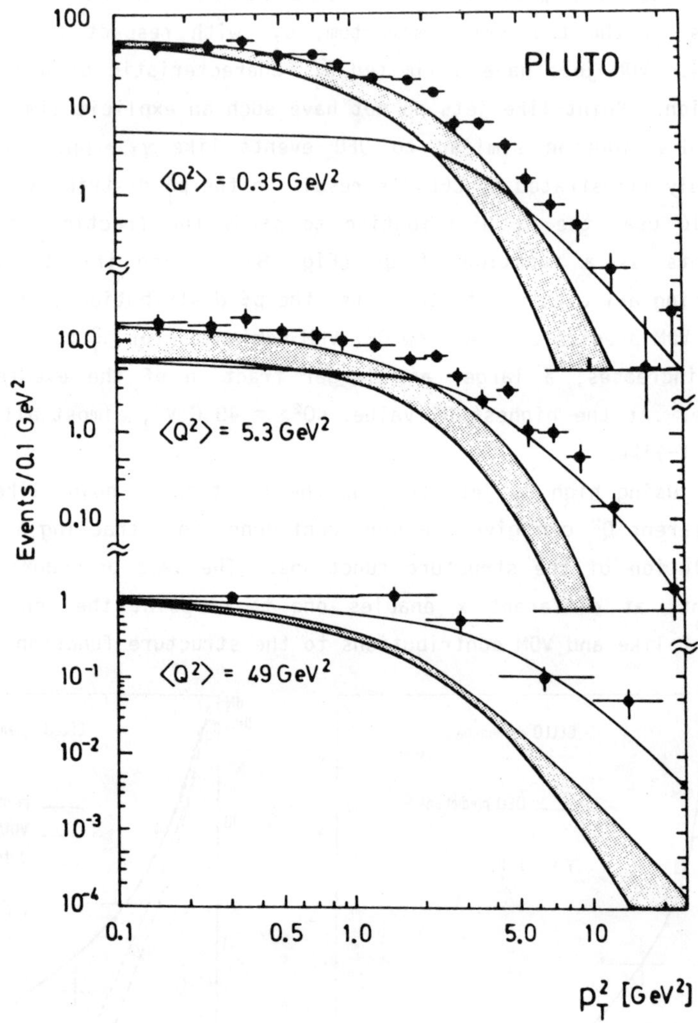

Figure 8. p_T^2 distribution at different Q^2 and comparison with VDM production. The shaded curve represents VDM and the solid line the point-like model.

V. Resonance Formation

Now we will turn to exclusive processes. Although exclusive reactions are prominent only at low W, they do provide insight into the fundamental content of photons and C = +1 resonances. They may even give a cleaner test of QCD.

The radiative widths of pseudoscalars are listed in Table 1. Half of all the results are new contributions to this conference. They are shown in italics.

$$\Gamma_{\pi^\circ \gamma\gamma} \quad (eV)$$

Browman et al.	7.83 ± 0.56	Primakoff
Crystal Ball (DORIS)	*7.9 ± 1.4 ± 1.6*	γγ

$$\Gamma_{\eta\gamma\gamma} \quad (kev)$$

Browman et al.	0.324 ± 0.046	Primakoff
Crystal Ball (SPEAR)	0.56 ± 0.12 ± 0.09	γγ
JADE	*0.56 ± 0.05 ± 0.08*	γγ
DM1/2	*B(η→γγ)=(44 ± 2.5 ± 7)%*	

$$\Gamma_{\eta'\gamma\gamma} \quad (keV)$$

MARK II	5.8 ± 1.1 ± 1.2	ργ
CELLO	6.1 ± 1.1 ± 0.8	ργ
JADE	5.0 ± 0.5 ± 0.9	ργ
TASSO	*5.1 ± 0.4 ± 0.7*	ργ
PLUTO	3.8 ± 0.26 ± 0.43	ργ
MARK II	*3.7 ± 1.0*	$\eta_{\pi\pi}$
DM1/2	*B(η'→γγ)=(2.8 ± 0.3 ± 0.4)%*	$J/\psi\rightarrow\gamma\eta$

Table 1. Pseudo-Scalar Radiative Widths

To illustrate these measurements, I will just show two mass spectra. Figure 9 shows the preliminary result from the Crystal ball at DORIS. A coplanar photon trigger was employed. The data are extremely clean, with signals for π^o, η, and even a couple of η' clearly seen. Figure 10 shows the Mark II measurement of $\eta' \rightarrow \eta\pi\pi$. All other measurements of η' radiative width are from the decay mode $\eta' \rightarrow \rho\gamma$ and are extremely sensitive to γ detection efficiency. This Mark II measurement was achieved only because a burn-out coil enable the very low momentum pions to be measured.

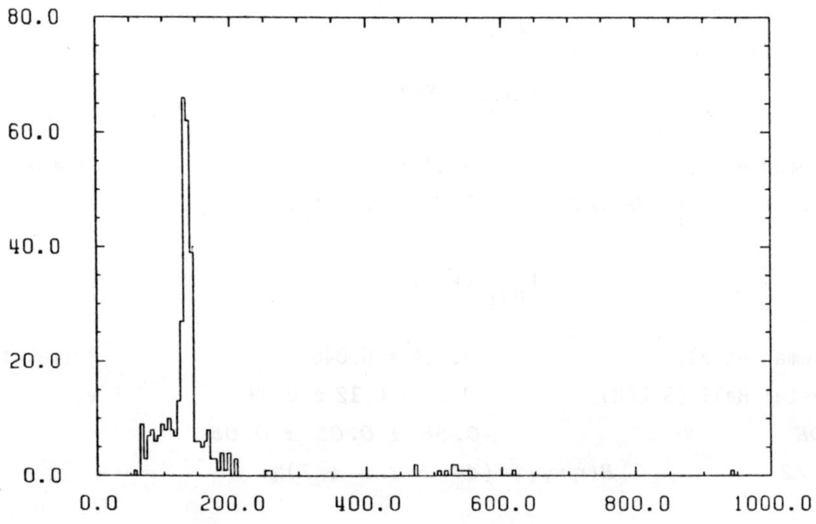

Figure 9.   Mass spectrum m($\gamma\gamma$) from coplanar $\gamma\gamma$ trigger of Crystal Ball at DORIS.

A deviation from the SU(3) expectation the radiative width could indicate that the resonance has a certain gluonium admixture. One can formulate a state as $|\psi\rangle = \sqrt{2} (|u\bar{u} + d\bar{d}\rangle) + y |s\bar{s}\rangle + z |G\rangle$, with $x^2+y^2+z^2 = 1$. Thus $x^2 + y^2 < 1$ is an indication of gluonium GS admixture. For R=η and η', an SU(3) calculation gives:

$$\frac{\Gamma_{R\gamma\gamma}}{\Gamma_{\pi^o\gamma\gamma}} = \frac{1}{9} \left(\frac{m_R}{m_{\pi^o}}\right)^3 (5\, x_R + \sqrt{2}\, y_\eta)^2 \; .$$

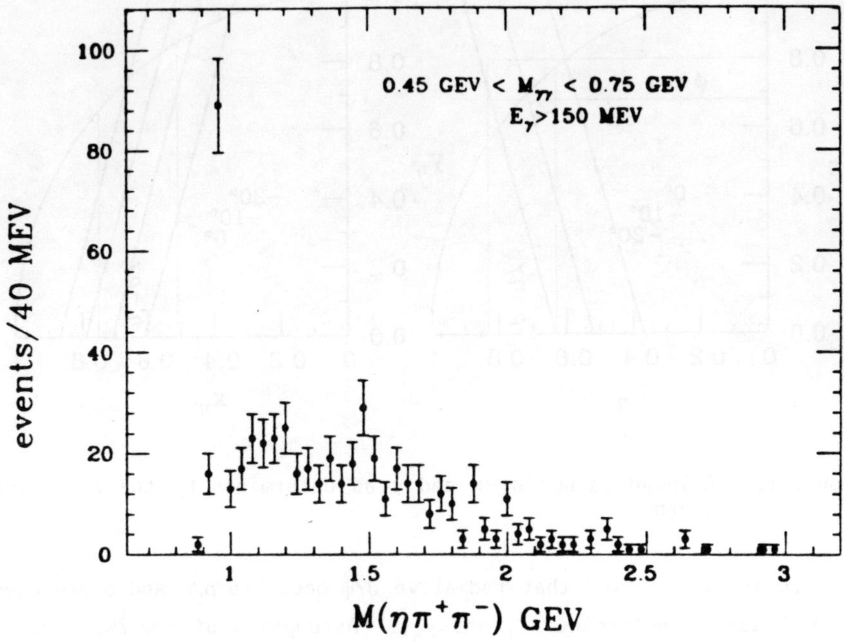

Figure 10. Mark II measure of $\eta\pi^+\pi^-$ mass.

The values x_η and y_η are also constrained by $\Gamma(\rho\to\eta\gamma)$ and $\Gamma(\phi\to\eta\gamma)$ to be 0.73 ± 0.10 and 0.74 ± 0.06 respectively. We see from Fig. 11 that the measured values of $\Gamma_{\eta\gamma\gamma}$ are consistent with $x_\eta^2+y_\eta^2$ being 1. In the η' case, only $x_{\eta'}$ is constrained by $\Gamma(\eta'\to\rho\gamma)$ to be $x_{\eta'}$=0.62±0.12. If we use the mixing angle given by the $\Gamma_{\eta\gamma\gamma}$ results (between -5° and -15°) then $\Gamma_{\eta'\gamma\gamma}$ is bounded by 3.5 and 6.5 keV. With the mixing angle restricted, $\Gamma_{\eta'\gamma\gamma}$ becomes a direct measure of the gluonium admixture. For $\Gamma_{\eta'\gamma\gamma}$ = 5.1 keV, $z_{\eta'}^2$ is between $\frac{1}{4}$ and $\frac{1}{3}$. But for $\Gamma_{\eta'\gamma\gamma}$=3.87 keV, $z_{\eta'}^2$ is between $\frac{1}{3}$ and $\frac{1}{2}$!

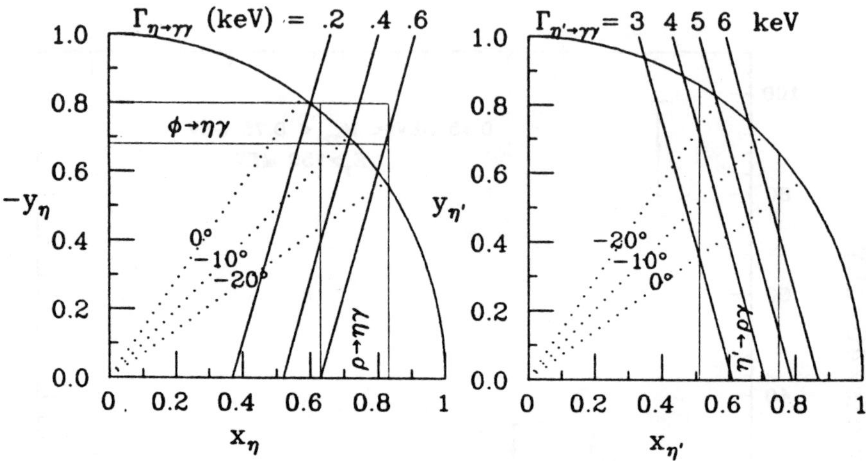

Figure 11.   Allowed values of x and y as determined by the radiative
widths.

It is interesting that radiative J/ψ decay to ρ/ω and ϕ are used
by Mark III to determine $x_{\eta'}$ and $y_{\eta'}$.   Assumption of the Zweig rule,
however is necessary.

The Q^2 dependence of η' formation is shown in Fig. 12.   In
principle the relative amount of $|u\bar{u}+d\bar{d}\rangle$ and $|s\bar{s}\rangle$ can be determined by
whether the Q^2 dependence is more like ρ-pole or ϕ-pole.   The present
statistics is far from adequate for this.

As to the new results on tensor ($J^{PC} = 2^{++}$) mesons, measurements
of $\Gamma_{f^\circ\gamma\gamma}$ have now been made by DELCO and PEP4/9 which have good
electron identification capabilities.   The final state, eeee, was
measured and used to normalize a QED Monte Carlo.   The QED Monte Carlo
for $ee\mu\mu$ was then subtracted from the data to obtain a sample of $ee\pi\pi$
events.   The subtracted distributions are shown in Fig. 13.   The data
are consistent with a Born term plus f° Breit-Wigner with $\Gamma_{f^\circ\gamma\gamma}$ being
the only free parameter. $\Gamma_{f^\circ\gamma\gamma}$ measured to be $2.39 \pm 0.06 \pm 0.30$ keV
by PEP4/9 and $2.70 \pm 0.05 \pm 0.20$ keV by DELCO.

PLUTO used a special trigger in the forward spectrometer to study
the $\pi\pi$ spectrum down to 400 MeV.   The boost provided by the $\gamma\gamma$ center

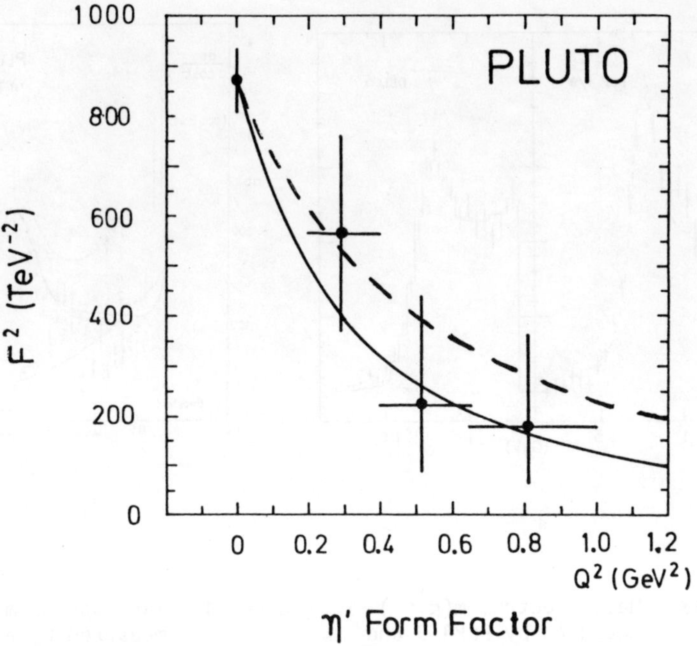

Figure 12. Q^2 dependence of η' formation as compared with ρ-pole form factor (solid line) and ϕ-pole form factor (dashed line).

of mass system allows particles to have sufficient momentum to be studied in the forward spectrometer; muons can be identified and removed. Electrons are removed by the Cerenkov counters. The result (Fig. 14) shows that the data fall significantly below the Born term (line 2) for $0.5 < W < 0.7$ GeV.

The radiative width of the f° can also be used to determine its quark content. In the case of ideal mixing of tensor particles, f and A_2 are composed of $|u\bar{u}>$ and $|d\bar{d}>$, while f' is composed of $|s\bar{s}>$. The reported values for the A_2 radiative width are $0.77 \pm 0.18 \pm 0.72$ keV and $0.81 \pm 0.19 \pm 0.27$ keV by Crystal Ball (SPEAR) and CELLO, respectively. PLUTO has reported a new measurement of the width to be $1.06\pm0.18\pm0.19$ keV from $A_2\to\pi^+\pi^-\pi^\circ$. This value fits very well in the ideal mixing scheme with the above f° radiative width.

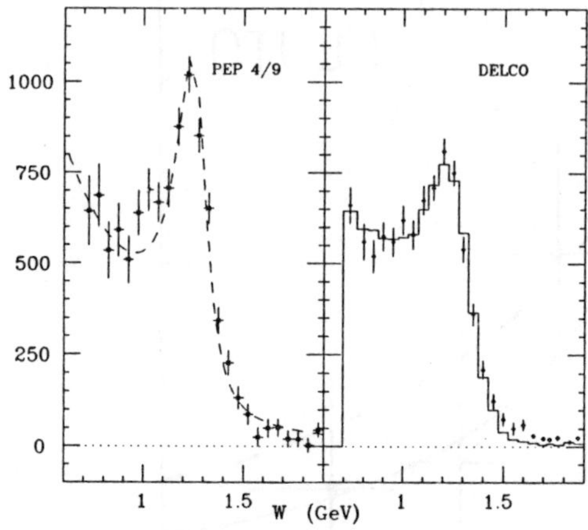

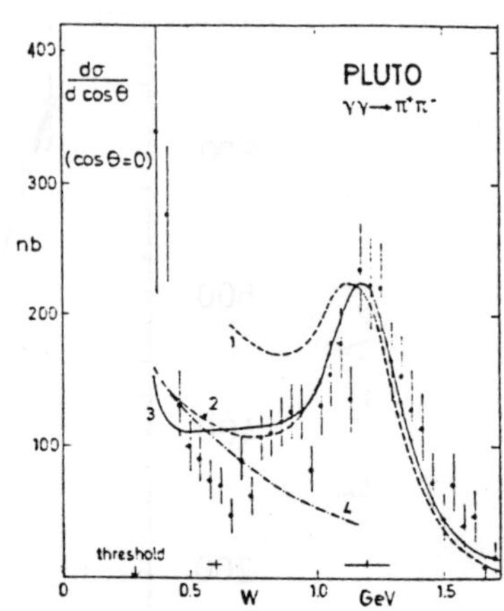

Figure 13. Mass spectrum m($\pi^+\pi^-$) Figure 14. Mass spectrum m($\pi^+\pi^-$)
measured by PEP4/9 and measured by a special
DELCO. Continous curves forward trigger of
are given by f° Breit-Wigner PLUTO.
plus the Born term.

VI. $\pi^+\pi^-\pi^+\pi^-$ and $K^+K^-\pi^+\pi^-$ Productions

The most dominant feature for $\pi^+\pi^-\pi^+\pi^-$ production in γγ interaction is the enhancement near $\rho^0\rho^0$ threshold. In the last two-photon
conference at Aachen, a preliminary result by JADE on $\pi^+\pi^-\pi^0\pi^0$
indicates a very small cross-section for $\rho^+\rho^-$ production. This
immediately rules out a simple $q\bar{q}$ resonance interpretation by isospin
relation. Unfortunately this important result has not yet been
published and there is no further information at this conference. One
model to explain the dilemma of large $\rho^0\rho^0$ cross-section and small $\rho^+\rho^-$
cross-section is that there are two states, one isospin 0 and the other

isospin 2. The two resonances interfere constructively in $\rho^\circ\rho^\circ$ case and destructively in $\rho^+\rho^-$ case. Since there is no I = 2 multiplet in $q\bar{q}$ mesons, these resonances are thought to be $qq\bar{q}\bar{q}$ states. This model further predicts suppressed $K^*\bar{K}^*$ production due to the same destructive interference, but reasonably large $\rho\omega$ and $\rho\phi$ cross-sections.

PEP4/9 and TASSO reported their result on $KK\pi\pi$ production. They both found the ϕ signal in the K^+K^- spectrum (PEP4/9 result in Fig. 15).

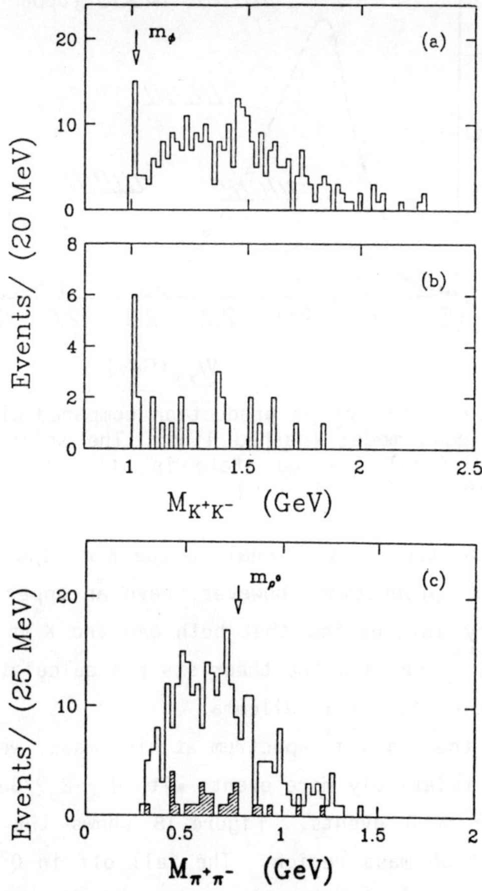

Figure 15. K^+K^- mass distribution for untagged (a) and tagged (b) $\gamma\gamma \to K^+K^-\pi^+\pi^-$ events, (c) $\pi^+\pi^-$ masses for events without ϕ mesons, i.e., $M(K^+K^-) < 1.04$ GeV. This plot includes both tagged and untagged events.

392

There is no ρ^o signal, however, in the $\pi^+\pi^-$ for all events or events associated with ϕ. The upper limit on $\phi\rho$ production is below the prediction of the four quark model. The cross-section prediction, however, can be brought below the upper limits by introducing yet another $\bar{q}q$ state with $I = 1$ (Fig. 16).

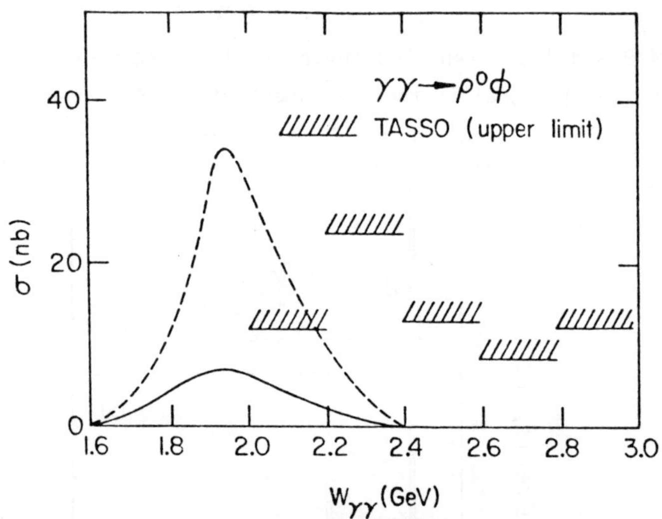

Figure 16. Upper limit of $\rho\phi$ production compared with the prediction of qqqq model (dashed line). The solid line is when an additional I = qq state is allowed. [Theoretical contribution by K.F. Liu]

PEP4/9 also sees a K^* signal in the $K^{\pm}\pi^+$ spectrum. The cross-section for $K^{*-}\bar{K}^{*}$ production, however, have an upper limit of 5.7 nb.

It is very interesting that both $\phi\pi\pi$ and $K^*\pi\pi$ are observed but not $\phi\rho$ and $K^{*-}\bar{K}^{*}$. Perhaps the theorists can calculate the final state interaction and explain this dilemma.

Examining the $\pi^+\pi^-\pi^+\pi^-$ spectrum at high mass, we see from Fig. 17 that there are relatively more events with $W \gtrsim 2.9$ GeV for non-zero Q^2 events than $Q^2 = 0$ events. Figure 18 shows the Q^2 dependence of events in the high mass region. The fall off in Q^2 is quite gentle. A ee$\mu\mu$ curve (therefore point-like behavior) fits the data quite well (dashed line). Rho pole does not fit the data at all (dotted line). It is interesting to note that the Q^2 dependence also fits a ψ pole (solid line). In the latter case one might want to interpret the excess near 3 GeV at large Q^2 to be the charmonium states.

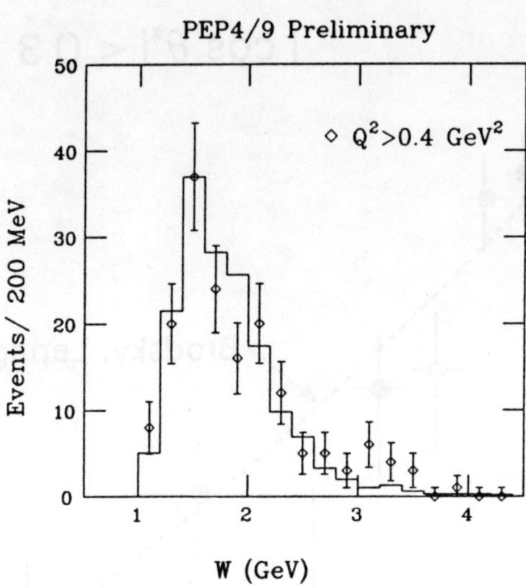

Figure 17. Comparison of $m(\pi^+\pi^-\pi^+\pi^-)$ at $Q^2=0$ and $Q^2>0.4$ GeV^2.

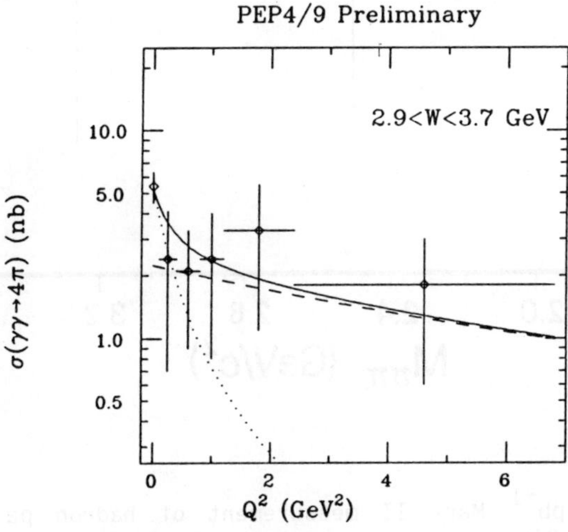

Figure 18. Q^2 dependence of the $\pi^+\pi^-\pi^+\pi^-$ production from $m(\pi^+\pi^-\pi^+\pi^-)$ between 2.9 and 3.7 GeV. The solid line is the ψ-pole form factor, dotted line is the ρ-pole form factor and dashed line is for $ee \to ee\mu\mu$.

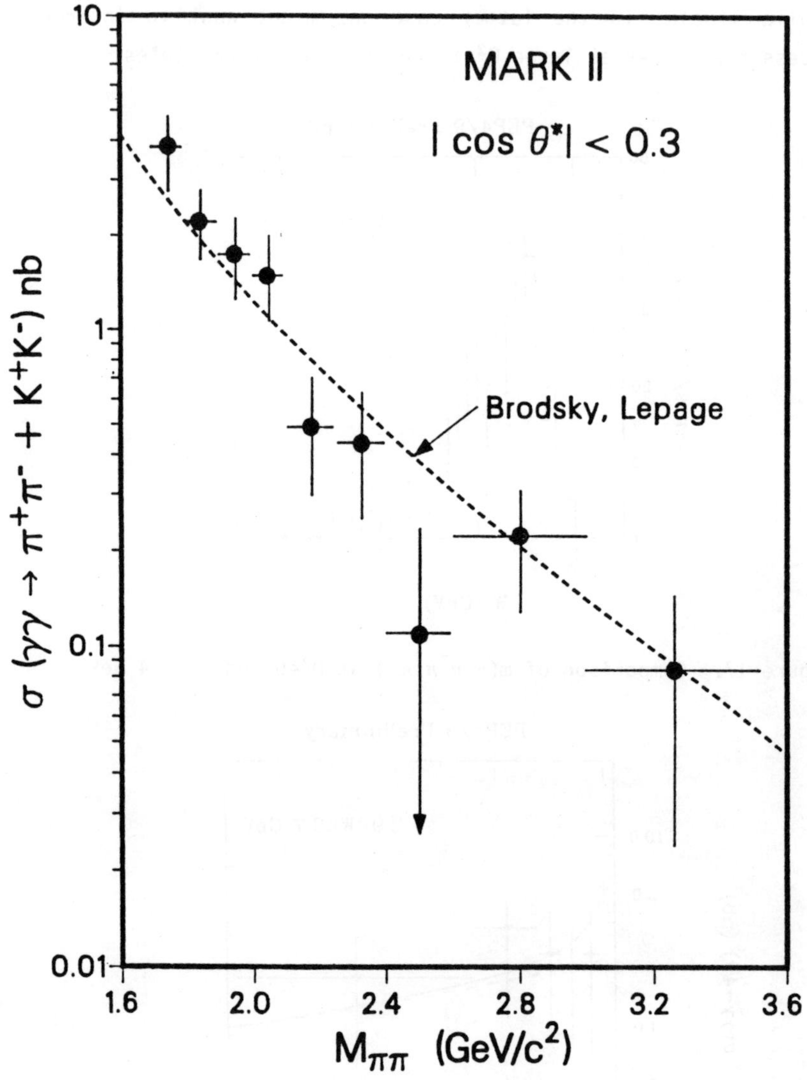

Figure 19. The 230 pb^{-1} Mark II measurement of hadron pair production. Pion masses are assumed. Comparison with theory is made with the assumption that the K-π ratio is 2.1 using kaon and pion form factors.

VII. High Mass Pair Productions

The point-like behavior, of which we saw a glimpse in the high mass 4π spectrum in the last section, is expected for high mass (high W) pair production. The characteristic of point-like behavior is

$$\frac{d\sigma}{d\Omega^{*}} = \frac{f(\cos\theta^{*})}{(W^2)^{n_s+1}}$$

where n_s is the spectator involved for the point-like scattering (n_s=2 for meson pairs and 4 for baryon pairs), and $f(\cos\theta^{*})$ is given by the details of fragmentation. After the power law W dependence is checked, $f(\cos\theta^{*})$ becomes a direction measurement of the fragmentation function. The classical example is the Mark II hadron pair mass spectrum above W = 1.6 GeV, exhibiting a nice power law fall off. In this conference Mark II presented an update with 230 pb^{-1} (Fig. 19). The impressive luminosity gives one sensitivity at higher mass. The data covers only $\cos\theta^* <0.3$ and there is no particle identification.

High mass $\bar{p}p$ production is studied by JADE (Fig. 20) and TASSO in the no tag case and PEP4/9 in the single tag case. PEP4/9, with its forward spectrometer is able to cover the whole $\cos\theta^{*}$ region. The W dependence measured by all experiments is much weaker than $\frac{1}{W^{10}}$ expected from point-like scattering. Experiment on $\bar{p}p$ production is ahead of theory, for which many Feynman diagrams have to be included in the calculation. Compared with the latest calculation using some simple proton fragmentation function, the $\bar{p}p$ cross-section measured is much higher than expected. The magnitude and W dependence for $\bar{p}p$ both suggest that we are not seeing the point-like behavior, but perhaps the tails of the resonance that we discussed in the last section.

High mass pair production may very well be a clean process to study fragmentation of quarks. It will be very interesting to separate $\pi^+\pi^-$ pairs from K^+K^- pairs and thereby study pion and Kaon fragmentation functions. Because of the sharp W dependence, high luminosity is needed, especially for $\bar{p}p$ production in which the present data did not appear to have the point-like region. More results on this subject are certainly expected by the next two-photon conference.

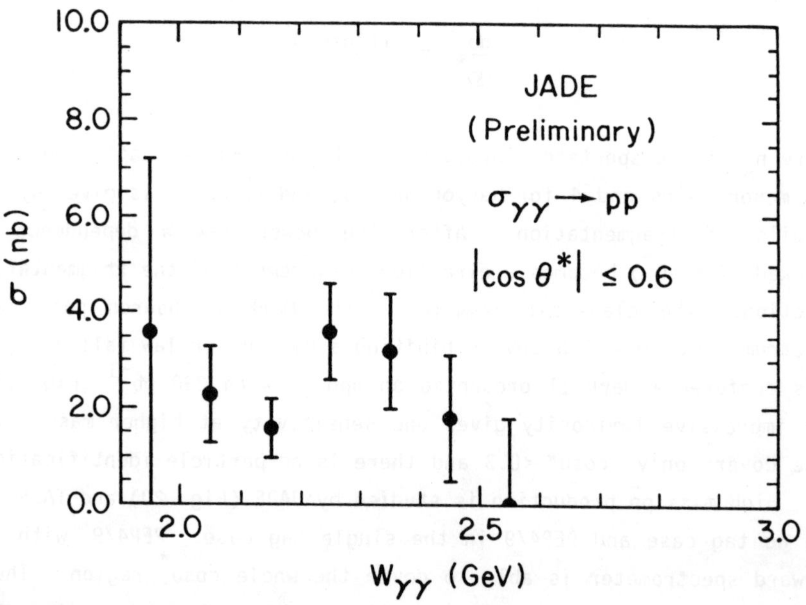

Figure 20. JADE measurement of p̄p production spectrum.

ACKNOWLEDGMENT

I like to thank all those who worked hard in contributing their experimental results. These contributions made this conference exciting and fruitful. I would like to thank all the speakers for their cooperation, and the student-helpers for their assistance in the preparation of this summary talk.

PHOTON-PHOTON COLLISIONS*

STANLEY J. BRODSKY

Stanford Linear Accelerator Center
Stanford University, Stanford, California 94305

1. Introduction

Over the past decade the field of photon-photon collisions[1] has emerged as an important laboratory for testing both perturbative and nonperturbative properties of quantum chromodynamics.

At this meeting a huge array of new and high quality experimental results for exclusive and inclusive two-photon channels has been reported, both at high and low transverse momentum and at high and low virtual photon mass.[2] In many theoretical areas the precision of QCD predictions has now been sharpened, allowing quantitative measures of the running coupling constant, $\alpha_s(Q^2)$, determinations of basic features of hadronic wavefunctions, as well as tests of specific QCD scaling laws and spin selection rules. The striking resonance structure measured[3],[4] in the $\gamma\gamma \to \rho^0\rho^0$ cross section suggests an interpretation in terms of $(qq\overline{q}\overline{q})$ bound states, which if confirmed, represents a manifestation of novel degrees of freedom of QCD. In the case of the photon structure function, we are now beginning to understand the interplay between point-like and hadron-like interactions of on-shell photons and how to meaningfully extract a high precision value for the QCD scale $\Lambda_{\overline{MS}}$.

In photon-photon collisions one studies in e^+e^- storage rings the production of even charge conjugation states by two elementary probes of variable mass and polarization. (See Fig. 1a.) The conventional viewpoint until the early 1970's had been that photon interactions were mediated by intermediate vector meson states (current field identity, generalized vector mesons dominance, etc.). Our viewpoint from the perspective of QCD is just the reverse: even on-mass-shell

---

* Work supported by the Department of Energy, contract $DE - AC03 - 76SF00515$.

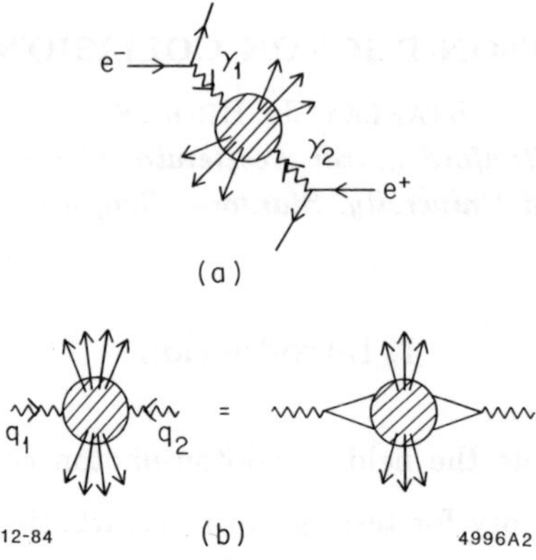

Fig. 1. (a) Photon-photon collisions in e^+e^- storage rings. (b)
The direct coupling of photons to
$q\bar{q}$ currents in QCD.

photons couple directly to local quark currents. (See Fig. 1b.) The observation[5]
of two-jet events: $\gamma\gamma \to$ jet + jet at high p_T is in dramatic conflict with the
VMD description since hadronic collisions nearly always produce four or more
final state jets. At an even more basic level, the confirmation of the QCD scaling
law[6]

$$F_{2\gamma}(x, Q^2) \cong f(x)\, \ell n\, \frac{Q^2}{\Lambda^2} \qquad (1.1)$$

for deep inelastic scattering on a photon target $(e\gamma \to e'X)$ over the range
of $1 \lesssim Q^2 \lesssim 100$ GeV^2 is in striking contrast to the observed scaling pattern
of hadron structure functions. In a related area of real photon physics, the
charge asymmetry reported at this meeting by the MAC group[7] in the reaction
$e^+e^- \to \gamma+$ jet automatically measures the direct coupling of a real photon to
the outgoing quark jet currents[8] and agrees with the QCD fractional charge
assignment for Σe_q^3.

The study of photon-photon collisions in the dynamical range accessible
at e^+e^- storage rings such as PEP and PETRA is well matched to the basic
energy scales of QCD: for processes in which the quarks and gluon propaga-

tor are far off-shell ($Q^2 \gg \Lambda^2_{\overline{MS}}$ and $Q^2 \gg \langle k_\perp^2 \rangle$), higher-twist power-law and non-perturbative corrections become small and perturbative calculations become justified. Experimentally, the specific QCD scaling of the photon structure function, the $p_T^{-4} f(x_T, \theta_{cm})$ scaling of the invariant jet production cross section, and the scaling of the $\gamma\gamma \to \pi^+\pi^-, K^+K^-$ cross sections at large momentum transfer confirm the basic validity of QCD perturbative predictions in this domain. Conversely, at lower momentum transfer one can study in the simplest channels non-perturbative dynamics, including multi-quark and gluon resonance formation, prebinding effects, and other fundamental aspects of hadronization. The total $\gamma\gamma$ cross section has now been measured over a large dynamic range—up to the maximum PEP and PETRA energies.[9] Here there is the outstanding theoretical question whether the box graph which is special to the $\gamma\gamma \to \gamma\gamma$ forward amplitude gives a local $1/W^2$ contibution separate from the conventional Reggeon parameterization.

In this summary, I shall discuss only a few of the many interesting topics presented at this workshop. My main emphasis will be on the use of photon-photon collisions as a primary tool for investigating QCD.

2. Two-body Production Processes

One of the most important areas where two-photon physics will have a critical impact in QCD is in the study of exclusive channels. The exclusive two-body processes $\gamma\gamma \to H\overline{H}$ at large $W_{\gamma\gamma}^2 = (q_1 + q_2)^2$ and fixed $\theta_{\text{c.m.}}^{\gamma\gamma}$ provide a particularly important laboratory for testing QCD, since the large momentum-transfer behavior, helicity structure, and often even the absolute normalization can be rigorously predicted.[10] Conversely, the angular dependence of $\gamma\gamma \to H\overline{H}$ cross sections can be used to determine the shape of the hadron distribution amplitudes[11] $\phi_H(x_i, Q)$—the process-independent probability amplitudes for finding valence quarks in the hadron, each carrying (light-cone) fraction x_i of the hadron's momentum collinear up to the momentum transfer scale Q of the process. The $\gamma_\lambda \gamma_{\lambda'} \to H\overline{H}$ amplitude can be written as a factorized form[10]

$$\mathcal{M}_{\lambda\lambda'}(W_{\gamma\gamma}, \theta_{\text{c.m.}}) = \int_0^1 [dy_i]\, \phi_H^*(x_i, Q)\, \phi_{\overline{H}}^*(y_i, Q)\, T_{\lambda\lambda'}(x, y; W_{\gamma\gamma}, \theta_{\text{c.m.}}) \qquad (2.1)$$

where $T_{\lambda\lambda'}$ is the hard scattering helicity amplitude for scattering the clusters of valence quarks in each hadron. $T_{\lambda\lambda'}$ can be computed in perturbation theory and scales according to the dimensional counting rules:[12] to leading order $T \propto \alpha(\alpha_s/W_{\gamma\gamma}^2)^{1,2}$ and $d\sigma/dt \sim W_{\gamma\gamma}^{-4,-6} f(\theta_{\text{c.m.}})$ for meson and baryon pairs, respectively. The distribution amplitudes $\phi_H(x_i, Q)$ require input from non-perturbative bound state physics, but their logarithmic dependence in Q^2 is determined by evolution equations.[11] Detailed predictions for pseudo-scalar and vector-meson pairs for each helicity amplitude are given in Ref. 10. The helicities of the hadron pairs are predicted to be equal and opposite to leading order in $1/W^2$. The QCD predictions have now been extended to mesons containing $|gg\rangle$ Fock states by Atkinson, Sucher and Tsokos,[13] to $\gamma\gamma \to p\overline{p}$ by Damgaard,[14] and to all $B\overline{B}$ octet and decouplet states by Farrar, Maina and Neri.[15] The normalization of the $\gamma\gamma \to p\overline{p}$ amplitude is constrained by the $\psi \to p\overline{p}$ rate.[11] The arduous calculation of 280 $\gamma\gamma \to qqq\overline{q}\overline{q}\overline{q}$ diagrams in T_H required for calculating $\gamma\gamma \to B\overline{B}$ is greatly simplified by using two-component spinor techniques.[15] However, since there is a disagreement between the calculations of Refs. 14 and 15, a third calculation is necessary.

The basic gauge-invariant measure of a hadron's wavefunction is the distribution amplitude $\phi_H(x_i, Q)$. Using the factorization theorem for exclusive scattering amplitudes, one can show that $\phi_H(x_i, Q)$ is the only non-perturbative input required to normalize and compute any exclusive hadronic scattering processes in QCD at high momentum transfer. Eventually one can hope to actually calculate distribution amplitudes from first principles in QCD, e.g. by solving the QCD light-cone equation of motion[10] or from numerical constraints obtained from lattice gauge theory.[16] At this point, we can utilize model distribution amplitudes for mesons and baryons as candidate forms for the nonperturbative dynamical input.

Candidate hadronic wavefunctions have been recently derived[17] using QCD sum rules which relate the first few x-moments of the meson and baryon distribution amplitudes to the QCD vacuum condensates $\langle 0| G^2_{\mu\nu} |0\rangle$ and $\langle 0| m\bar{\psi}\psi |0\rangle$. The resulting form for the nucleon distribution amplitude leads to a number of non-trivial predictions: the sign and magnitude of $G^P_M(Q^2)$ at high Q^2, the ratio G_{M_p}/G_{M_n} and the normalization of $\psi \to p\bar{p}$ are all correctly determined.[17] In the pion case, the normalization of the weak decay constant and high Q^2 form factor are consistent. All of this is contrary to the conclusions of Isgur and Llewellyn Smith[18] who had argued that QCD exclusive scattering formalism could not account for the normalization of the pion and nucleon form factors at presently available momentum transfer.[19]

Although there are a number of assumptions involved in applying QCD sum rules as wave function constraints, the distribution amplitudes derived by Chernyak and Zhnitskii serve as very useful Gedanken forms for making predictions for photon-photon exclusive cross sections. The postulated shapes differ significantly from the SU(6)-symmetric asymptotic solution to the distribution amplitude evolution equation: $\phi \propto x_1 x_2 x_3$, or the weak binding form $\phi \propto \delta(x_1 - \frac{1}{3})\delta(x_2 - \frac{1}{3})$. In particular, the proton quark distribution is strongly skewed: the u-quark with helicity parallel to that of the nucleon carries 65% of the nucleon's momentum. This asymmetry also implies that the hadron scattering amplitude is sensitive to the near-endpoint region of integration as well the dependence of the running coupling constant on the exchanged momentum in the hard scattering amplitude. Another special feature of the QCD sum-rule analysis is the strong sensitivity to the hadron helicity. This effect is induced due to the fact that the coupling of a pair of quarks through gluon exchange to the gluon condensate is strongest when the quark spins are antiparallel. The distribution amplitude for pions and rho-mesons with helicity zero is predicted to be double-humped, with a local minimum at $x = \frac{1}{2}$. The distribution amplitude of a rho-meson with helicity ± 1 is, however, peaked at equal momentum. This implies a strong dependence of the $\gamma\gamma \to \rho\rho$ amplitude on a non-perturbative vacuum condensate effect in the ρ wavefunction. In analogy one also expects that the Δ distribution amplitude has a very different shape for helicity $S_z = \frac{3}{2}$ and $S_z = \frac{1}{2}$ states. This dependence could have a striking effect on the relative normalization of the $\Delta\overline{\Delta}$ and $p\overline{p}$ production cross sections and could very possibly diminish the ratio of 60 predicted by Farrar *et al.* on the basis of symmetric helicity-independent nucleon and isobar wavefunctions. It is clearly important to repeat the $\gamma\gamma \to p\overline{p}$ calculations assuming the asymmetric form of the proton distribution amplitude derived from the ITEP QCD sum rules by Chernyak and Zhnitskii, since their model can readily account for the magnitude and sign of the proton and neutron form factors.

The normalization and angular dependence of the $\gamma\gamma \to \pi^+\pi^-$ predictions turn out to be insensitive to the precise form of the pion distribution amplitude since the results[10] can be written directly in terms of the pion form factor taken from experiment. The reason for this is that, for meson distribution amplitudes which are symmetric in x and $(1-x)$, the same quantity

$$\int_0^1 dx \, \frac{\phi_\pi(x,Q)}{(1-x)} \tag{2.2}$$

controls the x-integration for both $F_\pi(Q^2)$ and to high accuracy $M(\gamma\gamma \to \pi^+\pi^-)$. Thus we find the relation:

$$\frac{\frac{d\sigma}{dt}(\gamma\gamma \to \pi^+\pi^-)}{\frac{d\sigma}{dt}(\gamma\gamma \to \mu^+\mu^-)} \cong \frac{4|F_\pi(s)|^2}{1 - \cos^4\theta_{cm}} \, . \tag{2.3}$$

The scaling behavior, angular behavior, and normalization of Eq. (2.3) are all non-trivial predictions of QCD. Recent Mark II data[20] for $\pi^+\pi^-$ and K^+K^- production in the range $1.6 < W_{\gamma\gamma} < 2.4$ GeV near $90°$ are in excellent agreement with the normalization and energy dependence predicted by QCD (see Fig. 2). As reported by Gidal[21] at this meeting, the Mark II results have now been extended to pair mass beyond 3 GeV, again in agreement with the QCD predictions. It is clearly very important to test the angular dependence of the

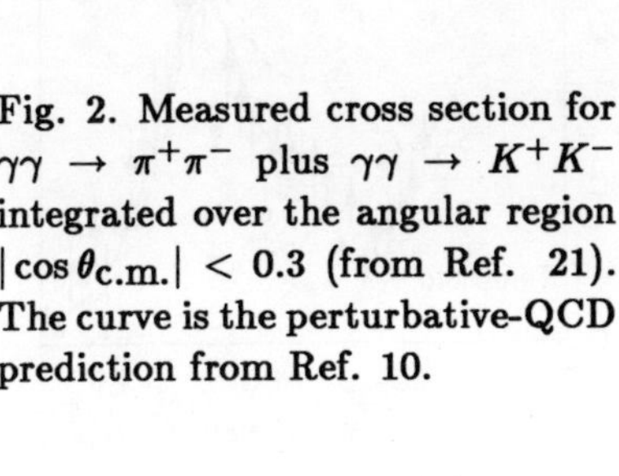

Fig. 2. Measured cross section for $\gamma\gamma \to \pi^+\pi^-$ plus $\gamma\gamma \to K^+K^-$ integrated over the angular region $|\cos\theta_{\mathrm{c.m.}}| < 0.3$ (from Ref. 21). The curve is the perturbative-QCD prediction from Ref. 10.

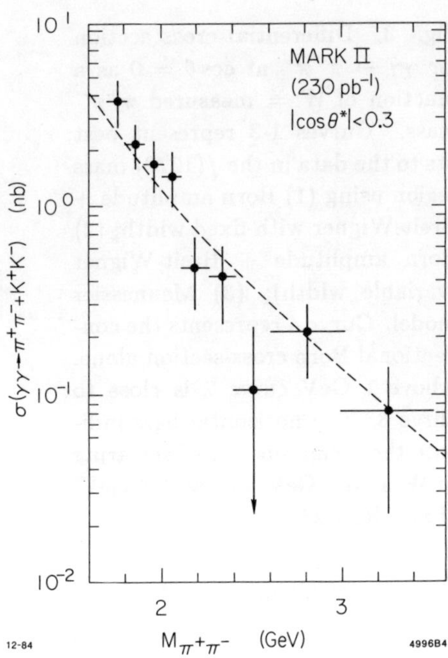

cross sections and separate the $\pi^+\pi^-$ and K^+K^- contributions. The onset of scaling at this range of momentum transfer for meson pair production is reasonable since the off-shell quark propagators in the diagrams for T_H carry momenta large compared to the relevant QCD scales: quark masses, intrinsic transverse momentum, and $\Lambda_{\mathrm{QCD}}^{\overline{MS}}$. However, just as in $e^+e^- \to H\overline{H}$, the scaling behavior of the Born cross sections can be distorted by resonance production; the leading order predictions are only be valid well above particle production thresholds and where low relative-velocity final-state corrections become unimportant. [Here we have in mind the QCD analogue of Coulomb interactions between attractive charged particles which, in the non-relativistic regime, give singular distortion factors[22] of the form $\varsigma/(1 - e^{-\varsigma})$ where $\varsigma = 2\pi\,\alpha/v$ ($\Rightarrow 8\pi\,\alpha_s/3v$ in QCD).]

It is also important to understand the $\gamma\gamma \to \pi^+\pi^-$ amplitude in the threshold region since this is the simplest two-body scattering amplitude in QCD. The amplitude is rigorously determined below threshold at $W = 0$ by the low energy

Fig. 3. Differential cross section for $\gamma\gamma \to \pi^+\pi^-$ at $\cos\theta = 0$ as a function of W = measured $\pi^+\pi^-$ mass. Curves 1-3 represent best fits to the data in the $f(1270)$ mass region using (1) Born amplitude + Breit-Wigner with fixed width; (2) Born amplitude + Breit-Wigner (variable width); (3) Mennessier model. Curve 4 represents the conventional Born cross-section alone. Above 1 GeV curve 2 is close to curve 3. The horizontal bars indicate the mean amount of smearing in W at 0.6 GeV and at 1.2 GeV. (From Ref. 24.)

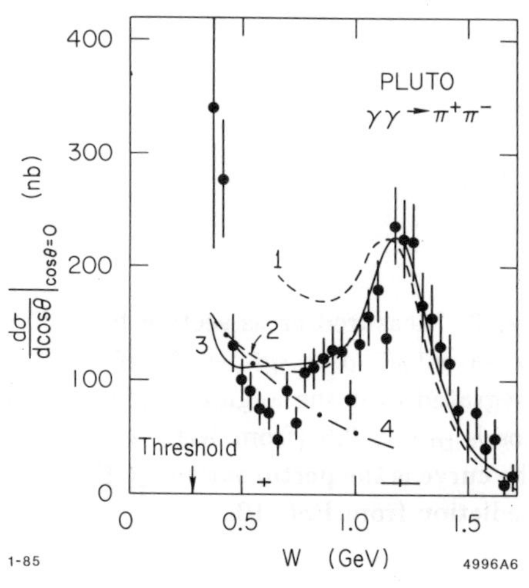

theorem for Compton scattering and crossing. The phase of each $\gamma\gamma \to \pi^+\pi^-$ spherical wave is related by Watson's theorem to the phase of the corresponding $\pi^+\pi^- \to \pi^+\pi^-$ scattering amplitude. The most detailed predictions employing these constraints, which are obtained by modifying the $\gamma\gamma \to \pi^+\pi^-$ point-like Born approximation, have been given by Menessier.[23] As shown in Fig. 3, the Pluto data[24] appears to differ significantly from the model predictions at energies below the f^0 contribution. If these results are confirmed, this would signal a large threshold enhancement in the $\gamma\gamma \to \pi^+\pi^-$ amplitude, possibly indicating low velocity distortion effects[22] as discussed above or a new resonance near or below the $\pi^+\pi^-$ threshold. Either possibility has important implications for QCD.

The data[3],[4] for $\gamma\gamma \to \rho^0\rho^0$ from PETRA and PEP are much larger than predicted by QCD in the region $1.2 < W_{\gamma\gamma} < 2.4$ GeV and are clearly suggestive of resonance enhancement near $M \sim 1.4$ GeV. (See Fig. 4.) The absence of a comparable signal in $\rho^+\rho^-$ precludes an explanation in terms of a single isoscalar resonance such as a glueball state. A possible, if not compelling, interpretation has been suggested by Achasov $et\ al.$,[25] and Li and Liu[26] in terms of two interfering $I = 0$ and $I = 2$, $J^{PC} = 2^{++}$, $q q \overline{q} \overline{q}$ resonances with masses 1.3

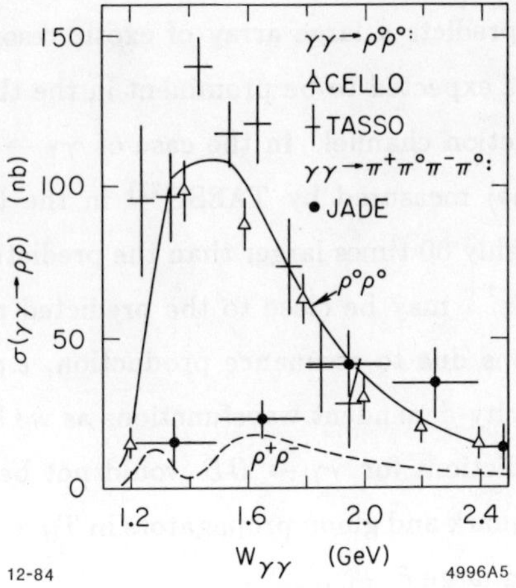

Fig. 4. Comparison of the $\gamma\gamma \to \rho^0\rho^0$ and $\rho^+\rho^-$ data[3),4)] with the mesonium ($qq\overline{q}\overline{q}$) resonance model of Achasov *et al.*[25)] See also Ref. 26.

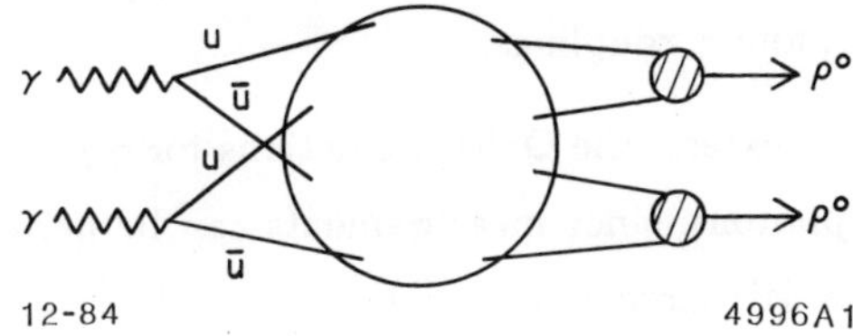

Fig. 5. Coupling of two photons to $q\overline{q}q\overline{q}$ systems decaying to the $\rho^0\rho^0$ final state. The $u\overline{u}u\overline{u}$ coupling is dominant.

and 1.6 GeV, respectively. Two photons couple naturally to such "mesonium" S-wave states since each photon is likely to produce a $q\overline{q}$ system. (See Fig. 5.) Since the $I = 0$ and $I = 2$ amplitudes add constructively in $\gamma\gamma \to \rho^0\rho^0$, they interfere destructively[25)] in $\gamma\gamma \to \rho^+\rho^-$. Identification of these resonances with the predicted couplings in $\psi \to \gamma 4\pi$ as well as other $\gamma\gamma \to V\overline{V}$ channels is crucial for a check of this hypothesis. At the high end of the experimental range, $W_{\gamma\gamma} \gtrsim 2$ GeV, the data could be approaching the magnitude predicted by perturbation theory.

In general, QCD predicts a large array of exotic resonances $q\bar{q}g$, gg, $q\bar{q}q\bar{q}$, $qqq\overline{qqq}$, *etc.*, which are expected to be prominent in the threshold region of the appropriate $\gamma\gamma$ production channel. In the case of $\gamma\gamma \to p\bar{p}$, the cross section $(d\sigma/d\cos\theta = 3 \pm 1\,nb)$ measured by TASSO[27] in the threshold region $2 < W_{\gamma\gamma} < 2.4$ GeV is roughly 60 times larger than the prediction of Farrar *et al.*,[15] although $\gamma\gamma \to \Delta^{++}\overline{\Delta}^{++}$ may be close to the predicted normalization. Again this suggests distortions due to resonance production, *e.g.*, $qqq\overline{qqq}$ baryonium states or strongly helicity-dependent wavefunctions as we have discussed above. The perturbative predictions for $\gamma\gamma \to B\overline{B}$ would not be expected to become valid unless all of the quark and gluon propagators in T_H are reasonably off-shell, *i.e.*, $W_{\gamma\gamma} \gtrsim 5$ GeV and large $\theta_{\mathrm{c.m.}}$.

An essential feature of the QCD predictions for baryon pair production is the fall-off of the cross section at large momentum transfer, reflecting the quark compositeness of the hadrons. One can compare these predictions with the large, rapidly increasing cross sections predicted[29] from effective Lagrangian models with *point-like p*, Δ, and γ couplings.

It is important to extend the QCD predictions for $\gamma\gamma \to H\overline{H}$ to the case of one or two virtual photons, since measurements can be performed with tagged electrons. In fact, for W^2 large and fixed $\theta_{\mathrm{c.m.}}$, the q_1^2 and q_2^2 dependence of the $\gamma\gamma \to H\overline{H}$ amplitude for transversely polarized photons must be minimal.[29] in QCD since the off-shell quark and gluon propagators in T_H already transfer hard momenta; *i.e.*, the 2γ coupling is effectively local for $|q_1^2|$, $|q_2^2| \ll p_T^2$.

The study of resonance production in exclusive two-photon reactions is particularly advantageous because of the variety of new and exotic channels, the absence of complications from spectator hadrons, and the fact that the continuum can be computed or estimated from perturbative QCD. The onset of open charm is particularly interesting since the sum of the exclusive channel cross section should saturate the $\gamma\gamma \to c\bar{c}$ plus $\gamma\gamma \to c\bar{c}q\bar{q}$ contributions. The channels with maximal spin and charge such as $\gamma\gamma \to B_{3/2}(cuu)\,\overline{B}_{3/2}(\overline{cuu})$ are likely to be dominant due to charge coherence and multiple helicity states.

We also note that photon-photon collisions provide a way to measure the running coupling constant in an exclusive channel, independent of the form of hadronic distribution amplitudes. The photon-meson transition form factors $F_{\gamma \to M}(Q^2)$, $M = \pi^0, \eta^0, f$, etc. are measurable in tagged $e\gamma \to e'M$ reactions. QCD predicts[10]

$$\alpha_s(Q^2) = \frac{1}{4\pi} \frac{F_\pi(Q^2)}{Q^2 |F_{\pi\gamma}(Q^2)|^2} \tag{2.4}$$

where to leading order the pion distribution amplitude enters both numerator and denominator in the same manner. The higher order corrections can be calculated using the methods of Ref. 30.

In the regime $s \gg p_T^2 \gg \mu^2$ the cross sections for $\gamma\gamma \to V\overline{V}$ and $\gamma\gamma \to \gamma V$ can be computed from $n \geq 2$ multiple gluon exchange diagrams by summing a series in $\alpha_s(p_T^2)\, \ell n\, s/p_T^2$. As shown by Ginzburg, Panfil, and Serbo,[31] the exponentiation of this series leads to large enhancement factors of order of 100 over Born contributions. The cross sections dominate over the lower-order quark exchange contributions at forward angles. Estimates are also given for $\gamma\gamma \to V q\bar{q}$, although in this case soft gluon radiation needs to be included.

3. The Photon Structure Function

A key physical quantity in QCD is the set of photon structure functions[32] $F_i^\gamma(x, Q^2)$ measured in $e\gamma \to e'X$:

$$\frac{2\pi\, d\sigma}{dx\, dy\, d\phi} = \frac{4\pi\, \alpha^2 s_{e\gamma}}{Q^4} \left[(1-y)F_2 + y^2 x F_1 + \epsilon(1-y) \cos 2\phi F_3 \right] \tag{3.1}$$

with $q^2 = -Q^2$, $x = Q^2/2k \cdot q$, $k^2 = 0$, $y = q \cdot k/q \cdot p_{e_1}$, and $\epsilon = 2(1 - \varsigma)/(1 + (1 - \varsigma)^2)$, where $\varsigma = q \cdot k/q \cdot p_{e_2}$ is the energy fraction transferred from the lepton beam to the real photon. As first shown by Witten,[6] QCD predicts, unlike hadron structure functions, the normalization, shape, and evolution of the F_i^γ to second order in $\alpha_s(Q^2)$. The basic scaling behavior, $F_2^\gamma(x, Q^2) \sim \ell n\, Q^2 f(x)$

408

predicted by QCD has now been confirmed for $0.3 < x < 0.8$ by PLUTO, JADE, TASSO and PEP4-PEP9 measurements[2] for Q^2 below 2 GeV^2 to beyond 100 GeV^2. The quark and gluon distributions[33],[34] in the photon obey (in leading order) the extended evolution equations ($t = \ell n\, Q^2/\Lambda^2$)

$$\frac{dq_i(x,t)}{dt} = \frac{3\alpha\, Q_i^2}{2\pi}\left[x^2 + (1-x)^2\right] + \frac{\alpha_S(t)}{2\pi}\int_x^1 \frac{dy}{y}$$

$$\left[p_{qq}\left(\frac{x}{y}\right) q_i(x,t) + p_{qG}\left(\frac{x}{y}\right) G(y,t)\right] \tag{3.2}$$

$$\frac{dG(x,t)}{dt} = \frac{\alpha_S(t)}{2\pi}\int_x^1 \frac{dy}{y}$$

$$\left[p_{Gq}\left(\frac{x}{y}\right) \sum_i q_i(y,t) + p_{GG}\left(\frac{x}{y}\right) G(y,t)\right] \tag{3.3}$$

where the inhomogeneous term is induced by the direct $\gamma\gamma \to q\bar{q}$ box diagram. It has been conventional to parametrize the QCD prediction in terms of a regular hadronic (vector meson dominance) piece plus the asymptotic solution to (Eq. (2)) of the form $q_i^\gamma(x_1 Q^2) = [(4\pi)/(\alpha_S(Q^2))]a_i(x) + b_i(x)$. However, in lowest order, this gives an artificial singularity in the photon structure function: $F_{2\gamma} = xq_2^\gamma \sim x^{-0.5964}$ at $x \to 0$. In higher order, $b_i(x) \propto x^{-2}$ implying a *negative* cross section for $x \to 0$ at fixed Q^2. These difficulties show that a straightforward separation of regular hadronic and pointlike contributions is *invalid*; diagrammatically both horizontal and vertical gluon exchange corrections to the box diagram must be taken into account.[35]

As emphasized by Glück *et al.*,[36] rigorous QCD predictions can be made by construction of quark and gluon distributions in the photon to agree with experiment at a given scale Q_0^2, and then using the evolution Eq. (2) to make predictions at large Q^2. The differences between higher and leading order predictions are found to be small. The fundamental prediction of QCD, $F_{2\gamma}(x, Q^2) \sim \log Q^2$ at fixed x and large Q^2, remains. The disadvantage of this procedure is that the

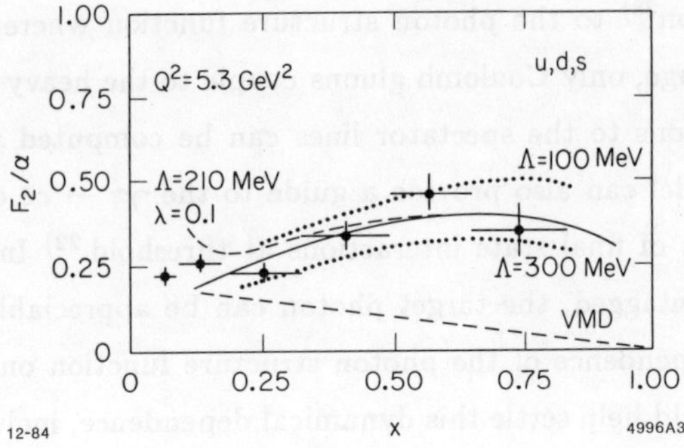

Fig. 6. Analysis of the photon structure function in the Antoniadis-Grunberg scheme.[37] See the reviews of A. Deuter and W. Wagner, these proceedings.

possibility of determining $\Lambda_{QCD}^{\overline{MS}}$ and making *a priori* predictions for the shape of the structure functions is lost.

A more convenient method has recently been proposed by Antoniadias and Grunberg.[37] They parametrize the photon structure moments in the form

$$\frac{4\pi}{\alpha} M_n^\gamma = \frac{4\pi}{\beta_0} \frac{a_n}{\alpha_s(Q^2)} + b_n^r + \frac{b}{n-2} [1 - (\lambda\alpha_s(Q^2))]^{d_n} + \ldots \qquad (3.4)$$

The parameter $\lambda \neq 0$ represents hadronic contributions and eliminates the potential singularity at $n = 2$. Thus at the expense of one extra parameter, one can make detailed predictions for QCD; for $x > x_0 \cong 0.25$ there is apparently little sensitivity to the hadronic input. A PLUTO analysis based on this procedure gives $\Lambda_{\overline{MS}} = 160 \pm 45 \ MeV$. (See Fig. 6.) Theoretical uncertainties still remain, however, concerning whether a background VMD contribution should still be included as in conventional fits and the manner in which charm quark contributions should be identified.

It clearly would be useful to test the accuracy of these methods in an example where the photon interactions and gluonic radiative corrections could be systematically computed. One such theoretical laboratory is the $\gamma^*\gamma \to Q\overline{Q}$ heavy quark contribution[38] to the photon structure function where, for $v^2/c^2 \ll 1$ and Coulomb gauge, only Coulomb gluons couple to the heavy quarks, and the radiative corrections to the spectator lines can be computed as an expansion in v/c. This model can also provide a guide to the $\gamma\gamma \to c\bar{c}$ contribution, including the effect of final state interactions at threshold.[22] In the case where one electron is untagged, the target photon can be appreciably off shell, thus obscuring the dependence of the photon structure function on $\Lambda_{QCD}^{\overline{MS}}$. Heavy quark models could help settle this dynamical dependence, including the degree of quenching of the hadronic contribution as $|k^2|$ increases.

The photon structure function plays a pivotal role in perturbative QCD and further measurements are very much warranted. Higher luminosity measurements at PEP, PETRA, and higher energies possible at SLC, LEP and Tristan will allow more precise measurements at high $Q^2 \geq 100\ GeV^2$, the separation of F_1^γ, F_2^γ, F_3^γ and checks of specific QCD predictions for F_L^γ, separation of heavy quark contributions, checks on the jet topological structure, moment analyses, etc. We emphasize the need to check the photon-off-mass dependence and the need for real photon target measurements, since even at moderate k^2 the sensitivity of F_2^γ to $\Lambda_{\overline{MS}}$ drops out.

In the future, it may be possible to measure real photon structure functions in $e\gamma \to e'X$ reactions where the photon target is obtained from a laser or wiggler beam back-scattering on a linac beam.[39] In addition to potentially high luminosity, one also has the advantage that the photon beam is polarized and can have an energy spectrum peaked at high energy, reducing the need for reconstruction of the hadron production energy W.

4. High Transverse Momentum Inclusive Processes

One area of considerable theoretical and experimental uncertainty in photon photon collisions is jet and hadron production at high p_T. As reported by the JADE collaboration, the predicted QCD scaling law[40)]

$$p_T^4 \, E \, d\sigma/d^3 p(\gamma\gamma \to Jet + X) = f(x_T, \theta_{cm})$$

appears to set in at p_T as low as 3 GeV. This is in remarkable contrast to the very high p_T $(p_T \gtrsim 20\,GeV))$ required before any semblance of scale-invariance is seen in pp collisions. The precocious scaling of $\gamma\gamma$ reactions could be due to a number of factors:

1. There are no logarithmic modifications predicted for the $\gamma\gamma \to 2$ jet, 3 jet and 4 jet processes in leading order. This is due to the fact that the scale-violation due to the running coupling constant in the $\gamma q \to gq$ and $qq \to qq$ subprocess contribution is compensated by the evolution of the quark distribution in the photon. The subprocess can be distinguished by the power of $(1 - x_T)^n$ at threshold $x_T = 2p_T/\sqrt{s_{\gamma\gamma}} \to 1$ or from jet topology.

2. Higher twist contributions receive less trigger bias in $\gamma\gamma$ compared to hadron-induced reactions.

The normalization of the JADE jet production cross section appears higher than QCD.[5)] As discussed in this meeting, it seems unlikely that this discrepancy could be due to an integrally-charged quark model. Within the context of QCD, there are other possible explanations: anomalous K-factors for $\gamma\gamma \to q\bar{q}$, etc., anomalous threshold corrections for $\gamma\gamma \to c\bar{c}$; mis-estimate of higher jet nucleon contributions, etc. Measurements of $\gamma\gamma \to \pi X$, and $\gamma\gamma \to \gamma X$ at high p_T could help to resolve these questions. One is also interested in understanding the photon mass dependence of the inclusive cross sections, backgrounds due to $ee \to ee\gamma^*$ single photon radiation contributions, jet coherence effects[43)] in 3-jet and 4-jet reactions, etc. On the theoretical side, we need to compute higher order corrections, relate the $\gamma\gamma \to Jet\,X$ cross sections to the measured photon structure function, analyse heavy quark production, etc. Aside from e^+e^- reactions, photon-photon collisions provide the simplest environment to understand jet production and hadronization.

5. Conclusions

The study of photon-photon collisions has progressed enormously, stimulated by new data and new calculational tools for QCD. In the future we can expect precise determinations of α_S and $\Lambda_{\mathrm{QCD}}^{\overline{MS}}$ from the $\gamma^*\gamma \to \pi^0$ form factor and the photon structure function, as well as detailed checks of QCD, determination of the shape of the hadron distribution amplitudes from $\gamma\gamma \to H\overline{H}$, reconstruction of $\sigma_{\gamma\gamma}$ from exclusive channels at low $W_{\gamma\gamma}$, definitive studies of high p_T hadron and jet production, and studies of threshold production of charmed systems. Photon-photon collisions, along with radiative decays of the ψ and Υ, are ideal for the study of multiquark and gluonic resonances. We have emphasized the potential for resonance formation near threshold in virtually every hadronic exclusive channel, including heavy quark states $c\bar{c}c\bar{c}$, $c\bar{c}u\bar{u}$, etc.

At higher energies (SLC, LEP, ...) parity-violating electroweak effects and Higgs production due to "equivalent" Z^0 and $W^\pm$ beams from $e \to eZ^0$ and $e \to \nu W$ will become important.[43] The basic form for the virtual (transversely polarized) Z^0 beam in an electron is $(x = (k_Z^0 + k_Z^z)/(p_e^0 + p_e^z))$

$$\frac{dN}{dx\,dk_\perp^2} = \frac{\alpha_Z}{2\pi}\,\frac{k_\perp^2}{(k_\perp^2 + M_Z^2)^2}\,\left[1 + (1-x)^2\right].$$

Asymptotic $\log s/M_Z^2$ scaling for dN/dx becomes relevant at $s \gg M_Z^2$, where $k_\perp^2 \cong M_Z^2$.

Many of the most important $\gamma\gamma$ studies are severely limited by counting rate, emphasizing the need for increasing detector acceptance and photon-photon luminosity. New accelerator developments,[39] such as backscattered lasers on linear collider beams or other coherent methods[44] which can generate intense beams of photons, could lead to dramatic increases in $\mathcal{L}_{\gamma\gamma}$. We note that many of the most interesting QCD tests require only modest photon energies $W_{\gamma\gamma} \lesssim$ 5 to 10 GeV, but high photon-photon luminosity.

ACKNOWLEDGMENTS

All of us, at this very productive workshop, are grateful to the U.C. Davis organization committee members, and the chairman Dick Lander for their outstanding efforts and assistance.

REFERENCES

1. For recent reviews of photon-photon collisions, see H. Kolanoski, Bonn-HE-84-06 (1984), to be published in Springer Tracts in Modern Physics; J. H. Field, LPNHE 84-04 (1984); and the *Proceedings of the Vth International Colloquium on $\gamma\gamma$-Interactions, Aachen,* ed. Ch. Berger, in *Lecture Notes in Physics*, Vol. 191, Springer Verlag (1983).

2. See also the reviews of P. M. Zerwas and W. Ko, in these proceedings.

3. R. Brandelik *et al.*, (TASSO collaboration) Phys. Lett. <u>97B</u>, 448 (1980); M. Althoff *et al.*, Z. Phys. <u>C16</u>, 13 (1982).

4. H.-J. Behrend *et al.*, (CELLO collaboration), Z. Phys. <u>C21</u>, 205 (1984). D. L. Burke *et al.*, (MARK II collaboration), Phys. Lett. <u>103B</u>, 153 (1980). J. Dainton, (JADE collaboration), *Proc. of the European Physical Soc. Conf. 1983.*

5. See F. Foster, these proceedings.

6. E. Witten, Nucl. Phys. <u>B120</u>, 189 (1977). The experimental data are reviewed by W. Wagner, these proceedings.

7. R. Prepost, these proceedings; E. Fernandez *et al.*, SLAC-PUB-3410 (1984).

8. S. J. Brodsky, C. E. Carlson and R. Suaya, Phys. Rev. $\underline{D14}$, 2264 (1976).

9. For a recent review see W. Ko, UCD-840915 (1984), to appear in the proceedings of the 22nd International Conference.

10. S. J. Brodsky and G. P. Lepage, Phys. Rev. $\underline{D24}$, 1808 (1981); G. P. Lepage, S. J. Brodsky, T. Huang, P. B. Mackenzie, in *Particles and Fields 2*, eds. A. Z. Capri and A. N. Kamal (Plenum, New York, 1983), p 83; S. J. Brodsky, T. Huang, and G. P. Lepage, *ibid.*, p. 143.

11. G. P. Lepage and S. J. Brodsky, Phys. Rev. $\underline{D22}$, 2157 (1980).

12. S. J. Brodsky and G. R. Farrar, Phys. Rev. Lett. $\underline{31}$, 1153 (1973). V. A. Matveev, R. M. Muradyan, A. V. Tavkheldize, Lett. Nuovo Cimento $\underline{7}$, 710 (1973).

13. G. W. Atkinson, J. Sucher, K. Tsokos, Phys. Lett. $\underline{137B}$, 407 (1984).

14. P. H. Damgaard, Nucl. Phys. $\underline{B211}$, 435 (1983).

15. G. R. Farrar, E. Maina, F. Neri, Rutgers preprints RU-83-33 (1983), and RU-84-13 (1984), and G. Farrar, these proceedings.

16. See, *e.g.*, B. Velikson and D. Weingarten, IBM/Brown University preprint (1984) and S. Gottlieb, P. B. Mackenzie, H. B. Thacker and D. Weingarten, Phys. Lett. $\underline{134B}$, 346-350 (1984).

17. V. L. Chernyak and A. R. Zhnitskii, Novosibirsk preprint I4F-83-108 (1983); A. R. Zhitnitskii, I. R. Zhitnitskii and V. L. Chernyak, Yad. Fiz. $\underline{38}$, 1074-1086 (1983); V. L. Chernyak and A. R. Zhitnitskii, IYF-83-108, 1983, p. 37; T. Huang, private communication.

18. N. Isgur and C. H. Llewellyn Smith, Phys. Rev. Lett. $\underline{52}$, 1080 (1984).

19. See also V. M. Belyaev and B. L. Ioffe, Sov. Phys. JETP $\underline{56}$, 493 (1982). B. L. Ioffe, A. V. Smilga, Phys. Lett. $\underline{114B}$, 353 (1983); V. A. Nesterenko and A. V. Radyuskhin, Phys. Lett. $\underline{115B}$, 410 (1980); $\underline{123B}$, 439 (1983).

20. J. R. Smith *et al.*, (Mark II collaboration), Phys. Rev. $\underline{D30}$, 851 (1984).

21. G. Gidal, these proceedings.

22. See, *e.g.*, *Quantum Electrodynamics* by A. I. Akhiezev and V. B. Berestetski, National Tech. Trans. Service, Washington D.C. (1953). Applications to the threshold behavior of heavy quark production is given in S. J. Brodsky and J. F. Gunion, SLAC-PUB-3527 (1984).

23. C. Mennessier, Z. Phys. $\underline{C16}$, 241 (1983).

24. Ch. Berger *et al.*, DESY 84-074 (1984).

25. N. N. Achasov, S. A. Devyanin, G. N. Shestakov, Phys. Lett. $\underline{108B}$, 134 (1982); Z. Phys. $\underline{C16}$, 55 (1982), and Novosibirsk preprints, TF-65-141 (1984), TF-56-137 (1984).

26. Bing-an Li and K. F. Liu, Phys. Lett. $\underline{118B}$, 435 (1982), $\underline{124B}$, 550(E) (1983), Phys. Rev. Lett. $\underline{51}$, 1510 (1983).

27. R. Brandelik *et al.*, Phys. Lett. $\underline{108B}$, 67 (1982). M. Althoff *et al.*, Phys. Lett. $\underline{130B}$, 449 (1983).

28. E. Bagán, A. Bramon and F. Cornet, Barcelona preprint UAB-FT-99 (1983).

29. S. J. Brodsky and G. P. Lepage, Ref. 2. S. J. Brodsky, F. E. Close, J. F. Gunion, Phys. Rev. $\underline{D6}$, 177 (1972).

30. M. H. Sarmadi, PITT-82-10, 1982, p. 10; F. M. Dittes and A. V. Radyushkin, Phys. Lett. $\underline{134B}$, 359-362 (1984); F. M. Dittes and A. V. Radyushkin, Yad. Fiz. $\underline{34}$, 529-540 (1981); S. J. Brodsky, P. Damgaard, Y. Frishman and G. P. Lepage, SLAC-PUB-3295 (1984).

31. I. F. Ginzburg, S. L. Panfil, V. G. Serbo, contributed to the 22nd International Conference, Leipzig; V. L. Chernyak and A. R. Zhitnitskii, Nucl. Phys. $\underline{B222}$, 382-388 (1983), and Novosibirsk Inst. Nucl. Phys. Acad. Sci. 82-044, 1982, p. 14.

32. S. J. Brodsky, T. Kinoshita, H. Terazara, Phys. Rev. Lett. $\underline{27}$, 280 (1971). T. F. Walsh, Phys. Lett. $\underline{36B}$, 121 (1971). T. F. Walsh and P. M. Zerwas,

Phys. Lett. <u>44B</u>, 195 (1973).

33. R. J. DeWitt, L. M. Jones, J. D. Sullivan, D. E. Willen, H. W. Wyld, Phys. Rev. <u>D19</u>, 2046 (1979), <u>D20</u>, 1751(E) (1979); W. A. Bardeen and A. J. Buras, Phys. Rev. <u>D20</u>, 166 (1979). S. Brodsky, SLAC-PUB-2447, *Proceedings of the SLAC Summer Institute (1979)*. D. W. Duke and J. F. Owens, Phys. Rev. <u>D22</u>, 2280 (1980).

34. W. R. Frazer and J. F. Gunion, Phys. Rev. <u>D20</u>, 147 (1979).

35. G. Rossi, Phys. Lett. <u>130B</u>, 105 (1983). W. Frazer, *Proceedings of the 4th International Colloquium on $\gamma\gamma$-Interactions, Paris (1981)*, ed. G. W. London.

36. M. Glück and F. Reya, Phys. Rev. <u>D28</u>, 2749 (1983). M. Drees, M. Glück, K. Grassie, F. Reya, Dortmund preprint DO-TH-84/12 (1984), and contributions to this conference.

37. I. Antoniadis and G. Grunberg, Nucl. Phys. <u>B213</u>, 445 (1983); I. Antoniadis, these proceedings.

38. See T. Uematsu and T. F. Walsh, Phys. Lett. <u>101B</u>, 263 (1981); Nucl. Phys. <u>B199</u>, 93 (1983).

39. See, *e.g.*, I. F. Ginzburg, G. L. Kotkin, V. G. Serbo, V. I. Telnov, Nucl. Instr. Methods <u>219</u>, 5 (1984); Yad. Fiz. <u>38</u>, 372 (1983).

40. S. J. Brodsky, T. A. DeGrand, J. F. Gunion and J. H. Weis, Phys. Rev. <u>D19</u>, 1418 (1979) and Phys. Rev. Lett. <u>41</u>, 672 (1978); C. H. Llewellyn Smith, Phys. Lett. <u>79B</u>, 83 (1978).

41. See, *e.g.*, A. Janah, these proceedings; A. Janah, UMI 82-26472-mc (1982), Ph.D. Thesis.

42. S. J. Brodsky, SLAC-PUB-3429 (1984).

43. For a related idea in hadron collisions, see R. N. Cahn and S. Dawson, Phys. Lett. <u>136B</u>, 196 (1984); Erratum-ibid <u>138B</u>, 464 (1984).

44. J. Spencer, SLAC-PUB-2677 (1981), and private communication.

ABSTRACTS OF CONTRIBUTIONS

TO

PARALLEL SESSIONS

Measurement of the reactions $\gamma\gamma \to K^+K^-\pi^+\pi^-$ and $\gamma\gamma \to \phi\pi^+\pi^-$

TASSO Collaboration

Presented by Uri Karshon
Department of Nuclear Physics, Weizmann Institute of Science, Rehovot, ISRAEL.

ABSTRACT

The reaction $\gamma\gamma \to K^+K^-\pi^+\pi^-$ has been studied in the $W_{\gamma\gamma}$ range of 1.8 to 3.0 GeV. Kaon identification is done in the inner time-of-flight counters of the TASSO detector. A cut of $|\Sigma\vec{p}_T| < 0.12$ GeV/c is applied to remove events with missing particles. Preliminary cross-sections for $\gamma\gamma \to K^+K^-\pi^+\pi^-$ in the above $W_{\gamma\gamma}$ region are in the range 5-20 nb. Requiring at least one identified $K^{\pm}$, a clear ϕ signal is seen in the K^+K^- mass distribution when the other "kaon" is not an identified pion or proton (figure a). Preliminary cross-sections for $\gamma\gamma \to \phi\pi^+\pi^-$ are given (figure b). The values for $1.8 < W_{\gamma\gamma} < 2.2$ GeV are 95% C.L. upper limits, including 30% systematic errors. Preliminary upper limits for the cross-section of the reaction $\gamma\gamma \to \phi\rho^0$ for $2.0 < W_{\gamma\gamma} < 3.0$ GeV are between 10 to 25 nb. These results are not inconsistent with theoretical calculations assuming $q\bar{q}q\bar{q}$ states as the origin of the $\phi\rho^0$ final state.

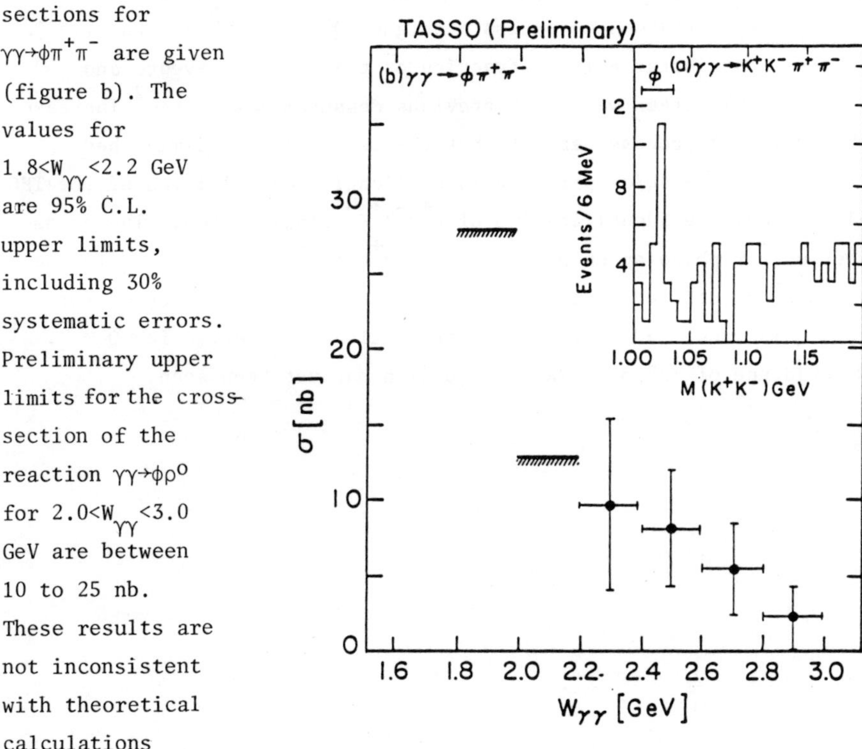

Exclusive Production of $\pi^+\pi^-\pi^+\pi^-$ and $\pi^+\pi^-K^+K^-$
in Photon Photon Collisions

Clark S. Lindsey

U.C. California-Riverside

Physics Dept.

Riverside, CA   92521

Using the PEP4/PEP9 TPC/Two-Gamma detector system, exclusive production of $\pi^+\pi^-\pi^+\pi^-$ and $\pi^+\pi^-K^+K^-$ final states in two-photon collisions is investigated at $Q^2{\sim}0$ and in the Q^2 range between 0.15 and 50 GeV^2. The low mass enhancement in the $\pi^+\pi^-\pi^+\pi^-$ system around 1.4 GeV, observed earlier at $Q^2{\sim}0$, is seen to persist at higher Q^2. The Q^2 dependence of the cross-section follows the ρ form factor out to 2.0 GeV^2 but shows an excess at higher Q^2. For $Q^2{\sim}0$, the $\gamma\gamma{\to}\rho^0\rho^0$ process contributes a large fraction of the $\pi^+\pi^-\pi^+\pi^-$ events and $\sigma(\gamma\gamma \to \rho^0\rho^0)$ agrees well with previous measurements. As Q^2 increases, the $\gamma\gamma \to \rho^0\rho^0$ process persists but the cross-section diminishes.

Use of the dE/dx particle identification capabilities of the TPC allows positive identification of $\pi^+\pi^-K^+K^-$ final states. The cross-section for phase space production of $\pi^+\pi^-K^+K^-$ is measured. In the $\pi^+\pi^-K^+K^-$ events, evidence for production of $\phi\pi^+\pi^-$ is observed for both $Q^2{\sim}0$ and $Q^2>0$. Also $K^{*0}K\pi$ production is observed for $Q^2{\sim}0$. No evidence of $\phi\rho^0$ or $K^{*0}\overline{K^{*0}}$ production has yet been seen.

RESONANCE PRODUCTION IN PLUTO

M. Poppe

DESY

ABSTRACT

In the determination of the radiative widths of resonances, details of the q^2 fall off of the $\gamma\gamma$ cross section may introduce systematic errors up to 0 (10%). The PLUTO collaboration has overcome this difficulty by using p_t constraint kinematic fits. All other systematic errors are by now fully controlled through the use of data test samples.

The A_2 meson has been seen in the reaction $\gamma\gamma \to \pi^+\pi^-\pi^o$. A spin parity analysis excludes $J < 2$ for the $\pi^+\pi^-\pi^o$ system and favours 2^+ over 2^-. The data show a slight preference for helicity $\lambda = 2$. Using a Monte Carlo program which takes into account the interference between the ρ^+ and the ρ^-, we measure $\Gamma_{\gamma\gamma}(A_2) = 1.06 \pm 0.18 \pm 0.19$ keV. This value is insensitive to assumptions about the A_2 helicity.

The reaction $\gamma\gamma \to \pi^+\pi^-$ has been measured from the f^o region down to near threshold, using the particle identification features of the PLUTO forward spectrometer. We obtain $\Gamma_{\gamma\gamma}(f^o) = 2.85 \pm 0.25 \pm 0.5$ keV, assuming $\lambda = 2$ and modelling the interference with the $\pi^+\pi^-$ continuum according to the model of Mennessier. However, the low mass continuum is not well described by either this model or the Born term Ansatz. No further resonance is seen below the f^o.

The cross section for $\gamma\gamma \to \rho^o\omega$ is found to be less than 20 nb (preliminary!) for invariant masses above 1.6 GeV. In this mass region, the cross section for $\gamma\gamma \to \omega\omega$ is less than 24 nb (preliminary!). Both limits exclude some VDM type models.

PRODUCTION OF VECTOR MESON
PAIRS IN $\gamma\gamma$ REACTIONS[+]

K.F. Liu
Department of Physics & Astronomy
University of Kentucky
Lexington, Kentucky 40506 USA

ABSTRACT

The late upper limit on the $\gamma\gamma\to\rho^+\rho^-$ cross sections from the JADE detector[1] in the mass region of 1.3 - 2.0 GeV shows that they are greatly suppressed as compared to the $\gamma\gamma\to\rho^0\rho^0$ cross sections[2] in the same region. This essentially rules out the interpretation of isoscalar resonances such as $Q\bar{Q}$, glueball, and $Q\bar{Q}g$ states which would predict a $\rho^+\rho^-$ cross section twice that of $\rho^0\rho^0$. Both the $\rho^0\rho^0$ and the $\rho^+\rho^-$ data can be accounted for quantitatively[3] in terms of the unique flavor and color-spin structures of the I=0 and I=2 0^{++} and 2^{++} $Q^2\bar{Q}^2$ states which have been predicted to be in this same mass region.

Production of other vector meson pairs are also considered. The $\omega\omega$, $K^{*+}K^{*-}$, and $K^{*0}\bar{K}^{*0}$ cross sections are very small due to the cancellation between two $Q^2\bar{Q}^2$ states. This seems to agree with experiments. On the other hand, the $\rho^0\omega$, $\rho^0\phi$ and ρ^0J cross sections may be reasonably large since there are no cancellation for these I=1 states. However, mixture with $Q\bar{Q}$ mesons at this mass region will perhaps bring them down to a level of $\sim$ 10-20 nb. Production of these mesoniums in hadronic collisions[4,5] and J/ψ radiative decays will be discussed.

[+]Work supported in part by U.S. DOE Grant No. DE-FG05-84ER40154.

REFERENCES

1. J. Dainton, Proceedings of European Physical Society Conference, Brighton, England, 1983.
2. M. Althoff et al. (TASSO Collaboration), Z. Phys. C16, 13 (1982); D.L. Burke et al., Phys. Lett. 103B. 153 (1981).
3. B.A. Li and K.F. Liu, Phys. Rev. Lett. 51, 1510 (1983).
4. B.A. Li and K.F. Liu, Phys. Rev. D28, 1636 (1983).
5. B.A. Li and K.F. Liu, Phys. Rev. D29, 426 (1984).

Investigation of $\gamma\gamma \to \pi^+\pi^-\pi^+\pi^-$ at 2.9 − 3.7 GeV

A.Buijs

NIKHEF, Amsterdam, The Netherlands

We study the process $\gamma\gamma \to \pi^+\pi^-\pi^+\pi^-$ in the charmonium mass range 2.9-3.7 GeV by comparing untagged events with events where one of the final state leptons is detected in the forward spectrometer of the PEP4/9 detector at SLAC. This approach is motivated by the expectatation that pointlike $\gamma\gamma$ interactions will not show the strong Q^2 dependence that vector meson dominance (VDM) type reactions exhibit.[†] Selecting events with sufficiently high Q^2 will therefore permit us to suppress the background from the tail of the well-known enhancement around 1.4 GeV in the $\gamma\gamma \to 4\pi$ data.[‡] This is illustrated in Fig.1, where the Q^2 dependence of the $\gamma\gamma \to 4\pi$ cross section is shown for various ranges of W, the $\gamma\gamma$ invariant mass. The **lower three mass bins are well described** by a pure ρ-pole formfactor (solid line), whereas the data in the 2.9-3.7 GeV range are best described by a the sum of a ρ-pole formfactor and a formfactor with a pole at the J/ψ mass. The dashed line shows the contribution of the J/ψ-pole formfactor. The ratio of tagged to untagged events as function of W is shown in Fig.2, indicating an excess of events in the mass range where we expect contributions from η_c, χ_0, χ_2 and η_c'. The solid line gives the expected ratio from a phase space Monte Carlo with a VDM scaling. The present mass resolution does not allow us to separate η_c, χ_0, χ_2 or η_c' signals. We obtain a cross section of 2.6 ± 1.4 nb for the process $\gamma\gamma \to 4\pi$ for W between 2.9 and 3.7 GeV, when extrapolated to $Q^2 = 0$.

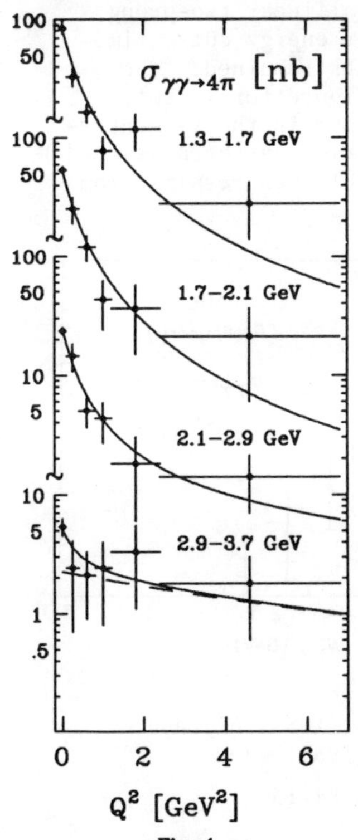

Fig. 1

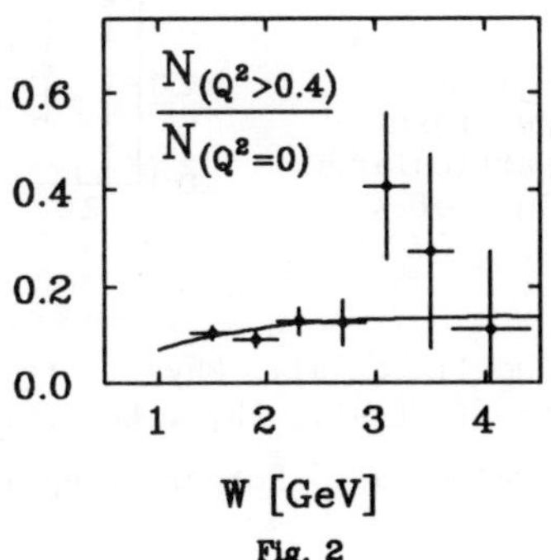

Fig. 2

† R.Kirschner, A.Schiller, Zeit. Phys. C16,141 (1982).
‡ D.L.Burke et al.,Phys. Lett. 103B,153 (1981).

JADE RESULTS ON $\gamma\gamma \rightarrow p\bar{p}$

John Arthur J. Skard

Department of Physics and Astronomy
University of Maryland
College Park, Maryland 20742

ABSTRACT

Preliminary results from the JADE collaboration on the exclusive production of proton-antiproton pairs in photon-photon interactions are presented. The data were taken at the PETRA e^+e^- storage ring at DESY, and represent an integrated luminosity of approximately 93 pb^{-1}. The protons were identified by their rate of energy loss (dE/dx) in the central detector. After cuts, which included a collinear two-prong trigger requirement, momentum cuts, shower energy cuts, time-of-flight cuts and a visual scan, 44 events remained. The resulting cross section is shown in the figure for $|\cos\theta^*| \leqslant 0.6$. ($\theta^*$ is the polar angle of the proton in the $\gamma\gamma$ center-of-mass system, and an isotropic distribution has been assumed.) The cross section agrees well with the published results from the TASSO collaboration[1]
and is ~ 200 times larger than that predicted by calculations using perturbative QCD[2]. Further analysis, particularly a study of the trigger efficiency for $p\bar{p}$ events, is expected to increase somewhat the JADE values for the $\gamma\gamma \rightarrow p\bar{p}$ cross section, particularly at low values of $W_{\gamma\gamma}$.

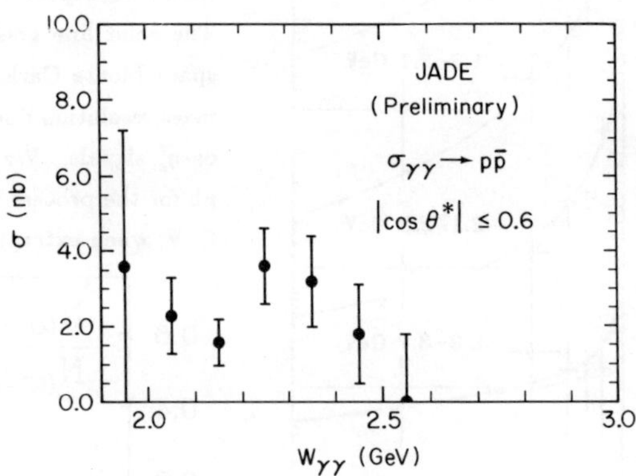

References:

1)   R. Brandelik et al., Phys. Lett. 108B (1982) 67, and
     M. Althoff et al., Phys. Lett. 130B (1983) 449.

2)   G. Farrar et al., Rutgers preprint RU-84-13.

MEASUREMENT OF THE PRODUCTION OF PROTON-ANTIPROTON PAIRS IN PHOTON-PHOTON COLLISIONS AT PEP

PEP-4 TPC/PEP-9 Two Gamma Collaboration

Roger McNeil
Physics Department
University of California
Davis, CA 95616

ABSTRACT

Exclusive production of proton-antiproton pairs in $\gamma\gamma$ collisions has been measured in the single tagged mode with the TPC-2γ detector at PEP. The final state particles are identified by dE/dX from the TPC in the central region, and by TOF in the forward/backwards spectrometer. The cross section of $\gamma\gamma \rightarrow p\bar{p}$ for $1.9 < M_{p\bar{p}} < 8.0$ GeV, $0.1 < Q^2 < 7.0$ GeV^2 and $0 < |\cos\theta^*| < 1$ is compared with predictions for a QCD continuum. The results are also discussed in light of recent evidence for exotic states in $\rho^\circ\rho^\circ$.

OBSERVATION OF HARD PROCESSES IN UNTAGGED $\gamma\gamma$ COLLISIONS

TASSO Collaboration

Presented by Ehud Duchovni
Department of Nuclear Physics, Weizmann Institute of Science
Rehovot, ISRAEL

The analysis of hard processes in $\gamma\gamma$ collisions was performed using high statistics data (70 pb^{-1}) without requiring an electron tag. By the use of a suitable set of cuts the background, due to the annihilation channel, was reduced to a level of 25% of the data (fig.1), with a systematic error of 10%. The P_t distribution of the final state particles is shown (fig.2) to deviate from the VDM expected exponential slope at P_t of about 1.2 GeV/c. A high yield of hadrons with P_t in the range of 1.5 to 3.0 GeV/c was observed. This yield exceeds the expectation of the Born term by a factor of four, however the shape of the P_t distribution is similar to the Born term expectations. Evidence for clustering of particles around the highest P_t track (fig.3) were found. This clustering is consistent, in shape, with the expectations of two-jet production via the Born term.

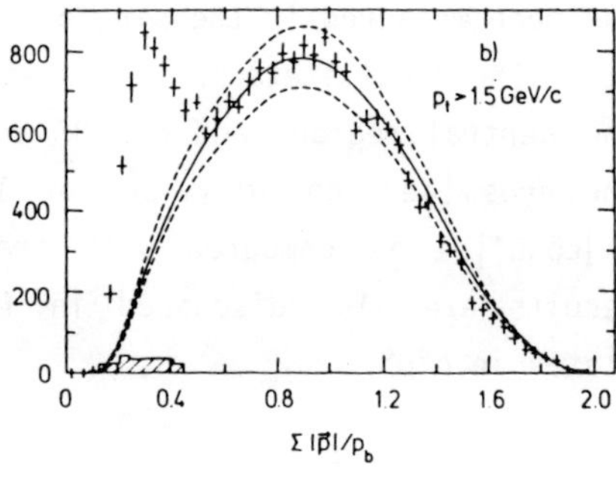

Fig 1.

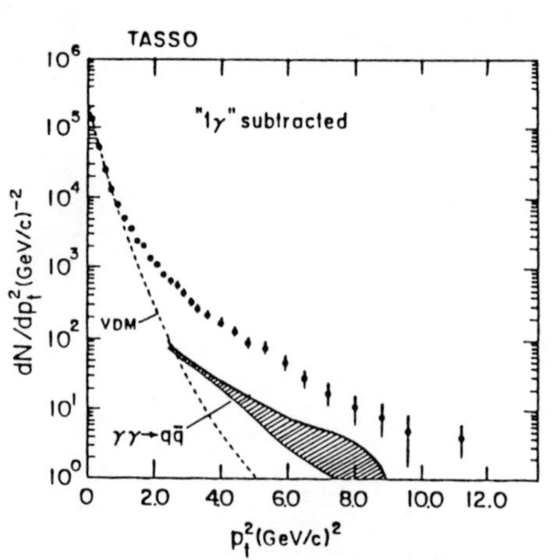

Fig 2.

Fig 3.

THE Q^2 AND TRANSVERSE MOMENTUM DEPENDENCE OF JET PRODUCTION IN PHOTON-PHOTON INTERACTIONS

D. Schmidt

DESY

ABSTRACT

We present an experimental study of jet-production in photon-photon interactions for $0.1 \lesssim Q^2 \lesssim 120$ GeV^2 and jet transverse momentum, p_T, up to 5 GeV/c. At all Q^2, the data show a high p_T tail, characteristic of a hard interaction. The jet production cross-section approaches the quark-parton model (QPM) expectation as either jet p_T or Q^2 increases. Overall, the data are well described in both total cross-section and event topology by the sum of a vector-dominance model and a point-like interaction, represented by the QPM.

MEASUREMENT OF THE TOTAL HADRONIC CROSS SECTION
IN PHOTON-PHOTON INTERACTIONS

D. Bintinger

University of California, San Diego
La Jolla, California   92093

The total hadronic cross section in photon-photon collisions, $\sigma\gamma\gamma\rightarrow$had, is measured for center-of-mass energy, W, between 2 and 20 GeV via the process e+e- $\rightarrow$ e+e- + hadrons. This measurement uses double-tagged events in which both scattered e+ and e-, the tags, are detected. The range of both incident photons' four-momentum squared, q^2, is -0.1 to -1.6 GeV^2. W is determined directly from the tags. The experiment was performed at PEP at SLAC using the PEP-4 TPC/PEP-9 2γ apparatus. The e+e- center-of-mass energy was 29 GeV. The tags were detected by two NaI shower detectors whose excellent energy resolution is essential for a good determination of W.

The q^2 dependence of the data agrees well with a Generalized Vector Meson Dominance (GVD) model with parameters chosen to describe inelastic ep scattering. In the figure the GVD model is used to inter- polate $\sigma\gamma\gamma\rightarrow$had to both q^2s= -0.3 GeV^2 and polarization parameters = 1, and to extrapolate it to both q^2s=0. The shaded band is a one standard deviation envelope for $\sigma\gamma\gamma\rightarrow$had assuming the form a+b/W. At both q^2s=0, a=360±60 nb, b=10±290 nb-GeV and the correlation coefficient = -0.92. Systematic uncertainties are 17% for W between 5 and 11 GeV and 23% elsewhere. Final states of two hadrons are excluded in the cross section reported. The photon helicity interfer- ence terms τ_{++} and τ_{+s} (where t and s refer to transverse and scalar photons respective- ly) are measured as $\tau_{++}/\sigma\gamma\gamma\rightarrow$ had = -0.49±0.24 and $\tau_{+s}/\sigma\gamma\gamma\rightarrow$ had = -0.02±0.04.

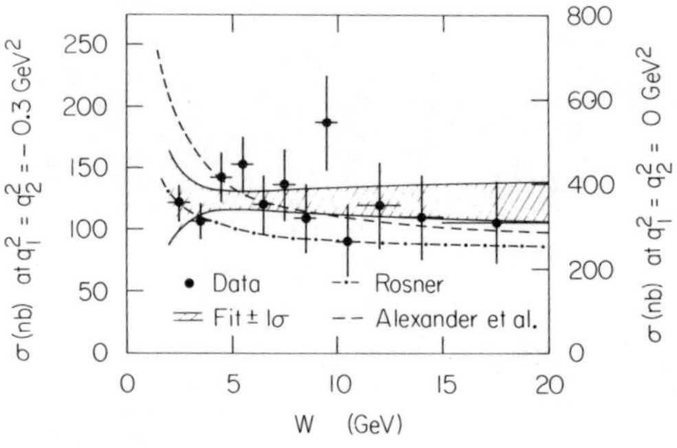

$\sigma\gamma\gamma\rightarrow$had vs W. Statistical errors only. Theoretical curves from J.L. Rosner, BNL Rep 17522, 316 (1972), and G. Alexander, U. Maor and C. Milstene, Phys. Letter 131B, 224 (1983).

RADIATIVE CORRECTIONS IN TWO PHOTON PHYSICS
INCLUDING NON-PERIPHERAL DIAGRAMS

W.L. van Neerven

Institut für Physik

Universität Dortmund

4600 Dortmund 50, West-Germany

J.A.M. Vermaseren

NIKHEF-H

Amsterdam

The Netherlands

Abstract

We have computed all first order radiative corrections to the two photon reaction $e^+e^- \rightarrow e^+e^-X$ where X is a point-like pseudo-scalar particle. The corrections consist out of two parts: A) The peripheral diagrams (σ_p) where the photon starts and ends on the same electron (positron) line. Here we calculated the soft plus virtual photon contribution σ_{pv} as well as the hard photon radiation σ_{pr}. B) The non-peripheral diagrams (σ_5) which consist out of all graphs where the photon starts from the electron (positron) line and ends on the positron (electron) one. The most prominent among them are the five point functions. Here we only took the soft plus virtual photon contribution σ_5 into account. For a C.M. energy $\sqrt{s} = 100$ GeV and a cut off of $m_e + \omega$ on the γ-electron invariant mass with $\omega = 5m_e$ we obtain the following results. For $M_X = 0.135$ GeV we get $\sigma_5 = -2.10^{-3}\%$, $\sigma_p = 1.55\%$ and for $M_X = 1.5$ GeV $\sigma_5 = -3.8 \cdot 10^{-3}\%$, $\sigma_p = 2.15\%$. Clearly the class B diagrams are completely negligible in the total cross-section. For the differential cross-section we find that $d\sigma_5 dt$ ($t = -10^4$ GeV^2) is of the order of -1% where the lowest order differential cross section is about $1-2$ pb/GeV^2. This is not observable. Notice that $d\sigma_p/dt$ ($t = -100$ GeV^2) is about 6%. Finally $d^2\sigma_5/dt_1 dt_2$ is of order of $5-10\%$ if both $|t_1|$ and $|t_2|$ exceed 10^3 GeV^2. In this region the cross section is $10^{-2}\%$ of the lowest order total cross-section so that this correction cannot be observed either. Summarizing we conclude that the contributions from the non-peripheral graphs (class-B) are negligible which implies that two photon processes are factorizable to a high degree of accuracy.

RADIATIVE CORRECTIONS TO THE PROCESS $e^+e^- \to e^+e^-\mu^+\mu^-$

F.A. Berends, P.H. Daverveldt and R. Kleiss,
Instituut-Lorentz, Leiden, The Netherlands.

We have designed a Monte Carlo simulation program which produces unweighted events for the process $e^+e^- \to e^+e^-\mu^+\mu^-(\gamma)$. All radiative corrections to the electron and positron line in the two t-channel (or multiperipheral) diagrams are calculated, except for the negligible contribution of 6-point functions and interferences between real photon emission from the electron and the positron line.

When constructing an event generator, we have to make sure that the cross section is positive at every part in phase-space. Therefore we must beware of infra-red divergent cancellations which yield a finite but negative result for the cross section with soft photon emission. This problem is solved by a rather unusual distinction between soft and hard photon phase-space. Hard photons have, by definition, an energy greater than an arbitrary parameter k_s and an invariant mass with the scattered beam paricle from which it was radiated, greater than an arbitrary parameter μ_o. Soft photons occupy the complementary part of phase-space. μ_o is chosen so that the maximum energy of soft photons equals 1 % of the beam energy. By tuning k_s we can force the soft photon cross section to be positive everywhere in phase-space.[1]

In the no-tagging case the radiative corrections turn out to be on the percent level. However in the large angle tagging case, when differential cross sections are studied which depend heavily on the energy of the scattered particles or when cuts are applied to the total amount of neutral energy, radiative corrections will become sizeable.[1]

1 F.A. Berends, P.H. Daverveldt and R. Kleiss, Leiden preprint
   (to be published in Nucl. Phys. B).

COMPLETE LOWEST ORDER CALCULATIONS FOR FOUR-LEPTON FINAL STATES IN ELECTRON POSITRON COLLISIONS

F.A. Berends, P.H. Daverveldt and R. Kleiss,
Instituut-Lorentz, Leiden, The Netherlands.

We present a Monte Carlo simulation program which generates unweighted events for the following processes : $e^+e^- \to e^+e^-\ell^+\ell^-$ ($\ell = e,\mu,\tau$) , $e^+e^- \to e^+e^-q\bar{q}$, $e^+e^- \to \mu^+\mu^-\mu^+\mu^-$, $e^+e^- \to \mu^+\mu^-\tau^+\tau^-$, $e^+e^- \to \mu^+\mu^-q\bar{q}$. All these processes are treated exactly, that is, all Feynman diagrams contributing in lowest order α^4, all interferences between them and all mass effects are properly taken into account. The design of these generators poses two major problems. Firstly, there is the calculation of the complete matrix element squared corresponding to a large number of Feynman graphs (36 in the eeee case). The method followed to overcome this difficulty resembles closely the procedure described in ref. 1 . Again we perform the calculation at the amplitude level by choosing suitable spinor definitions.[2] The second intricacy has to do with the generation of unweighted events according to a multi-differential cross section which has very many different peaks (657 in the eeee case) in phase-space. This problem is solved by the use of the superposition principle for the MC simulation of distributions.[2] The contributing Feynman graphs are divided into several groups. Next we construct for each group a separate subgenerator. Consequently the interference between groups of graphs is dealt with by a weight assignment. This scheme has the advantage that we are now allowed to choose in each subgenerator that set of integration variables that describes best the occurring peaks in the differential cross section. Moreover it opens perspectives for the accurate calculation of other processes of similar complexity such as double bremsstrahlung.

1 R. Kleiss, Nucl.Phys. B241 (1984) 64.

2 F.A. Berends, P.H. Daverveldt and R. Kleiss, Leiden preprint
   (to be published in Nucl. Phys. B)

Radiative corrections in a deep inelastic electron-photon scattering experiment

S. Ong, C. Carimalo and P. Kessler

Laboratoire de Physique Corpusculaire, Collège de France, Paris, France

Radiative corrections of order α have been computed for an experiment measuring the photon's hadronic or leptonic structure function $F_2(x,Q^2)$. For the hard-photon contribution the peaking approximation has been used. The total correction, for a typical experiment performed with existing high-energy electron-positron colliding-beam machines (PETRA, PEP), appears from our calculation to vary with x and Q^2, ranging from minus to plus a few percent. It tends to increase in absolute value with the beam energy.

DM2 RESULTS ON RADIATIVE DECAYS OF THE J/ψ*

Bernard MICHEL

Laboratoire de Physique Corpusculaire
Université de Clermont II
B.P. 45, 63170 AUBIERE
FRANCE

The DM2 experiment, installed on the interaction region of the Orsay storage rings DCI, has now recorded 8×10^6 events at the J/ψ energy. Results on decay modes of the J/ψ in 3 photons are presented here. They correspond to a sample of 6.4×10^6 analysed events. Candidates with only 3 showers and no charged tracks are selected. The conversion point of each photon is reconstructed with an accuracy better than 1°. Their energy is measured by the shower detector, using both the number of hit tubes and the energy deposition in scintillators. The knowledge of the 3 conversion points determines completely the kinematical parameters of the reaction. Loose cuts are applied on the difference between the reconstructed and the measured energy of each photon. $\pi°$, η and η' signals are clearly visible on the distribution of the minimum invariant mass combination of two photons among three. The $\pi°$ signal, after a visual scan of each candidate, is found to be 28 ± 5 events and corresponds to a branching ratio $B(J/\psi \to \gamma\pi°) = (2.8 \pm .7 \pm .4) \times 10^{-5}$. The η and η' signals (552 and 104 events respectively) are fitted on gaussian shapes with a mass resolution of 15 MeV. This fit leads to : $M_\eta = (547.9 \pm .7)$ MeV and $M_{\eta'} = (956.1 \pm 1.6)$ MeV.

The products of the branching ratios are measured to be :

$$B(J/\psi \to \gamma\eta) \times B(\eta \to \gamma\gamma) = (3.8 \pm .2 \pm .5) \times 10^{-4}$$

$$B(J/\psi \to \gamma\eta') \times B(\eta' \to \gamma\gamma) = (1.0 \pm .1 \pm .15) \times 10^{-4}$$

The first of the two uncertainties is statistical only, while the second one comes from a 15 % systematic error on the normalisation. The values of the branching ratios are in good agreement with previous measurements from other experiments. No other signal is visible in any invariant mass combination of two photons, especially at masses foreseen for ι, θ and η_c.

---

* DM2 Collaboration :
  Clermont-Ferrand - Orsay - Padova - Frascati.

CONNECTIONS BETWEEN CHARMONIUM PHYSICS AND TWO - PHOTON PHYSICS

Lutz Köpke

University of California, Santa Cruz, USA

The present status of η_c, x_0, and x_2 hadronic and $\gamma\gamma$ decays is reviewed including the first measurement of $\Gamma(\eta_c \to \gamma\gamma) = 4.5\,^{+3.8}_{-3.6}\,^{+8.2}_{-2.7}$ [KeV] by a CERN-JSR experiment.

The Mark III collaboration made a systematic study of $J/\psi \to 1^{--}0^{-+}$ decays. Assuming ideal mixing of the vector meson nonet and neglecting double disconnected diagrams, the quark content of pseudoscalar mesons can be deduced. While the η meson comes out to be saturated with u, d, s quarks, there is room for additional contributions (glue?) in the η'.

The production mechanisms for states excited in J/ψ radiative decays ("color charge") and $\gamma\gamma$-collision ("electric charge") are different. Thus the study of glueball candidates ($\theta(1670)$, $\iota(1440)$) in their radiative and $\gamma\gamma$ decays is important. While only upper limits on $B(\iota \to \gamma\gamma)$ and $B(\theta \to \gamma\gamma)$ exist, the $\iota(1440)$ might have shown up in the $\gamma\rho$-decay mode. An enhancement around 1.4 [GeV] has been observed in $J/\psi \to \gamma\gamma\rho^0$ by the Mark III, Crystal Ball and DM2 collaborations.

The reactions $J/\psi \to \gamma\,1^{--}1^{--}$ (a) and $\gamma\gamma \to 1^{--}1^{--}$ (b) show different patterns. While a low mass enhancement in $\rho^0\rho^0$ is common to both channels, a similar structure in $\rho^+\rho^-$ and $\omega\omega$ has only been seen in reaction (a). In addition, the preference of negative parity (0^-) in reaction (a) and for positive parity (0^+, 2^+) in (b) underline the different dynamic origin.

A FAST MONTE CARLO GENERATOR
FOR ee → eeX UNTAGGED EXPERIMENTS

André Courau

Laboratoire de l'Accélérateur Linéaire
Orsay 91405, France

ABSTRACT

We describe for untagged $\gamma\gamma$ experiments a specific and very fast Monte Carlo based on the Double-Equivalent-Photon Approximation. This generator takes into account the experimental constraints in order to perform approximations involving experimentally unobservable effects and to generate events only within the experimental acceptance. It allows a very fast simulation of the events of any cross section in any invariant-mass range and, in particular, a simultaneous fit to the data of the parameters in any model of the $\gamma\gamma$ hadronic cross section.

A TEST OF THE QUARK CHARGES IN A
QUARKONIUM GLUONIUM MIXING MODEL FOR
THE LIGHT PSEUDOSCALAR AND TENSOR MESONS

J.H.FIELD [*]

L.P.N.H.E. University Pierre et Marie Curie
4 Place Jussieu Tour 32 F-75230 Paris Cedex 05

ABSTRACT

A test is made of the quark charges using the measured 2γ widths of the pseudoscalar (π°,η,η') and tensor (A_2,f,f') mesons. The model used for $\Gamma_{\gamma\gamma}$ is non relativistic quarkonium annihilation with binding and SU(3) breaking corrections. The flavour mixing is given by linear mass matrices in the basis $|ns\rangle,|s\rangle, |g\rangle$ for the isoscalar mesons $(\eta,\eta', i(1450)$ and $(f,f',\theta(1720))$.

For the pseudoscalars a fit is made to 5 experimental measurements sensitive to flavour mixing, but independent of the quark charges. The best fit indicates significant gluonium amplitudes of 0.14 ± 0.08, $0.62^{+0.08}_{-0.07}$ in the η,η' respectively and is consistent with the experimental value of $\Gamma_{\gamma\gamma}(\eta')/\Gamma(\omega \to e^+e^-)$ for fractional charge quarks only. The ratio $\Gamma_{\gamma\gamma}(\eta')/\Gamma_{\gamma\gamma}(\eta)$ indicates $|\psi(0)|^2 \simeq M^{2\cdot i}$ and predicts $\Gamma_{\gamma\gamma}(\pi^\circ) = 7$ eV, $\Gamma_{\gamma\gamma}(i) = 5$ keV.

For the tensors, with fractional quark charges, the measured ratios $\Gamma_{\gamma\gamma}(f)/\Gamma_{\gamma\gamma}(A_2),\Gamma_{\gamma\gamma}(f')/\Gamma_{\gamma\gamma}(A_2)$ and the f,f',θ masses completely determine the mass matrix. The f,f' are found to be ideally mixed quarkonium states with only weak gluonium admixtures. For gauge integer charge quark models charged gluonium annihilation may give large contributions to $\Gamma_{\gamma\gamma}(f,f')$. Such models are excluded by constraints on the quarkonium gluonium mixing coming from the strong decays $f,f' \to \pi\pi$, $K\bar{K}$ and the measured upper limit on $\Gamma_{\gamma\gamma}(\theta)BR(\theta \to K\bar{K})$.

---

[*] On leave-of-absence from DESY Hamburg.

TAGGING AND INTEGER-CHARGED QUARK EFFECTS IN $\gamma\gamma$ PRODUCED JETS

Arjun Janah

Physics Department

Kansas State University

Manhattan, KA 66506

INTEGER CHARGE QUARKS AND JET PRODUCTION

T. Jayaraman
Theoretical Physics Department
University of Madras
Madras 600025, India

A background study for direct-photon pair production

C. Carimalo, M. Crozon, <u>P. Kessler</u> and J. Parisi

Laboratoire de Physique Corpusculaire, Collège de France, Paris, France

An estimation of the indirect-photon background (mainly due to π^0 and η decay, and to quark bremsstrahlung) is given for a measurement of direct-photon pair production under (pp or p$\bar{\text{p}}$) colliding-beam conditions at high energy ($\sqrt{s}$ = 60 or 540 GeV). Assuming both photons to be measured at 90° with equal and opposite momenta, and (within experimental limits) unaccompanied by any hadrons or additional photons, it is shown that the noise/signal ratio can be reduced to acceptable proportions. That calculation includes smearing effects of the partons involved in the hard collision, using a formalism proposed by Gunion and Petersson.

DIRECT PHOTON PRODUCTION IN e^+e^- ANNIHILATION INTO HADRONS

Takashi Maruyama
Department of Physics
University of Wisconsin, Madison, WI   53706

on behalf of the MAC Collaboration[*]

Direct photon production in hadronic events from e^+e^- annihilation has been studied at $\sqrt{s}$ = 29 GeV using the MAC detector at the PEP storage ring. A charge asymmetry A = (-12.3 ± 3.5)% is observed in the final state jets. The cross section and the charge asymmetry are in good agreement with the predictions of the fractionally charged quark-parton model. Both the charge asymmetry and total yield have been used to determine values of quark charges. Limits have been established for anomalous sources of direct photons.

------

[*]E. Fernandez, W. T. Ford, N. Qi, A. L. Read, Jr., and J. G. Smith, Dept. of Physics, University of Colorado, Boulder, CO 80309; T. Camporesi, R. De Sangro, A. Marini, I. Peruzzi, M. Piccolo, and F. Ronga, Laboratori Nazionali di Frascati dell'INFN, Frascati, Italy; H. T. Blume, R. B. Hurst, J. C. Sleeman, J. P. Venuti, H. B. Wald, and Roy Weinstein, Dept. of Physics, University of Houston, Houston, TX 77004; H. R. Band, M. W. Gettner, G. P. Goderre, O. A. Meyer, J. H. Moromisato, W. D. Shambroom, and E. von Goeler, Dept. of Physics, Northeastern University, Boston, MA 02115; W. W. Ash, G. B. Chadwick, S. H. Clearwater, R. W. Coombes, H. S. Kaye, K. H. Lau, R. E. Leedy, H. L. Lynch, R. L. Messner, L. J. Moss, F. Muller, H. N. Nelson, D. M. Ritson, L. J. Rosenberg, D. E. Wiser, and R. W. Zdarko, Dept. of Physics and Stanford Linear Accelerator Center, Stanford University, Stanford, CA 94305; D. E. Groom, and H. Y. Lee, Dept. of Physics, University of Utah, UT 84112; M. C. Delfino, B. K. Heltsley, J. R. Johnson, T. L. Lavine, T. Maruyama, and R. Prepost, Dept. of Physics, University of Wisconsin, Madison, WI 53706.

MEASUREMENT OF THE PHOTON STRUCTURE FUNCTION F_2

Armin Deuter*

1. Physik. Inst. der RWTH Aachen

ABSTRACT

The structure function F_2 for a quasi real photon has been measured in the Q^2 range 1.5 to 100 GeV^2 using $\sim$ 1600 multihadron events obtained with the PLUTO detector at PETRA. The x dependence of F_2 has been corrected for the effects of experimental resolution and incomplete acceptance. The events in the Q^2 range 1.5 to 16 GeV^2 were used to present F_2 as a function of x with high statistics at fixed Q^2 = 5.3 GeV^2. To study the Q^2 evolution in more detail the data were divided into 4 Q^2 bins (1.5-3, 3-6, 6-16, 18-100 GeV^2); within these intervals F_2 has been interpolated to fixed Q^2 = 2.4, 4.3, 9.2, 45 GeV^2.

Furthermore the central x-region (0.1-0.9) is plotted versus Q^2. The Charm contribution was estimated with the Box-diagram and statistically subtracted.

The x dependence and the Q^2 evolution are well described by higher order calculations (Bardeen and Buras) regularized by the scheme of Antioniadis and Grunberg.

A Λ fit in this scheme to the data in the Q^2 range 6-100 GeV yields $\Lambda = 240^{+40}_{-50},\ ^{+50}_{-60}$ MeV with the ansatz F_2^{had}/alpha = a(1-x), a = 0.2 for the hadronic component of the photon. If a is included in the fit the result is $\Lambda = 185^{+40}_{-60},\ ^{+50}_{-70}$ MeV and a $\sim$ 0.

*now at DESY

Single-Tagged Inclusive Production of Hadrons via $\gamma\gamma$ Interactions

J.R. SMITH

Physics Department
University of California, Davis
Davis, CA. 95616

ABSTRACT

We present results from the PEP4/PEP9 TPC/TWGM Collaboration on the production of inclusive hadronic two-photon events in terms of the structure function $F_2(x, Q^2)$. The hadronic final states are accompanied by a tag with Q^2 in the range $0.5\,\text{GeV}^2 \leq Q^2 \leq 7.0\,\text{GeV}^2$. The structure function $F_2(x, Q^2)$ is extracted by applying the unfolding procedure of V. Blobel to the data. The input of the unfolding program consists of the data and also a set of Monte Carlo events produced with a constant structure function. The fragmentation of the hadronic final state in the Monte Carlo events (i.e. the $p_\perp^2$ and multiplicity distributions) are matched to the data and then the transformation from the x-visible to x-true distributions is computed. The x-true distribution of the data is then compared directly with the x-true distribution of the Monte Carlo events produced with a constant structure function. The unfolding program then adjusts the Monte Carlo x-true distribution until a reasonable fit to the x-true distribution of the data is achieved (convergence subject to smoothness requirements). The structure function that results when convergence is achieved is the final $F_2(x, Q^2)$ measurement. The Q^2 evolution of the structure function was compared directly with Higher Order QCD in the x-true region $0.1 < x < 0.4$. The data agree with Higher Order QCD with $\Lambda = 200\,\text{MeV}$ over the entire Q^2 range of the data.

MEASUREMENT OF $F_2(x,Q^2)$ IN THE HIGH Q^2 REGION

TASSO Collaboration

Presented by Ehud Duchovni
Department of Nuclear Physics, Weizmann Institute of Science
Rehovot, ISRAEL

The process of $e^+e^- \rightarrow e^+e^- +$ hadrons, where one of the scattered electrons was detected in the TASSO liquid argon endcap shower counter, has been investigated. After background subtraction (82 events), 262 events were found. The Q^2 of these events ranged from 7 to 70 $(GeV/c)^2$ with an average value of 23 $(GeV/c)^2$. The data were analysed in terms of the photon structure function $F_2(x,Q^2)$. In order to obtain a model-independent measurement of F_2 (see figure), an unfolding procedure was applied to convert the measured $\gamma\gamma$ c.m. energy into the true energy. Results with and without the use of this procedure were compared to the quark-parton model prediction, to first order QCD calculations, and to QCD calculations which include higher order corrections (HOQCD). None of these models was excluded. Target photon mass corrections were done and found to be important in the determination of $\Lambda_{\overline{ms}}$. The value obtained for $\Lambda_{\overline{ms}}$ is $(120^{+80+180}_{-50-75})$ MeV or, using the unfolding procedure: $(150^{+100+210}_{-80-90})$ MeV.

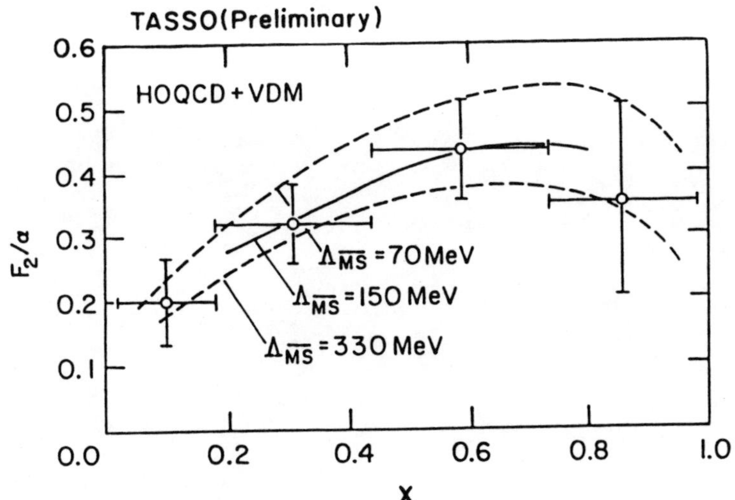

444

FINAL STATE ANALYSIS IN SINGLE-TAG
DEEP INELASTIC ELECTRON-PHOTON SCATTERING

CELLO Collaboration

Paper presented by D. Fournier (Orsay)

Data taken at 2x17 GeV (and already used for structure function analysis and Λ determination) are used to study in a preliminary analysis the hadrons of the final state. 215 events corresponding to 10 pb^{-1} are available.

By considering momenta transverse to the scattering plane (e_{in}, e_{tag}) we show that one is sensitive to the scattering angle θ^* in the CMS of the reaction $\gamma\gamma^* \to q\bar{q}$.

Using a simple algorithm to separate jets associated with each quark, we show that the observed transverse momentum distribution agrees with the point-like coupling of the photons (Vermaseren angular distribution) whereas a soft quasi-real photon (VDM) would predict softer transverse momentum than those observed.

As a consistency check it is shown that the similar distribution for $\mu^+\mu^-$ final state agrees well with the point-like prediction.

EXPERIMENTAL STUDY OF THE PHOTON STRUCTURE FUNCTION F_2 AT Q^2 VALUES FROM 10 TO 220 GeV^2

T. Nozaki (JADE)

Two photon initiated hadronic total cross section was measured under the single tag condition at Q^2 values from 10 to 220 GeV^2 using JADE detector at PETRA. The observed X distributions were analyzed in terms of photon structure functions F_2 and F_1 at two average Q^2 values of 24 and 100 GeV^2 and were found to be well described by asymptotic leading order QCD. The observed X distributions are also well described by regularized asymptotic higher order QCD (modified so as to be free from the singularities of F_2 at small X) plus VMD contribution. From the latter model the QCD scale parameter can be determined to be Λ_{MS} = 0.25 +0.1-0.1(statistical)+0.1-0.6 GeV(systematic), where the parameter is assumed to be equal to one and the radiative correction as well as the correction for the non zero target photon mass are taken into account. A good description of the data are also obtained by both the leading order and higher order non asymptotic QCD predictions including the hadronic component of F_2 calculated by solving the Q^2 evolution equation for the parton distribution in the photon. The F_2 functions were determined by unfolding the observed X and Q^2 distributions for the particle loss and the finite resolution of the JADE detector at two average Q^2 values of 24 and 100 GeV^2. The $F_2(X.Q^2)$ function averaged over the range X > 0.1 was observed to increase as a function of Q^2 with a slope being consistent with combined effect of the $\ln(Q^2)$ dependence of F_2 and Q^2 dependence of the charm contribution, although the data show a slightly steeper slope than the prediction.

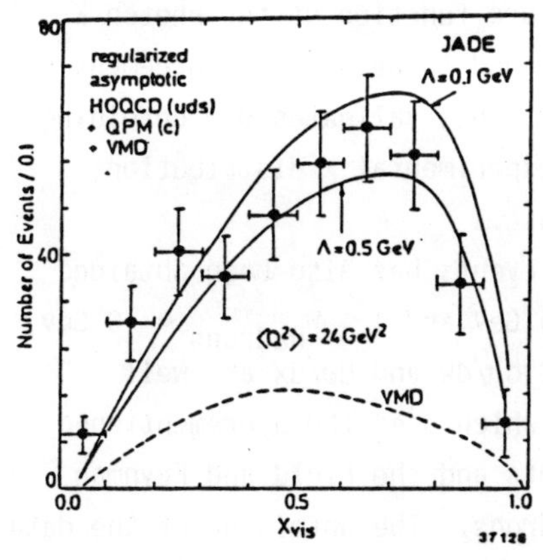

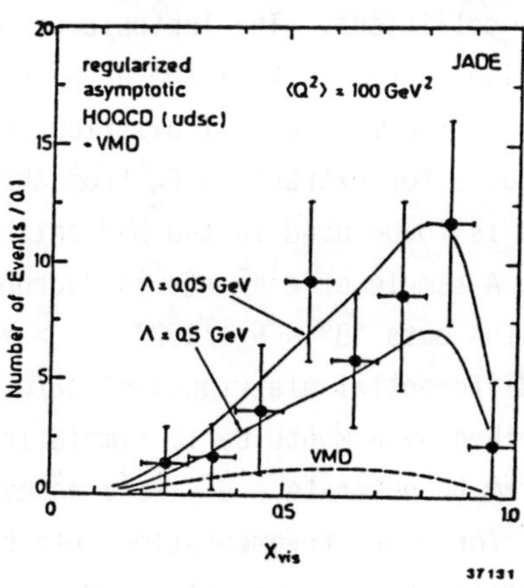

DEEP INELASTIC ELECTRON-PHOTON SCATTERING FROM MAC

Enrique Fernandez
Department of Physics
University of Colorado, Boulder, CO 80309
on behalf of the MAC Collaboration

Results on the single-tag two-photon reactions $e^+e^- \longrightarrow (e^{\pm})e^{\mp}\mu^+\mu^-$ and $e^+e^- \longrightarrow (e^{\pm})e^{\mp}$**hadrons** have been obtained with the MAC detector at PEP. The experimental conditions are such that the unseen electron is scattered at angles smaller than 11° from the beam while the tagged electron is required to have $\theta > 18°$ in order to be reconstructed in MAC's central drift chamber.

In the leptonic case the cuts imposed are $20° < \theta < 160°$, $E' > 5$ GeV, where E' is the energy of the tagged electron, and $W > 1$ GeV where W is the effective mass of the μ-pair system. The average Q^2 ($Q^2 = 4E_{beam}E'\sin^2(\theta/2)$) of the events in the sample is 40 Gev^2. The total cross section within our acceptance, as well as the differential cross sections $d\sigma/dQ^2$, $d\sigma/dW$ and $d\sigma/dx$, where x is the scaling variable $x = Q^2/(Q^2 + W^2)$, are in good agreement with a QED Monte Carlo simulation (J.A.M. Vermaseren, NIKHEF report NIKHEF-H/82-15, 1982) which includes the two basic diagrams for production of μ-pairs in $\gamma\gamma$ collisions. The leptonic structure function of the photon F_2 was also extracted from the data and found to be in good agreement with the expected theoretical distribution. This validates our unfolding procedure for extracting F_2 from the experimental x distribution, which is to be used in the hadronic case.

A sample of $e^+e^- \longrightarrow (e^{\pm})e^{\mp}$**hadrons** events has also been obtained with the cuts $18° < \theta < 162°$, $E' > 7.5$ GeV and $1 < W_{hadrons} < 10$ Gev. The differential distributions $d\sigma/dQ^2$, $d\sigma/dW$ and $d\sigma/dx$ are well described by a Monte Carlo simulation which uses the aforementioned program to generate $e^+e^- \longrightarrow e^+e^- q\bar{q}$ events and the Field and Feynman model for quark fragmentation into hadrons. The unfolding of the data to obtain the hadronic structure function F_2 is in progress.

COMPARISON OF LEADING AND HIGHER ORDER QCD-PREDICTIONS
FOR THE PHOTON STRUCTURE FUNCTION F_2^γ

M. Drees, M. Glück, K. Grassie and E. Reya
Institut für Physik, Universität Dortmund, 4600 Dortmund 50,
West Germany

The structure of the photon can be described in terms of quark and gluon distributions q^γ. The Q^2-evolution of these distributions is determined by a set of coupled inhomogeneous integro-differential equations whose form can be calculated from perturbative QED and QCD. In order to solve these equations one has to impose boundary conditions[1] of the form $q^\gamma(Q_0^2)=q_0^\gamma$; perturbative QCD then only predicts the evolution of q^γ with Q^2.[2,3,4] Neglecting the q_0^γ-dependent terms in the solution for $q^\gamma(Q^2)$ means to create unphysical $x \to 0$-divergences which have to be removed by hand. But even for medium values of Bjorken-x the hadronic terms provide corrections of $O(30\%)$ if $Q_0^2 \sim 1 \text{GeV}^2$ Therefore we used the full solutions throughout.

The only measurable quantity is F_2^γ. Since the relation between F_2^γ and q^γ is different in LO and HO the input distributions q_0^γ have to also be different; this fact has been long known in case of deep inelastic lepton-nucleon scattering.

Choosing appropriate LO- and HO-input functions q_0^γ such that F_2^γ fits the $Q^2 = 5.3 \text{ GeV}^2$ PLUTO data, and evolving them up according to the evolution equations we have shown that differences between LO- and HO-predictions lie below 10% for $1 \text{ GeV}^2 \le Q^2 \le 110 \text{ GeV}^2$, even if we use $\Lambda_{LO} = \Lambda_{HO,\overline{MS}}$. Therefore we conclude that these predictions are undistinguishable in the foreseeable future.[2,3]

[1] M. Glück and E. Reya, Phys. Rev. D28, 2749 (1983).
[2] M. Glück, K. Grassie and E. Reya, Phys. Rev. D30, 1447 (1984).
.[3] M. Drees, M. Glück, K. Grassie and E. Reya, to appear in Z. Phys. C.
[4] M. Drees, to appear in Z. Phys. C.

448

SECOND ORDER QCD PREDICTIONS FOR THE PHOTON STRUCTURE FUNCTIONS[*]

Ignatios Antoniadis

SLAC, Stanford University, Stanford, California 94305

In deep inelastic photon-photon scattering QCD predicts not only the Q^2-evolution, as in most other processes, but also the normalization and the shape of the photon structure function up to second order in the strong coupling constant $\alpha_s(Q)$. However, the naive inclusion of next to leading corrections produces a negative structure function at small x because the calculable ''point-like'' contribution has a simple pole at the $n = 2$ moment which produces a $-1/x$ singularity in the structure function. This singularity is cancelled by a term in the non-calculable ''hadronic'' part that is important only in the 2nd moment.

In Ref. 1, a singularity free parametrization of the photon structure function is derived. Its behavior at small x can be understood with the help of one additional parameter λ (besides $\Lambda_{\overline{MS}}$) related to the non-perturbatively calculable constant term in the second moment. The structure function is found to be sensitive at small x ($\lesssim 0.2$) to the value of this parameter and the problem of a negative structure function can be solved, the perturbative expansion being well behaved. The moments M_n^γ take the form:

$$\frac{4\pi}{\alpha} M_n^\gamma = \frac{4\pi}{\beta_0} \frac{a_n}{\alpha_s(Q)} + b_n^r + \frac{b}{n-2} \left\{ 1 - [\lambda \alpha_s(Q)]^{d_-^n} \right\} + \mathcal{O}(\alpha_s^{\delta_n}) ,$$

where α is the e.m. coupling constant, a_n, b_n^r, d_-^n, β_0 and b are calculable quantities and $\delta_n > 0$ for $n \geq 2$. For $\lambda \neq 0$ M_2^γ is well defined.

A comparison with experiment[2] shows that QCD describes the data quite well with $\Lambda_{\overline{MS}} = 160 \pm 45$ MeV and $\lambda \sim 10$. The above value of $\Lambda_{\overline{MS}}$ is in good agreement with the results from different processes.

A similar analysis has been made[3] for the decay of a heavy time-like photon into a quasi-real photon and hadrons. The time-like structure function is found to be ~ 10 times larger than the space-like one and to have no crossing relation with the latter. The next to leading corrections are important but the elimination of the $n = 2$ singularity is less spectacular. Thus, data are needed to further test QCD.

1. I. Antoniadis and G. Grunberg, Nucl. Phys. B213 (1983) 445.

2. A Deuter-PLUTO; W. Wagner, same proceedings.

3. I. Antoniadis, L. Baulieu and R. Lacaze, Phys. Lett. 125B (1983) 92.

---

[*] Work supported by the Department of Energy, contract DE−AC03−76SF00515.

Higher order corrections to $\gamma\gamma \rightarrow hX$ in Quantum Chromodynamics

Patrick Aurenche
LAPP, F-74019 Annecy-le-Vieux, France

The contribution of higher order diagrams is evaluated for inclusive hadron production in untagged e^+e^- reactions. We assume in the calculation real photons and massless quarks and gluons. The cross-section at the lowest-order (the Born cross-section) is given by the process $\gamma\gamma \rightarrow q\bar{q}$ calculated at order 0 in QCD. We consider the first order process $\gamma\gamma \rightarrow q\bar{q}g$ together with its associated virtual diagrams[1]. Such a calculation gives two types of terms:

a) the higher order corrections proper (denoted HO) of order α_s. We are more general than a previous calculation[2] since we take into account the contribution of the gluon fragmentation to the single inclusive cross-section; also, the result is obtained for the whole rapidity range of the observed hadron. Assuming a non universal definition of the fragmentation functions and a factorization scale $Q^2 = s_{\gamma\gamma}$, the corrections are negative and very small (see table).

b) terms of leading order (denoted ANO) where one of the photons couples to the hard subscattering process via its anomalous component. In our work we also consider the case where both photons interact through their anomalous components.

Besides the perturbative results above we estimate the Vector dominance model (VDM) contribution assuming the photon is an incoherent superposition of its perturbative and its hadronic component. Following ref. 3) we take $F_{2\gamma}(VDM)/F_{2proton} = (2/3)/(1-x_R)^2$ where x_R is the radial scaling variable. The data are taken from NA14 collaboration[4] and Donaldson et al.[5]. The results for the PEP/PETRA energy range ($p_e = 17$ GeV/c) are shown in the table which gives the contribution of various terms to the cross-section $d\sigma/d^2p_T$ (expressed in $10^{-37}cm^2GeV^{-2}$). In conclusion, the calculated corrections (perturbative and VDM) tend to reduce the disagreement between the theory and the experimental data of CELLO and TASSO collaborations[6].

$p_T \backslash \dfrac{d\sigma}{d^2p_T}$	BORN	HO	ANO	VDM
2 GeV	5.7	$-.54$	2.9	7.1
3 GeV	.61	$-.04$	.21	.37

1) P.Aurenche, R.Baier, A.Douiri, M.Fontannaz, D.Schiff, work in progress.
2) F.Khalafi, P.V.Landshoff, W.J.Stirling, Phys.Lett. 130B (1983) 215.
3) S.J. Brodsky et al., Phys. Rev. Lett. 41 (1978) 672.
4) I. Siotis, XV Symp. on Multiparticle Dynamics, Lund, 1984.
5) G. Donaldson et al., Phys.Lett. 73B (1978) 375.
6) CELLO and TASSO Collaborations, this conference.

PHOTON STRUCTURE AND SUPERSYMMETRIC PARTICLE PRODUCTION AT HERA

M. Drees, K. Grassie
Institut für Physik, Universität Dortmund
4600 Dortmund 50, West-Germany

Abstract

Q^2-dependent parametrizations for quark and gluon distributions within real photons are obtained from full solutions of the corresponding leading-order evolution equations. It is shown that the results differ appreciably from fits of the "asymptotic" solutions.

Based on our parametrizations applications for the production of heavy quarks as well as squarks and gluinos at the HERA collider are discussed. Results for total production rates in which the hadronic structure of the photon is involved are comparable to direct mechanisms, where the photon is treated to be pointlike. So the ratio of resolved to direct mechanism for the gluino-squark production is about 2 for $m_{\tilde{g}} = 25$ GeV and $m_{\tilde{q}} = 40$ GeV. Furthermore gluino pair production necessarily involves the hadronic content of the photon. If the gluino mass is 20 GeV and $m_{\tilde{q}} = 40$ GeV, one expects about 50 events/day for the reaction. This production rate is even comparable with selectron (sneutrino) squark production and might therefore lead to the first SUSY signal to be detected at HERA.

SEARCH FOR ANOMALOUS SINGLE PHOTONS

Kwong Lau
Stanford Linear Accelerator Center
Stanford, CA 94305
on behalf of the MAC Collaboration

Preliminary result for the search for single photon final
state in e^+e^- interactions at PEP using the MAC detector
is presented. One candidate event is observed with a
transverse energy above 3 GeV and is consistent with
expectation from radiative neutrino pair production. This
determines an upper limit of $6 \cdot 10^{-38} cm^2$ for the production
cross section in our search region. Using a model
calculation of radiative photino pair production reaction,
a lower limit of 33 GeV/c^2 for the mass of the selectron
is obtained.

The MAC detector has been used to search for production of single-
photon final state in e^+e^- interactions at PEP. The interest is to
detect production of non-interacting neutrals (NIN) accompanied by a
single photon which either comes from the initial-state radiation or is
a decay product of the NIN's. A copious signal would indicate the
existence of NIN's other than the three known neutrino species. Events
were triggered by requiring having EM energy in the shower chamber
>1.5 GeV. Computer programs together with human scanning were used
to reduce the data (100 pb^{-1}) to 158 final candidates. The bulk of
these events are consistent with $ee\gamma$ events with the electrons missing
down the beam pipe. One event was found to have a $p_t > 3$ GeV. This is
consistent with the expected signal of 0.54 event from the neutrino
production process.

Measurement of the Radiative Width $\Gamma_{\eta\gamma\gamma}$ at PETRA
JADE Collaboration, presented by J.Olsson

The reaction $e^+e^- \to e^+e^-\eta$, $\eta \to \gamma\gamma$ (1), was studied with the JADE detector. The scattered electrons were not observed. For triggering on the final state of two low-energy photons, the 2520 leadglass counters in the central cylinder were grouped into seven sectors and the energy sums were used. The threshold was ~180 MeV. The trigger was given by two sectors set in a coplanar fashion (1+4, 1+5, 2+5, etc.). A veto on any T.O.F. counter hit was applied. ~48000 triggers were obtained, corresponding to 14.6 pb^{-1} at a beam energy of 17.3 GeV. Cosmic background was removed using the large solid angle μ-filter. 2γ exclusive events were selected with $E_\gamma > 140$ MeV and $\Delta\varphi \epsilon [176°-180°]$, $\Delta\varphi$ being the opening angle between photons in the plane perpendicular to the beam. The events were then visually scanned. Remaining beam-gas background was assessed using separated beam data. The background subtracted $\gamma\gamma$-mass distribution is shown in Fig. 1, together with the distributions from Monte Carlo simulations. Excellent agreement is seen. Background from other exclusive reactions amounts to 5.7 events in the η region ($m(\gamma\gamma)<800$ MeV/c^2). The η signal contains 220.8 ± 17.6 events and with an overall detection efficiency of 2.4%, the radiative width was determined to be (preliminary):

$$\Gamma_{\eta\gamma\gamma} = 0.56 \pm 0.05 \pm 0.08 \text{ keV} \quad .$$

Errors are statistical and systematic, respectively. This value confirms the recent measurement by the Crystal Ball Coll., but contradicts earlier measurements, which used the Primakoff effect in photoproduction processes.

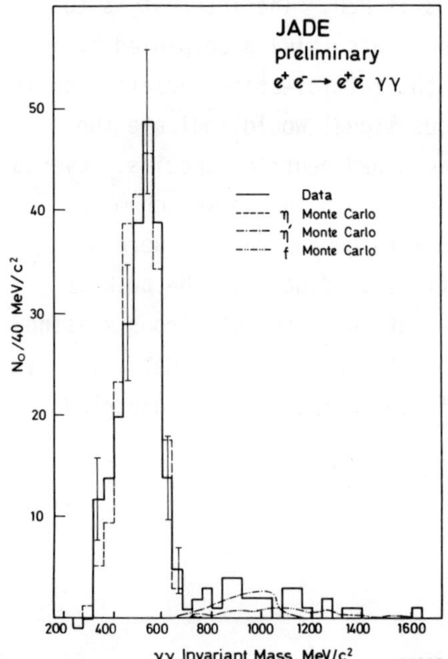

$\pi^0\eta$ PRODUCTION IN PHOTON–PHOTON COLLISIONS

Klaus Wacker

(Representing the Crystal Ball Collaboration)

Stanford Linear Accelerator Center

Stanford University, Stanford, California 94305

Abstract

The Crystal Ball collaboration has analyzed the reaction $\gamma\gamma \to \pi^0\eta$ in a data sample corresponding to $93pb^{-1}$ taken at DORIS II. Production of the resonances $A_2(1320)$ and $\delta(980)$ is observed. The A_2 is produced with a two-photon width compatible with previous measurements. The δ is observed for the first time in photon–photon collisions. A preliminary measurement yields $\Gamma(\gamma\gamma \to \delta) \times BR(\delta \to \pi\eta) = (0.10 \pm 0.04 \pm 0.06)keV$.

OBSERVATION OF PHOTON-PHOTON PRODUCTION OF π^o

David A. Williams
(Representing the Crystal Ball Collaboration)

Physics Department
Harvard University
Cambridge, Massachusetts   02138
USA

ABSTRACT

A run with a special trigger has enabled the Crystal Ball
Detector at DORIS II to study two-photon production of all
neutral final states with mass as small as 100 MeV. Photon-
photon production of a single π^o is observed for the first
time in e^+e^- collisions. An analysis of 6.8 pb^{-1} of e^+e^-
data yields a preliminary value of 7.9 $\pm$ 1.4 $\pm$ 1.6 eV for
the π^o full width.

Measurement of the reaction $\gamma\gamma \to \pi^+\pi^-\gamma$

TASSO Collaboration

Presented by Uri Karshon

Department of Nuclear Physics, Weizmann Institute of Science, Rehovot, Israel.

ABSTRACT

The reaction $e^+e^- \to e^+e^-\eta'(958)$ has been observed at beam energies of 7-18 GeV in the $\pi^+\pi^-\gamma$ final state, where the outgoing e^+ and e^- were not detected. Two different electromagnetic calorimeters were used to detect the low energy photon: the Liquid Argon Barrel Calorimeter (LABC) and the Hadron Arm Shower Counter (HASH).

The figure shows the $\pi^+\pi^-\gamma$ mass distributions for photons detected in the LABC(above) and HASH(below) with a cut of $|\Sigma\vec{p}_t|<0.07$ GeV/c to eliminate events with missing particles. Clear η' signals are observed. The shaded areas correspond to events where the $\pi^+\pi^-$ mass is in the ρ^0 region $(0.60 \leq M(\pi^+\pi^-) \leq 0.85$ GeV$)$. The solid lines are results of fits to a sum of the η' shape obtained from Monte-Carlo calculations, an A_2 contribution and a polynomial background (dashed lines). The $\gamma\gamma$ width of the η' has been measured to be $\Gamma(\eta' \to \gamma\gamma)=5.1\pm0.4$(stat.)$+0.7$(syst.) keV, in agreement with previous measurements and with fractionally charged quark models.

A search for the $\iota(1440)$ in the $\rho^0\gamma$ final state yields no signal in this mass region(shaded histograms). A 95% C.L. upper limit $\Gamma(\iota \to \gamma\gamma) \cdot B(\iota \to \rho^0\gamma) < 1.5$ keV has been obtained.

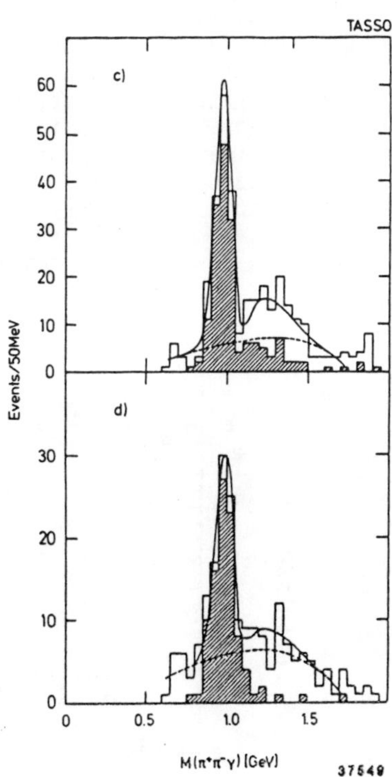

$$\gamma\gamma \rightarrow \eta' \rightarrow \eta\pi^+\pi^-$$

MARK II Collaboration

Lawrence Berkeley Laboratory and Department of Physics
University of California, Berkeley, California 94720

Stanford Linear Accelerator Center
Stanford University, Stanford, California 94305

Department of Physics
Harvard University, Cambridge, Massachusetts 02138

(presented by G. Gidal)

ABSTRACT

We present a measurement of the $\gamma\gamma$ decay width of the η' using the decay mode $\eta' \rightarrow \eta\pi^+\pi^-$, with the η detected via its $\gamma\gamma$ decay. The observation of this decay mode is made possible by the 2.3 kilogauss magnetic field in the MARK II, which allows us to trigger on relatively low p_T charged particles. The Liquid Argon Calorimeter is used to detect photons with fairly good efficiency down to energies of 150 MeV. The uncorrected $\eta\pi^+\pi^-$ spectrum is shown in Figure 1 and shows a clear η' peak, consistent with the expected resolution.

The preliminary result, based on 100 ± 13 events observed in 95 pb^{-1} of integrated luminosity at PEP is $\Gamma(\eta' \rightarrow \gamma\gamma) = 3.7 \pm 1.0$ keV. The error primarily reflects the statistical errors in the data sample and the Monte Carlo efficiency. No attempt has yet been made to estimate systematic errors.

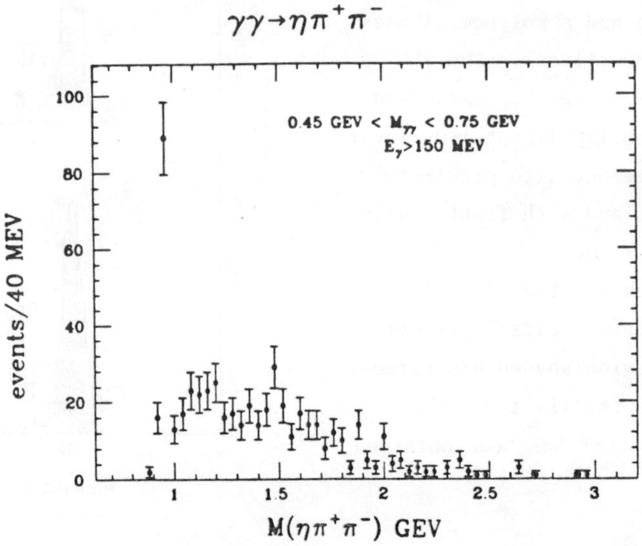

$\gamma\gamma \rightarrow$ High Mass Hadron Pairs

MARK II Collaboration

Lawrence Berkeley Laboratory and Department of Physics
University of California, Berkeley, California 94720

Stanford Linear Accelerator Center
Stanford University, Stanford, California 94305

Department of Physics
Harvard University, Cambridge, Massachusetts 02138

(presented by G. Gidal)

ABSTRACT

We present an updated[1] measurement of the cross section for $\gamma\gamma \rightarrow (\pi^+\pi^- + K^+K^-)$ for invariant masses between 1.6 and 3.5 GeV. The measurement is based on the full 232 pb^{-1} data sample obtained with the MARK II detector at PEP. The Liquid Argon and Muon detectors are used to reduce the ee and $\mu\mu$ backgrounds to a modest level (10-30% of the hadron pairs). The result for $|\cos\theta^*| < 0.3$ is shown in Fig. 1. The errors shown include both statistical and systematic contributions. The smooth curve is the perturbative QCD prediction of Brodsky and Lepage.[2]

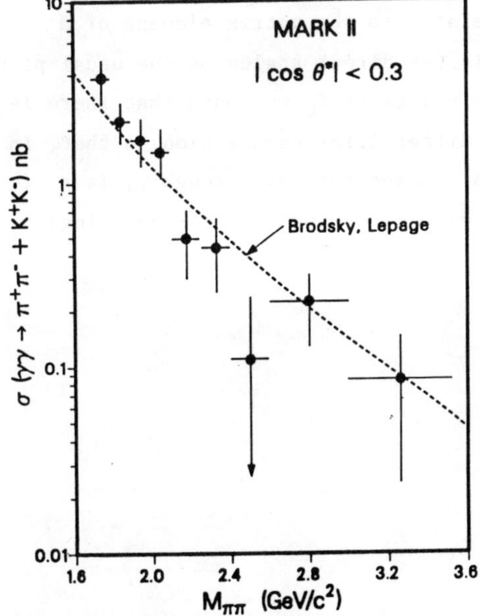

---

[1]For a more detailed description of the analysis, based on the first 35 pb^{-1} see J. R. Smith et al., Phys. Rev. D30, 851 (1984).

[2]S. J. Brodsky and G. P. Lepage, Phys. Rev. D24, 1808 (1981). Note that the K^+K^- contribution is calculated for pion masses.

EXCLUSIVE TWO-PHOTON PROCESSES: THE ROLE OF GLUONS[*]

G.W. Atkinson, J. Sucher and K. Tsokos
Dept. of Physics
University of Maryland
College Park, MD 20742
USA

ABSTRACT

We have computed the crossection for the exclusive process $\gamma\gamma \rightarrow M\bar{M}$ at high momentum transfer, with one or both mesons M, a flavor singlet state such as the η' . This calculation generalizes earlier work of Brodsky and Lepage. We find that an additional set of graphs has to be included due to the coupling of gluons to the flavor singlet state. This coupling, f_G, is related to the matrix element of a gluonic twist two operator and is the direct analog of the usual pion decay constant. With reasonable values of f_G we found that there is a substantial contribution to the differential crossection so that, in principle, a measurement of this crossection, even roughly, is a direct test of the existence of gluons and their vector couplings.

---

[*]Talk presented by K. Tsokos

LEPTON AND PION PAIR PRODUCTION
IN PHOTON-PHOTON COLLISIONS AT DCI[*]

Bernard MICHEL

Laboratoire de Physique Corpusculaire
Université de Clermont II
B.P. 45, 63170 AUBIERE
FRANCE

Lepton and pion pairs produced at low invariant mass (from 200 MeV to 1 GeV) are measured at large angles in a magnetic detector. One at least of the two photons is tagged by a zero degree tagging system, which provides a very efficient background rejection. Moreover, owing to angular and energy resolutions of both central detector and tagging device, we have been able to perform a kinematical identification of produced particles by measuring their mass squared μ^2. The observed μ^2 distribution is then fitted to Monte-Carlo predictions for e^+e^-, $\mu^+\mu^-$ and $\pi^+\pi^-$ production, including experimental acceptances and resolutions, as well as radiative effects. The best fit leads to event numbers for each of the three processes, and to the corresponding cross-sections within the experimental acceptance (electron pair production being chosen as a normalisation). Final results of the DM1 experiment are :

$$\sigma^e_{exp} = 803 \text{ pb} \quad ; \quad \sigma^\mu_{exp} = 437 \pm 40 \text{ pb} \quad ; \quad \sigma^\pi_{exp} = 74 \pm 16 \text{ pb}$$

The cross-section expected from QED for muon pair production, $\sigma^\mu_{th} = 395$ pb is in good agreement with our measurement. The observed cross-section for pion pair production is twice larger than the expected one from Born terms only $\sigma^\pi_{th} = 35$ pb.

This excess of 2.5 standard deviations, cannot be explained by S-wave final state interactions : unitarized Born terms, just as in the Mennessier's model, would give $\sigma^\pi_{th} = 33$ pb. This discrepancy could be explained by a direct coupling of a broad low mass scalar ε ($m_\varepsilon = 820$ MeV, $\Gamma_\varepsilon = 340$ MeV) to photon-photon states with a partial width $\Gamma_{\varepsilon\gamma\gamma} = 15 \pm 5$ keV. Data of DM2 experiment are analysed in a very similar way. Results are still somewhat preliminary. Nevertheless, they seem to indicate an excess of pion pairs, with respect to Born terms, quite consistent with DM1 results.

[*] DM1 and DM2 collaborations :
Clermont-Ferrand - Orsay - Padova - Frascati.

460

STUDY OF f^0 FORMATION AND IT'S DECAY INTO
$\pi^+\pi^-$ IN PHOTON-PHOTON COLLISIONS AT PEP

Kenneth Kwong

Physics Department
University of California, Riverside
Riverside, CA   92521
USA

ABSTRACT

Exclusive production of $\pi^+\pi^-$ in photon-photon collisions has been measured with the PEP-4/PEP-9 detector in both the untagged and the tagged modes. The electron component of the QED background has been subtracted using the dE/dx particle identification in the TPC. This observed electron distribution has been used to understand the detector acceptance and in turn provide reliable muon background subtraction. An alternative approach to muon background subtraction, using the statistical dE/dx separation of muons and pions, is also discussed. The 2γ width of the f^0 meson is fitted with a simple model in which the cross section can be expressed as a combination of the Born term, the f^0 resonance term, and interference term.   The fitting gives an explanation of the mass shift observed and a 2γ width of $2.39\pm0.06\pm$ 0.30 KeV. The Q^2 dependence of the 2γ width is shown to be consistent with GVDM.

PION PAIR PRODUCTION FROM $\gamma\gamma$ INTERACTIONS AT PEP AND THE $\gamma\gamma$ WIDTH OF THE f^o MESON

R.P. Johnson

Stanford Linear Accelerator Center and Physics Department

Stanford University, Stanford, California 94305

ABSTRACT

An untagged $e^+e^-\rightarrow e^+e^-+\pi^+\pi^-$ experiment has been performed at PEP with the DELCO detector. In the invariant-mass range, $0.7<W_{\pi\pi}<2.0 \text{GeV}/c^2$, the QED e^+e^- background has been identified and separated from the muons and charged hadrons by using the highly efficient Cerenkov counters. The kinematic distributions formed from the sample of electron pairs agree well with QED predictions based on the Double-Equivalent-Photon Approximation, as well as the Vermaseren Monte Carlo. Therefore, the $\gamma\gamma$ luminosity inferred from the e^+e^- measurement may be used as an accurate normalization for the subtraction of the $\mu^+\mu^-$ and K^+K^- backgrounds and for the $\pi^+\pi^-$ prediction. The resulting $\pi^+\pi^-$ spectrum agrees well with a simple model of superposition and inter-ference of the f^o (1270) resonance, produced with helicity 2, with a Born-term continuum. From a fit of the model to the data, the $\gamma\gamma$ width of the f^o is determined to be $\Gamma_{f^o\rightarrow\gamma\gamma}$ = 2.70 ± 0.21 keV, where the statistical contribution to the error is 0.05 keV.

Work supported by the Department of Energy, contract number DE-AC03-76SF00515.

BGO IN TWO-PHOTON PHYSICS

J.C. Sens

NIKHEF

The use of BGO as a tagger in two-photon physics is discussed. It is pointed out that:

1. The growth of single crystals out of the Bi_2O_3/GeO_2 binary mixture is complicated by the presence of several non-scintillating compounds in the phase diagram.

2. The measured energy resolution at $E \gtrsim 5$ GeV is comparable to that of NaI.

3. Radiation damage inflicted by 1000 RAD Co^{60} varies strongly from crystal to crystal. All crystals are observed to exhibit "SELF-HEALING", with widely differing effectiveness and time constants.

4. A 120 crystal BGO calorimeter is currently being installed in PEP-4/9 TPC to cover $5<\theta<9$ mrad. Preliminary data indicate a 1 s.d. resolution of 2.5% for 14.5 GeV Bhabha's.

Study of resonances decaying into $K\bar{K}$ in $\gamma\gamma$ collisions

TASSO Collaboration

Presented by Uri Karshon

Department of Nuclear Physics, Weizmann Institute of Science, Rehovot,
ISRAEL.

ABSTRACT

The reaction $\gamma\gamma \to K_S^0 K_S^0 \to \pi^+\pi^-\pi^+\pi^-$ has been studied in the $W_{\gamma\gamma}$ range between 1.0 and 2.5 GeV. A significant $K_S^0 K_S^0$ signal is seen when a cut of $|\Sigma\vec{p}_T| < 0.12$ GeV/c is applied to remove $\gamma\gamma$ events with missing particles.

The figure shows preliminary cross-sections $\sigma(\gamma\gamma \to K^0\bar{K}^0)$, corrected for all the undetected decay modes, as function of $W_{\gamma\gamma}$. The cross-section is obtained by comparing the data to Monte-Carlo events generated with isotropic production of the K^0(or $\bar{K}^0$) in the $\gamma\gamma$ rest frame. A clear f'(1515) signal is visible, and the cross-section near the $K^0\bar{K}^0$ threshold is large. The acceptance corrected angular distribution in the threshold $K^0\bar{K}^0$ region is consistent with isotropic production. No structure is seen in the regions of the $\Theta(1720)$ and $\xi(2220)$. Preliminary 95% C.L. upper limits for the product of the $\gamma\gamma$ partial width and the $K\bar{K}$ branching ratio are: $\Gamma(\Theta \to \gamma\gamma) \cdot B(\Theta \to K\bar{K}) < 0.14$ keV; $\Gamma(\xi \to \gamma\gamma) \cdot B(\xi \to K\bar{K}) < 0.6$ keV.

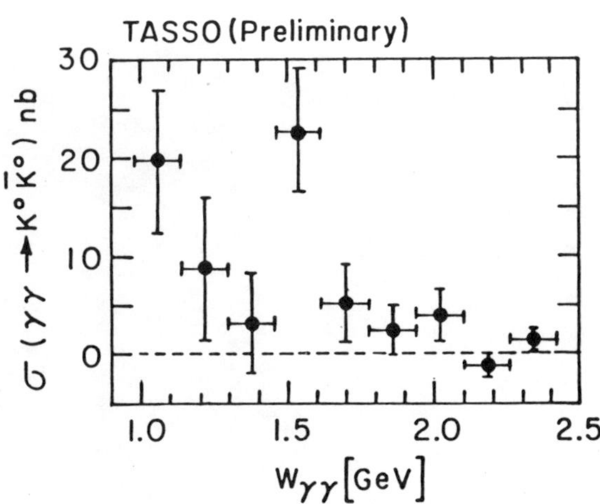

EXCITED FERMION SIGNATURES*

Y. Srivastava
LNF, Frascati, Italy
and
Northeastern U., Boston

Present evidence about excited leptons and quarks carrying definite hypercharge and weak I-spin is reviewed. The interpretation of the Collider ($\ell\bar{\ell}\gamma$) events as (60 ÷ 75) GeV excited leptons can be decisively tested by a factor (2 ÷ 3) improvement over the <u>present</u> data on $e^+e^- \to \gamma\gamma$. At higher energies, $e^+e^- \to$ energetic γ + ME can provide a clean test of excited neutrinos since there is no QED Born term here. Similarly in the two photon channel, $\gamma\gamma \to \nu\bar{\nu}$ can only proceed this way and we find $\dfrac{\sigma(\gamma\gamma \to \nu\bar{\nu})}{\sigma(e^+e^- \to \mu^+\mu^-)} \simeq \dfrac{1}{2} \dfrac{(S\gamma\gamma)^2}{(m^2)}$ which can be sizable. A promising channel to hunt for excited quarks is provided by $\gamma\gamma \to 2$ jets since it shows violations of scaling which grow <u>quadratically</u> with Sjj.

---

* Parts of this work were done in collaboration with N. Cabibbo, L. Maiani and G. Pancheri.

List of Participants

I. Antoniadis

P. Aurenche

A.H. Bäcker

W.A. Bardeen

A.R. Barker

F.A. Berends

D.A. Bauer

D.L. Bintinger

G.J. Bobbink

S.J. Brodsky

A. Buijs

P. Burt

F. Butler

D.O. Caldwell

Ch. Carimalo

M. Chanowitz

G. Cochard

A. Cordier

D. Cords

A. Courau

A. Deuter

M. Drees

E. Duchovni

A.M. Eisner

F.C. Erné

G.R. Farrar

E. Fernandez

J.H. Field

F. Foster

D. Fournier

G. Gidal

G. Godfrey

K. Grassie

J.F. Gunion

A. Janah

T. Jayaraman

R.P. Johnson

U. Karshon

P. Kessler

J. Kiskis

G. Knies

W. Ko

H. Kolanoski

L. Köpke

H. Kück

M. Kwan

K. Kwong

R.L. Lander

W.G.J. Langeveld

K.H. Lau

J. Layter

P. Lecoq

C.S. Lindsey

K.-F. Liu

T. Maruyama

R. McNeil

B. Michel

D. Millers

T. Nozaki

J. Olsson

D.E. Pellett

M. Poppe

W. Repko

M. Ronan

E. Ronat	K. Teshima
D. Schmidt	K. Tsokos
F. Schrempp	W.L. van Neerven
K. Schwitkis	B. van Uitert
J.C. Sens	K. Wacker
B. Shen	W. Wagner
J.A. Skard	Y. Wang
J.R. Smith	M. Wayne
J. Spencer	R. Wedemeyer
Y. Srivastava	A. Weinstein
B. Stella	D. Williams
D. Stork	S. Yellin
K. Strauch	P.M. Zerwas
C.R. Sun	

G
Granlibakken
at Lake Tahoe